T0261419

PERSPECTIVES IN COGNITIVE NEUROSCIENCE

Stephen M. Kosslyn
GENERAL EDITOR

THE BRAIN'S SENSE OF MOVEMENT

Alain Berthoz

Translated by Giselle Weiss

HARVARD UNIVERSITY PRESS

Cambridge, Massachusetts

London, England

Printed in the United States of America

First published in France as *Le Sens du Mouvement*

Copyright © Editions Odile Jacob, 1997

Published with the assistance of the French Ministry of Culture–CNL

Pages 327–328 constitute an extension of the copyright page

Library of Congress Cataloging-in-Publication Data

Berthoz, A.
[Sens du Mouvement. English]
The brain's sense of movement / Alain Berthoz ; translated by Giselle Weiss.
p. cm. — (Perspectives in cognitive neuroscience)
Includes bibliographical references and index.
ISBN 0-674-80109-1 (cloth)
ISBN 0-674-00980-0 (paper)
1. Motion perception (Vision)   2. Orientation (Physiology)
3. Proprioception.   4. Brain.   5. Neuropsychology.
I. Title.   II. Series.

QP493 .B47   2000
612.8′2—dc21   00-023758

FOREWORD TO THE AMERICAN EDITION

This book was published in 1997. Science moves fast, and many new ideas and observations have emerged in the intervening time. Yet the ideas proposed in this book have, if anything, been reinforced by recent knowledge about the brain. I did not think it useful to revise all the references, but I have included a number of papers published by our laboratory or close collaborators since 1997. I have also included a number of reviews from 1998 and 1999 in which the reader will find more general recent references.

ACKNOWLEDGMENTS

I would like to thank Odile Jacob, who agreed to publish the French edition of this book at the suggestion of Jean-Pierre Changeux and who saw it through production with the combination of keen intuition and professionalism that is her trademark.

Mountain climbers need a guide. Writing a book is a little like climbing a mountain. The guide knows the way to the top, avoiding loose rocks and hazardous trails. He knows the moods of the mountain and its traditions. He keeps everyone on course, makes detours when needed, sets the pace, and transforms the climb step by step into a rewarding human adventure. My editorial guide for this book was Gérard Jorland. The breadth of his knowledge as a philosopher and historian, his generosity, his trust, and his critical comments helped me, encouraged me, and touched me.

Maya, you accompanied this book. In form as well as in substance, you helped me to make it not only readable but understandable.

The work of our laboratory would not have been possible without the support of the Centre National de la Recherche Scientifique (CNRS), the Centre National d'Etudes Spatiales (CNES), and the Collège de France, which managed simultaneously to have faith in us, to rigorously evaluate our work, and to tolerate the gambles that are necessary for new ideas to emerge.

I would also like to thank the people who willingly read parts of the manuscript, my friends and colleagues Pierre Buisseret, Valérie Cornilleau-Pérès, Jacques Droulez, Jean-René Duhamel, Werner Graf, Alexej Grantyn, Isabelle Israël, Joseph McIntyre, Edmund Rolls, Jean-Michel Roy, Jean-Jacques Slotine, Yves Trotter, Pierre-Paul Vidal, Paolo Viviani, Sidney Wiener, as well as our entire laboratory team, who carried out the experiments described here.

I am grateful to the journal *La Recherche* for permission to reproduce a portion of the text and illustrations that appeared in a special issue on the brain in 1996.

I am also grateful to Solange Fanjat de Saint Font for editorial help and for assembling the bibliography. Her expertise was critical in bringing this difficult project to completion. And last but not least, I thank Frédéric Lacloche, whose computer graphics skill I admire. He adapted the illustrations.

Finally, I am most grateful to Giselle Weiss who translated this book in English with remarkable speed, exceptional insight into what I wanted to say, and a wide and deep culture, which has earned my admiration. My gratitude also extends to the team at Harvard University Press for the thorough professionalism with which they handled this project.

The translator is deeply indebted to Greer Ilene Gilman and Barbara Breasted Whitesides for editorial and research assistance.

CONTENTS

THE BRAIN'S SENSE OF MOVEMENT

INTRODUCTION

In the beginning was the Deed.

—*Faust*

Irish storytellers always begin this way: "It was not in my time, not in your time, not in anybody's time." The story I am about to tell also belongs to no time because it relates to the greatest mystery of all time: the brain.

This book is a reflection on how the brain works, based on the idea that the brain is used to predict the future, to anticipate the consequences of action (its own or that of others), and to save time. To this end, very diverse biological mechanisms developed over the course of evolution: the architecture of the skeleton, the subtle properties of sensory receptors, and the marvelous complexity of the central nervous system. These mechanisms endowed the brain with internal models of the world and of the body—not just any models, but models that reflect the laws of nature, the *Umwelt* of each species, to use Uexküll's term, and that ensure the survival of each animal.[1] The brain is not a reactive machine; it is a proactive machine that investigates the world. To become a ski champion, it is not enough for the skier to continuously process sensory cues and correct his trajectory; he must go over the run in his mind, anticipate its stages and the state of his sensory receptors, foresee possible solutions to every error, take chances and make decisions before he makes a move.

Our ability to understand the brain and cognition is of course limited, but it is not for lack of trying. First we must consider the question of how the brain relates to the mind. At a recent meeting on the subject of cognition and geometry that brought together mathematicians, physicists, physiologists, and philosophers, one participant, a philosopher, suddenly lost his temper, exclaiming angrily: "I don't even want to hear what physiologists have to say, because brains are just bloody things that I see at the local butcher shop!" This outburst would be a suitable epigraph for an anthology on militant dualism. It shows that we still have a long way to go to persuade even the most eminent thinkers that cognition is a property arising from the amazing complexity of the brain.

I think that we are partially responsible for prejudices like those the philosopher and mathematician Châtelet exposes when he writes:

> Typically it seems as though the work of science is limited to maintaining knowledge, ignoring the contributions of those who further that knowledge and disseminate it. But these are the very people who rescue science from the burden of endless accumulation and stratification, the fatuity of established ideas, the convenience of expediency, and, finally, the temptation to seek refuge in rules. They demonstrate how urgently we need to construct a genuine theory of the way information is processed and to learn about learning. Such reasoning is of course the antithesis of a *neural barbarism* that is stubbornly dedicated to hunting down the neuronal basis of thought and that confuses the activity of learning with the fruits of learning, that is, knowledge. Schelling saw this more clearly: he knew that, in any event, thought was not encapsulated in a brain, that it could be anywhere . . . outside . . . in the morning dew.[2]

We may never succeed in combating these attitudes if we do not describe the *complexity* of the brain. Now, complexity can be an excuse for not explaining anything, and I know that this book may well appear very complex. But I am convinced that the image often portrayed of the brain is too simplistic and especially too static to allow a glimpse of its mechanisms. It is not enough to say that there are millions of neurons. Anyone who believes in the soul will conclude the need for a higher principle to organize this intricacy. The dynamic, flexible, and adaptable character of biological mechanisms must be demonstrated. And physiology would appear to be the right discipline to do this, because it brings together discoveries in anatomy and in cellular biology, mathematical and physical models, and experiments in cognitive psychology to frame its explanations.

The most essential properties of human thought and sensibility are dynamic processes—ever changing, ever adapting relationships among the brain, the body, and the environment. *"Panta chōrei, ouden menei,"* said Heraclitus of Ephesus. "Everything flows and nothing stays." Thought and sensibility are nothing more than states of cerebral activity induced by certain relationships among the physical world, the body, the hormonal and neuronal brain, and its memory of thousands of years of culture.

The anger and incredulity that this idea evokes are rooted in our often oversimple vision of what the brain actually is. Like Leonardo da Vinci, we once thought that it was composed of cavities, then, like Descartes, that it was full of animal spirits. Today, it is known to be populated with little creatures called neurons. But some are not convinced that these neurons are really the basis of the subtle capacities that produce music and mathematics. At the dawn of this century, when the heart turned out to be just a pump, people had to make the best of it. Poets kept on singing about love nonetheless.

I must show the complexity, but I must explain it in simple terms. And there is the challenge. Reader, we must meet each other halfway. I will try to be simple, but you must seek complexity in what I write; and in the shortcuts I take, accept that behind each tree hides a forest. The great misunderstanding between physics and biology stems from the fact that the first is good at describing reality with simple formulas, whereas the second is wary of formulas and aims instead to build a genuine theory of dynamic complexity.

Why, given this climate, write a book about movement? First, because I think that the most refined cognitive abilities of the brain are a product of the need to carry out many difficult tasks. The species that passed the test of natural selection are those that figured out how to save a few milliseconds in capturing prey and anticipating the actions of predators, those whose brains were able to simulate the elements of the environment and choose the best way home, those able to memorize great quantities of information from past experience and use them in the heat of action. Relationships between perception and action are the model of choice for studying the functions of the nervous system. Unlike language, they lend themselves to analysis of human and animal behavior as well as to exploration of the neural mechanisms that underlie them, across the multitude of species that evolution has produced.

In fact, the main problem posed by the command and control of movements is that of inertia and the considerable forces that oppose these movements in the water, in the air, and on land. In general we only have one chance to survive—a single shot, as it were—but one that commits our muscles and our bodily mass to movement. To catch prey that is moving at thirty-six kilometers per hour, that is, ten meters per second, a predator must anticipate its position in less than one hundred milliseconds and head for where the prey will be in a moment's time. It must also prepare the gesture of capture as well as that needed by the muscles to compensate for the weight of the prey and overcome its resistance. To anticipate the behavior of the prey, the predator must make a guess, take a gamble; it must construct a theory of mind by predicting what dodges the prey might attempt in that context. These extremely fast and essentially dynamic processes, and everything that goes on during them, happen in a few dozen milliseconds. The brain is above all a biological machine for moving quickly while anticipating.

In the chapters that follow I will return to all these aspects. They have been neglected by researchers whose study of cognitive function gave priority to language and reasoning. And yet, even poets, those guardians of the language, have recognized the importance of action. Thus, in Goethe's drama, Faust reflects:

> It is written: "In the beginning was the Word!"
> Already I have to stop! Who'll help me on?
> It's impossible to put such trust in the Word!
> I must translate some other way
> If I am truly enlightened by the spirit.
> It is written: "In the beginning was the Thought!"

He hesitates again, then continues:

> It should be: "In the beginning was the Power!"
> Yet even as I write it down,
> Already something warns me not to keep it.
> The spirit helps me! All at once I see the answer
> And write confidently: "In the beginning was the Deed!"[3]

Despite Faust's warning that "in the beginning was the Deed," many read only the first line of Goethe's poem and still believe that "in the beginning was the word." Through an understandable fascination, it was actually believed for twenty years that understanding language, an attribute specific to humans, would enable us to understand cognitive function. This belief, reinforced by the influence of several groups in the eastern United States who have promoted the idea that the brain is a computer, led to a symbolic and computational conceptualization of the nervous system.

The American functionalist school and some of its European adherents defend a cognitive psychology that maintains in principle that the higher functions of the brain must be studied without any reference whatsoever to their neural underpinnings. These functions may be emergent or dissociated, but in the end they are viewed merely as superstructure. We have probably not heard the last of the dualist tendencies.

This risk was brilliantly exposed by Kant:

> Misled . . . by such a proof of the power of reason, the demand for the extension of knowledge recognises no limits. The light dove, cleaving the air in her free flight, and feeling its resistance, might imagine that its flight would be still easier in empty space. It was thus that Plato left the world of the senses, as setting too narrow limits to . . . the understanding, and ventured out beyond it on the wings of the ideas, in the empty space of the pure understanding. He did not observe that with all his efforts he made no advance—meeting no resistance that might, as it were, serve as a support upon which he could take a stand, to which he could apply his powers, and so set his understanding in motion. It is, indeed, the common fate of human reason to complete its speculative structures as speedily as may be, and only afterwards to enquire whether the foundations are reliable.[4]

Plato forgot the body. This book is an apology for the body.

I will proceed in the following way. First I will lead you on a walk through the land of the senses. In addition to the receptors that make up the usual five senses—vision, smell, hearing, touch, and taste—we must identify several others, those of the muscles, joints, and inner ear. In fact, we do not have only five senses, but eight or nine. Is there any point in listing them?

Not at all, because the brain does not process sensory cues independently. Each time it commits to an action, it makes assumptions about the state of certain receptors as the action unfolds. The ski champion cannot constantly be checking the state of all his sensory receptors; he mentally simulates the course of his run down the slope, and it is only from time to time, intermittently, that his brain checks to see whether the state of certain sensory receptors is in accordance with its prediction of the angle of the knees, the distance from the ski poles, and so on. These groupings of receptors are called configurations, and it appears that the brain checks configurations of specific receptors as it plans movement.

With that in mind, I will define all the sensory receptors that enable us to analyze movement in space. These receptors are collectively responsible for what is called the sense of movement, or kinesthesia. Kinesthesia is the result of cooperation among several sensors, and it requires the brain to coherently reconstruct movement in the body and in the environment. When this coherence cannot be achieved, perceptual and motor disturbances result, as well as illusions, which are actually solutions the brain devises to deal with discrepancies between sensory information and its internal preperceptions.

Next I will examine how the brain uses memory to predict the consequences of action. An especially interesting example is that of episodic memory and working memory. Various mechanisms enable the brain to preserve the traces of recent events, to combine motor and sensory signals, and to represent the required process for accomplishing a move or attaining a goal.

Having completed this ramble through the land of the senses and the mechanisms of perception, I will ask the reader to make a special effort to consider a fundamental aspect of the relationships between perception and action: the mechanical properties of bodily masses. Indeed, it is impossible to understand anything about how the brain works without conceding that its main problem is to put mass into motion. Instead of saying "mass," one might say "inertial moment," or the considerable, complex forces that arise as soon as a mass is in motion, like the Coriolis forces created by three-dimensional movements of the head, the resultant of angular accelerations in several planes. The Russian physiologist Bernstein drew attention to the fact that in animals and in man, the limbs have enormous "degrees of freedom," and consequently nature has had to find tricks to simplify the work of the brain. Study of the geometry of movements helps to grasp the natural solutions devised by the nervous system to solve this problem. Skeletal anatomy can be explained in this context; moreover, prewiring motor synergies is another way of simplifying neurocomputation, and simple kinematic relationships connect the geometry of movement and dynamics.

After examining these aspects of perception and movement, I will offer a few concrete examples of motor organization in connection with locomotion, gaze orientation, and control of balance. I will consider how neural mechanisms and internal models enable prediction, how preselection of sensory messages is carried out, and especially what the fundamental role of synaptic inhibition and simultaneous parallel and hierarchical processing is. I will examine how the same structures can be activated in both executed and imagined movement and, in the case of damage or of sensory conflict, how the brain invents new solutions to restore functional plasticity.

In short, I present a theory one step ahead of those that consider the brain to be a simple proactive or representational organ. I suggest that the brain is a biological simulator that predicts by drawing on memory and making assumptions. Flight simulators neither predict nor invent. The brain has a need to create; it is an inventive simulator that forecasts future events. It also acts to emulate reality.

I finish with a tirade against architects who have abandoned the brain's instincts and preferences in favor of their new idols: uniformity, profitability, and the tyranny of the right angle. They have forsaken all sense of finesse for a geometric approach of unparalleled mediocrity.

My scientific tool will be physiology. It is necessarily multidisciplinary, because it involves both structure and function. In the compelling words of Bergson, whose seductive dualism will be challenged several times in these pages,

> We shall find that wherever it is a question of this idea[,] Claude Bernard attacks those who refuse to see in physiology a special science distinct from physics and chemistry. The qualities, or rather the dispositions of the mind, which make the physiologist are not the same, according to him, as those which make the chemist and physicist. He is not a physiologist who has not the organizing sense, that is to say, the sense of that special coordination of the parts to the whole characteristic of the vital phenomenon . . . Claude Bernard defends physiology both against those who believe the physiological fact to be too elusive to lend itself to experimentation and against those who, while judging it to be accessible to our experiments, would not distinguish these experiments from those of physics or chemistry. To the first group he answers that the physiological fact is governed by an absolute determinism and that physiology is consequently a rigorous science; to the second, that physiology has its proper laws and proper methods, distinct from those of physics and chemistry, and that physiology is in consequence an independent science.[5]

A final word: I will not talk much about emotion in this book, except in the chapter where I briefly examine the problem of decision making in the context of a theory of Damasio. And yet, there is no perception of space or movement, no vertigo or loss of balance, no caress given or received, no sound heard or uttered, no gesture of capture or grasping that is not accompanied by emotion or induced by it. But it is necessary to construct a physiology of relationships between movement and emotion, as Ribot suggested:

> What are called agreeable or painful states only constitute the superficial part of the life of feeling, of which the deep element consists in tendencies, appetites, needs, desires, translated into movements. Most classical treatises (and even some others) say that sensibility is the faculty of experiencing pleasure and pain. I should say, using the same ter-

minology, that sensibility is the faculty of tending or desiring, and *conse-quently* of experiencing pleasure and pain. There is nothing mysterious in the tendency; it is a movement or an arrest of movement in the na-scent stage. I employ this word "tendency" as synonymous with needs, appetites, instincts, inclinations, desires; it is the generic term of which the others are varieties; it has the advantage over them of embracing at the same time both the psychological and physiological aspects of the phenomenon.[6]

This book blends analytical considerations of the components of percep-tion and action with synthetic concepts borrowed from experimental and cog-nitive psychology. One of the major developments of the last ten years has been the reconciliation of psychology and the neurosciences, which made it possible to revive study of cerebral function according to complex paradigms.[7] Despite the rear guard battle being waged by some to keep the soul and the body apart, a fruitful, irreversible collaboration has taken root among psychia-trists, neuropsychologists, philosophers, psychologists, and neurobiologists. I hope to add my modest contribution to it. Pardon the philosophical quota-tions, for they are intended neither as passing references nor as pedantry. We all make implicit assumptions, and these citations are one way of clarifying them—perhaps inappropriate, but it gives me pleasure. And is not pleasure the stimulus for all knowledge?

1

PERCEPTION IS SIMULATED ACTION

Our sensations are purely passive, while all our perceptions or ideas are born out of an active principle which judges.

—*J.-J. Rousseau*

The Motor Theory of Perception

A major theme of this book is that perception is more than just the interpretation of sensory messages. Perception is constrained by action; it is an internal simulation of action. It is judgment and decision making, and it is anticipation of the consequences of action. This is not an entirely new concept, as a glimpse at some of the ideas that prefigured it will show. There are so many theories of perception that I will not try to cover them all here. I wish simply to provide a taste of what is in store.

A history of the motor theory of perception was recently published by Viviani, who recalls that it was very fashionable before 1940.[1] One of the first modern works to signal the importance of movement in perception was that of Lotze in 1852.[2] His theory affirmed that spatial organization of visual sensations results from their integration with a muscular sense. The idea that the information that triggers a motor command is used by the brain to recognize movement was proposed by Helmholtz.[3] For him, motor control operates by comparing sensations with predictions based on the motor command. In 1890, William James also described a neural circuit that anticipates the sensory consequences of movement (Figure 1.1).[4] A simple way of conceptualizing the perception of movement, he proposed, would be to assume that a sensory cell

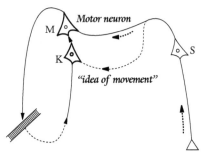

Figure 1.1. William James's concept of an anticipatory neuronal pathway.

S is excited and activates a motor neuron *M* that induces a muscle contraction. A kinesthetic cell *K* detects the movement and modifies the motor neuron *M*. James imagined an additional circuit: a collateral axon of cell *S* activates the kinesthetic cell *K* at the same time that the motor neuron *M* is activated. *K* is thus activated even before it receives information about the muscle movement, which allows it to anticipate the consequences of movement.

In France, the ideas of Janet were very close to those of the pioneers of the motor theory of perception. He established a hierarchy of actions. The most elementary are reflex actions, followed by actions that are "perceptual, social, simple intellectual; and at the level of verbal and volitional action, thoughtful, rational, experimental, progressive."[5] He wrote: "Reflex action is impulsive action as opposed to the considered action that characterizes perceptual behaviors."[6] In other words, perception is interrupted action, and especially goal-oriented action. "In all these perceptual actions, the starting point is a complex object, like prey or its burrow, and the action leads to a transformation and use of the object: the action adapts to the object, rather than simply following from superficial stimuli."[7] But Janet went further and suggested that perceptual action was predictive: "The action that is triggered by the initial stimulus adapts not only to this stimulus but to all the others that the object induces successively; it adapts to stimuli that do not yet exist, but that will only occur later owing to the action itself. This adaptation to an aggregate of future and potential stimuli is characteristic of perceptual behaviors."[8]

Thus the dissociation between perception and action must be discarded. *Perception is simulated action.* A further example, also borrowed from Janet, illustrates this point in a way that is very close to my own interpretation: "When we perceive an object, an armchair, for example, we do not see ourselves as acting at that instant, because we are still standing, unmoving, perceiving the armchair. This is an illusion. In reality, we already have within us the action we associate with the armchair, which we call a perceptual schema—here, the action of sitting in a particular way in the armchair."[9] Merleau-Ponty put it very well: "Vision is the brain's way of touching."[10]

Research on the physiological bases of the relationship between perception and action nevertheless remained limited until just recently. Viviani attributes the post-1940 eclipse of the motor theory of perception to the appearance of the analytical neurophysiology of Sherrington, the influence of Gestalt, and even to the constructivism of Piaget. The compartmentalization of disciplines whose efforts should converge, such as biomechanics, experimental and cognitive psychology, psychophysics, functional neurobiology, and so on, were for a long time another obstacle that today the cognitive sciences are attempting to overcome. After 1950, the motor theory of perception was revived. For example, consider the work of Lashley, Gibson, the Teuber school, and more particularly the work of Held and Hein on the role of action (not just activity) in the development of the visual system. Or consider the work of the psychologist Johansson in Scandinavia and that of Fessard and Piéron in France—Piéron maintaining that perception is awareness of external objects and events that give rise to sensations. In the 1970s the research groups of Imbert and Jeannerod made major discoveries concerning the development and functioning of the visual system and its relationship to the vestibular system, the control of ocular movements and posture.

The Concept of Acceptor of the Results of Action

Though underrated in the West, the Russian Anokhin's seminal ideas made him a trailblazer.[11] In his day, Pavlov's theories and experiments on conditioned reflexes, which profoundly influenced our century, were predominant. But Anokhin found Pavlov's definition of reflexes too limiting. He also criticized Descartes for having sidestepped the question of the relevance of the reflex response to the higher brain and thereby influenced future studies of the complex adaptive actions of animals and man for years to come. Anokhin introduced a key idea to the theory; that at the end of the reflex arc there is a reflex action. Although Pavlov's idea was already old, Western physiologists ignored subsequent contributions of Russian physiologists for fifty years, except to plagiarize their ideas without citing them—intellectual petty thievery that the isolation of the Russians helped to disguise. Western physiologists used the word "response" to refer to the effect of a stimulus in a reflex arc, whereas the Soviet literature emphasized and still emphasizes the concept of reflex action. The crucial difference is that if every reflex requires a complete action, this involves the entire organism and consequently its faculties of invention, creativity, and adaptation.

Anokhin provided no experimental proof, but he constructed a theory he called "acceptor of the results of action" that is worthy of comment. He began with the following line of questioning: If the result of reflex activity is an action, doesn't execution of this action need to be approved one way or another by some configuration of sensory afferent information? And doesn't execution of the action also need to be compared with a predicted configuration?

Say that we wish to pick up a cup from a table cluttered with dishes but that, just when we are about to grasp the cup, we are distracted and pick up a pitcher in its place. As we all know from personal experience, in general we correct an error like this one immediately. But what physiological mechanism permits us to notice our error and to correct it? The appearance of the pitcher and the gripping of its handle as well as the appearance of the cup and the gripping of *its* handle are just an aggregate of afferent signals that differ by a few constituents. Why is it, then, that the latter afferentation is precisely the one we choose to sanction our action?

According to Anokhin, our move with the pitcher was initially satisfactory because the set of sensory signals constituting its grasp matched a configuration that was predicted, expected, specified before making the gesture. Mathematicians would say that the sets of afferents corresponding to the cup and the pitcher contain an intersection sufficient to be allowed by Anokhin's acceptor of action. He states that this ready-made excitatory complex, which precedes the reflex action, must in some way be an afferent control apparatus that determines how closely the return afferentation from the central nervous system corresponds to it.

Thus, Anokhin developed a concept equivalent to what is now called an internal model of a group of preselected elements. He was careful not to use the word "representation." He referred to Pavlov's earlier observation that the chemical composition of saliva from a conditioned dog is related precisely to the kind of food used for reinforcement and thus to the character of salivation.

Anokhin next turned to the neural basis of this concept. He used the expression "acceptor of the results of action" to refer to a cortical system specialized in the analysis of complex afferents (sensory information) resulting from reflex action. This analyzer determines how the afferent inputs it receives relate to the planned action as a function of the past experience of the animal. Noting that he could as well have called his apparatus "the acceptor of the afferent results of a completed reflex action," Anokhin explained that he had chosen the word "acceptor," from the Latin *acceptare*, because it combined two ideas: accept and approve. Consequently, he incorporated a clear role for the acceptor of the result of action in decision processes. For example, if a person sitting in the living room decides for one reason or another to go into the dining room, at the precise moment of the decision, the set of afferents from all the stimulatory cues he has received in the dining room in the past (acceptor of action) is reproduced in his cerebral cortex. If, after the person has entered the dining room, the cues coincide perfectly with what the acceptor of the result of action predicts, the person then moves on to the next element of intended behavior. But if the acceptor of the result of action detects an error, an incongruity with what is predicted, the brain produces an orientation reaction, as described in the Soviet literature—that is, it reacts by analyzing new events.

Bernstein's Comparator

Another master of modern physiology who had an even more profound influence on our generation was the Russian physiologist Bernstein. Persecuted for his overly original ideas in the time of Pavlovian hegemony, he studied natural movement and inferred general rules of cerebral functioning from it.[12] To avoid a naïve linear description of the regulation of the coordination of movement as a succession of phases—prediction, preparation, execution, and control—Bernstein proposed a circular schema (Figure 1.2) that introduced the concept of the action-perception cycle. The basic element is a "comparator" that establishes the "required value."

This required value fulfils at least three different functions, all equally important. The first is detecting an error between accomplished movement and predicted movement that triggers a correction (from a cybernetic vantage). The second is recognizing that an action has been accomplished, which allows progression to the next action in sequence. "This aspect of function," says Bernstein, "is mainly reminiscent of what Anokhin has termed 'sanctional afferentation.'" The third function is the adaptation itself. Indeed, when action encounters surprises, it is impossible, or irrelevant, for corrective impulses to reestablish the initial plan of action. In this case, the receptor of information acts as an initiator (and not as a regulator) of adaptive changes in the program being executed. It does this by introducing either small technical changes into

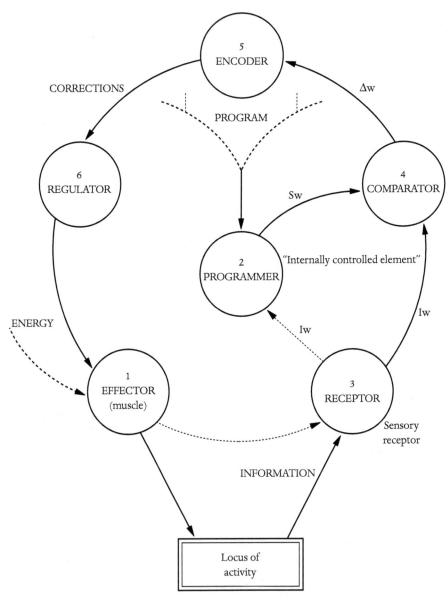

Figure 1.2. Schematic diagram showing the cerebral organization for control of movement proposed by Bernstein.

the movement or taking an adjacent trajectory. And it does it until the program has been completely reorganized, even down to the repertoire of consecutive elements and the staging of motor action—in other words, it adopts a new tactical approach to the task.

I think that the highest cognitive functions are the result of an evolutionary thrust toward developing this ability to reorganize action according to unforeseen events. This ability requires developing a memory of the past, the faculty for predicting and simulating the future, and the metafaculty, in a way, to mobilize all these capabilities rapidly, because they must integrate with a perception-action cycle that sometimes lasts only a tenth or twentieth of a second.

For Bernstein, these corrective processes depend heavily on what he calls the comparator. This neural device occupies a strategic position between the information supplied by the receptors and the elements that will effect the necessary corrections or reorganization. It does not function between two successive or simultaneous receptors to compare two distinct events, but between flowing, continuous reception and an internal guide.

An important property of the comparator is its capacity to detect *variations* in sensory information owing to the central nervous system's use of fresh traces. Our bodies, says Bernstein, have no receptive apparatus capable of perceiving velocity directly. This task is resolved in the central nervous system by the comparator. It instantaneously compares the cues about the position of the moving organ with the fresh trace of its position approximately 0.1 second before. The brain thus recognizes two positions with a certain interval of time between them. So it can easily reconstruct the velocity, which is displacement (the difference from one position to the other) divided by time.

Bernstein saw that the control of movement is not continuous but discrete, under the control of the comparator. He arrived at this conclusion from contemporary work that linked motor action with brain waves having a frequency of 8 to 14 cycles per second—that is, α waves measured by electroencephalogram.[13] This is, at least in part, the frequency of rhythmic oscillations of excitability for the principal elements of the reflex circuitry of our motor apparatus. Bernstein regarded the intervals between each cycle at the α rhythm as units of an internal physiological clock, or pacemaker.

Bernstein's vision was prophetic, since it was not until 1990 that Llinás suggested that movement is subtended by neural activity oscillating at 10 Hz (ten oscillations per second).[14] This suggestion remains viable, as it is possible to observe synchronizations of 10 Hz in the neck muscles of an alert cat. Movement may therefore be subtended by coupled oscillators that function at this frequency, as Bernstein foresaw. The work of Sokolov and his school in Russia on the hippocampus,[15] that of Rougeul-Buser in Paris, and of olfaction and sleep specialists has also demonstrated the importance of these central rhythms during various tasks that require alertness or organized movements.[16] It seems clear now that several oscillatory processes with frequencies whose values are 8 to 12 Hz, 16 Hz, 40 Hz, 70 to 90 Hz, and so on subtend the internal workings of perception and movement. Llinás summed up his ideas by saying that "we think at 40 Hz and move at 10 Hz," asserting laconically that the minimum time to process mental data is of the order of 25 milliseconds, and the minimum time for a motor control operation around 100 milliseconds. He proposed a theory that explains how the brain uses internal circuits that oscillate forty times per second (40 Hz) to develop multisensory perception and guarantee its coherence (Figure 1.3).

Bernstein completes his analysis of the microfunction of the perception-action cycle with more general considerations of the role of anticipation. He observes that in fact a wide range of movements anticipate or extrapolate by analyzing fresh traces of the past. I will offer many examples of these neural memories, within the cell itself or in complex networks, which maintain sensory elements for a certain time, the new idea being that they allow prediction of the future. If I am not mistaken, Husserl's concept of simultaneity of "protention" and "retention" is quite close to this idea.

A final basic conceptual contribution by Bernstein concerns the very principle of motor control. Suppose that I wish to move my arm from one position to another: the motor command an engineer might use would employ a potentiometer to measure the angle, then construct a movement command that controls the motors, which in this case are the muscles. A force command, or rather a coupling command (force multiplied by the lever arm), would be the *controlling variable,* and the resulting displacement the *variable detected* by the muscle receptors that are the biological equivalent of the potentiometer.

Bernstein reasoned that the brain controls another variable, the *point of equilibrium* between the two muscles that are in opposition around a joint—the biceps and triceps in the arm, for example. He started from the idea that when the forces exerted by these two muscles—which depend on the activation of motor neurons and their mechanical properties—are equal, the arm will be in a given position. To define a position in space or to maintain a certain relationship among muscular forces is equivalent. Thus, space is perhaps not coded explicitly, but the trajectory of movement results from this dynamic equilibrium. Movement is simply a progressive shifting of postures.

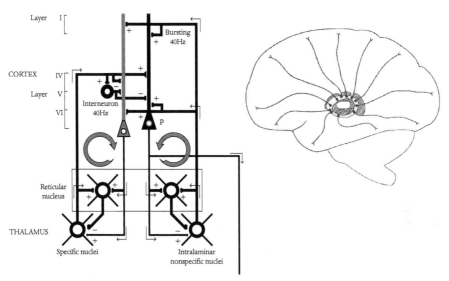

Figure 1.3. Thalamo-cortical circuits accounting for perceptual coherence through synchronization of neuronal oscillators. *(A):* On the left, sensory information received from vision, proprioception, and so on converge on specific thalamic nuclei. The neurons in these nuclei project to the cerebral cortex (to the pyramidal neurons) *(P)* in layer VI. Activated by various interneurons that oscillate at 40 Hz, the pyramidal neurons of the cortex project in turn (two projections, one inhibitory, one excitatory) to the thalamus and thus constitute a circuit swept by activity that oscillates at 40 Hz. To the right, a second circuit comprises the intralaminar nonspecific thalamic nuclei that project to the more superficial layers of the cortex and toward the reticular nuclei. The pyramidal neurons activated at 40 Hz also project back to the thalamus. *(B):* Sketch showing the neurons of the intralaminar nucleus of the thalamus, which projects in a starlike formation toward the cortex. Note that these neurons are arranged in a circular ring. The oscillation at 40 Hz is reproduced in the intralaminar nucleus along this ring. Consequently, the neurons of the cerebral cortex are activated sequentially. The other nuclei of the thalamus are indicated by two shaded regions.

Memory Predicts the Consequences of Action

A concept proposed a long time ago to link perception, action, and memory is that of the motor schema. According to Schmidt, movement structures, which he calls "schemas," are stored in the brain.[17] These schemas are not sensory or motor elements, but memorized relationships—topological links, mathematicians would say—between several sensory or motor components of action (like the position of limbs, the state of a target in space, and so on). Schmidt

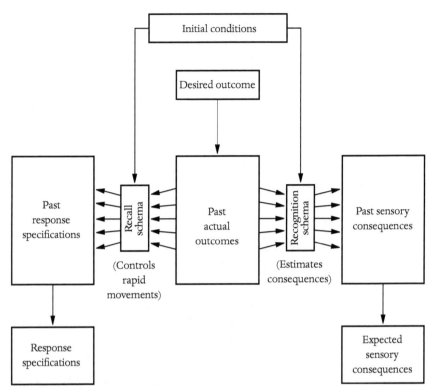

Figure 1.4. The brain anticipates the consequences of movement from past action to prepare and initiate a movement. Reading from top to bottom, anticipated movements begin with a set of initial conditions and a plan of action that will lead to the desired outputs. These are compared with the results of past actions recalled by two types of memory: left, memory of motor commands, and right, memory of sensory data associated with past movements and their effects on the environment. It is therefore possible to recall the expected sensory consequences, or the messages detected by the receptors during and after action over the course of movement. Control of movement involves estimating discrepancies between anticipated and actual sensory data.

proposes the following definition of a schema (Figure 1.4). When a person executes a move to reach a goal, he stores up four types of elements: the initial conditions, provided by the senses; the program for the motor command, which Schmidt calls "specification of past responses"; the sensory consequences of the movement; and other consequences of the movement, like outcome. The schema is not the set of these data, but their *relationships*.

What is interesting about this theoretical work is that it links prediction of the consequences of action with the memory of past consequences, without reducing movement either to a simple chain of packaged reflexes or to the simple execution of a centralized motor program. Schmidt's model also shows how relationships between the elements of the schema can be modified.

This concept turns up again in the work of Neisser, who insisted on the inseparable character of what he calls the "perception-action cycle."[18] In this cycle, exploration of the visual world by a subject is directed by defined anticipatory schemas, like blueprints, for *perceptual action*—perception is thus conceived as active.

Mental Nodes

The idea that the brain does not simply detect the physical parameters that excite the senses is a very old idea. One interesting (unfortunately not very well known) expression of it was offered by MacKay in several theoretical essays dating from 1970 to 1980. His theory, called "mental nodes," aimed at integrating interactions between perception and action. Others before him had tried this approach: "Not all theoretical thinking in psychology has adopted the assumption that components for perception and action are completely separate or unshared. In particular, Lashley proposed that speech comprehension and production make use of common components and mechanisms because 'the processes of comprehension and production of speech have too much in common to depend on wholly different mechanisms.' . . . I conclude that perception and production share some components but not others."[19] But, as he observed at the time, the dominance of research on perception isolated from action derived as well from the philosophical view of researchers that "action [was] functionally, temporally, and evaluatively subordinate to perception; functionally subordinate because they considered perception the sole means by which knowledge is acquired (empiricism); temporally subordinate because they considered perception a necessary precursor to action (paleobehaviorism); and evaluatively subordinate because they viewed the contemplative life as superior to a life of action (see Plato)."[20]

According to MacKay, two major mechanisms underlie most mental operations in sensorimotor systems. The first processes information from the sensory receptors that the system forwards to a comparator where it is subjected to criteria, such as a desired position. Error signals for checking corrections are sent from the comparator—Bernstein's classical loop—but MacKay adds a second circuit based on sensory information. Sensory data in this second circuit are not processed by the nervous system to extract velocity, force, volume, dilation, and so on; they are interpreted as *attributes,* that is, configurations of pertinent information that have a category-specific meaning. These data are directed to an operator that organizes or decides or chooses or finally executes a projective *feedforward* under the control of the supervisor. This operator recognizes pertinent cues for the task and evaluates the sensory context to anticipate the movement; it controls motor activity.

What is missing in this model is direct action by the organizer on the receptors; that is, an influence of active perception on sensation, although extensive proof exists for such action as I have indicated here, and as I will show again later. In particular, what is missing is a conception of the brain as a repertoire of sensorimotor schemas that are just so many possible actions and that organize perception even before sensory stimuli are processed. I will develop this idea further, but it is worth examining a clear example at the outset.

Mirror Neurons

Support for the idea that the brain contains schemas in its neuronal organization that constitute veritable behavioral actions was recently provided by Rizzolatti.[21] He discovered neurons in area F5 of monkeys, later termed "mirror neurons," which fire each time the animal makes a particular gesture, for example, lifting a peanut to its mouth, turning a handle, and so on. But the same neurons also fire when the monkey sees the experimenter make that same gesture. In other words, these neurons are activated both when the animal makes the gesture and when it sees the gesture being made. From that it can readily be deduced that the network of neurons, which includes this particular neuron, codes for a schema of the behavioral repertoire of the monkey. Do we make that deduction? I will discuss it; but whatever the outcome of the debate, this discovery argues for the existence of a repertoire of *preperceptions* linked to a repertoire of actions. Consequently, the brain can simulate actions to predict their consequences and to choose the most appropriate among them.

A word in passing about the term "encoding." I will use this term to designate the message contained in the firing of neurons whose properties I will describe. This usage is defensible even though it makes a major assumption, which is that the *temporal or frequency characteristics* of the discharge contain signals relevant to the functioning of the nervous system. The frequency of neural discharge is not the only code possible. The coincidence between the firing of several neurons or the temporal organization of their firing, the average activity of a population of neurons, synaptic modifications at the molecular level, are just so many different codes that we still do not understand. Substantial research efforts are ongoing to understand the respective contributions of temporal encoding and frequency encoding. I can only touch on this question, though it is fundamental.

Simulation, Emulation, or Representation?

A final word about the philosophical framework of this book. I will rarely use the word "representation," and when I do, I will use it reluctantly. Indeed, I think that it is very trendy and very convenient—for hiding our ignorance. Under the guise of seemingly centralist theories, it very often conceals subtle forms of dualism that I reject. In a recent review of theories of perception, in particular that of Helmholtz, Bouveresse articulated the dangers inherent in the concept of representation:

> Theories that describe perception as the construction of internal representations of the external world commit themselves at the outset to a path that may perhaps lead somewhere, but not where they claim to go. If we assume that our concept of mental representation is inspired by the idea of a material image for which the object itself is lacking, and that this image we have is distinct from the object itself, it is unlikely that we can thereby achieve a satisfactory theory of how objects are perceived.[22]

Bouveresse suspects Helmholtz of dualist tendencies:

> If the transformation of sensory excitation into psychic representations of objects in the external world is the result of neurological processes, which remain entirely below the threshold of consciousness, it seems there are good reasons for thinking that in the final analysis it is no less physiological than such activity that results in the production of sensations. The risk in dividing things up the way Helmholtz does is that of returning to an implicit form of dualism that accounts for both the raw data coming from stimulation of our sensory receptors by external objects and the activity that results in the nerves, like a sort of material, which the psychic apparatus (in plain language, the soul) must then process and transform according to its own particular principles or methods. It is probably this risk that Mach denounces when he makes the point that "everything psychical is physically based (fundiert), and not just one part of the psyche. Consequently, explanation solely by means of judgment and inferences is unacceptable. The apparatus for hearing does not imply that the soul proceeds independently, apart from hear-

ing. Cerebral phenomena are in effect *themselves* the psychic life." And Mach concludes that "the eye must not make a detour through the intellect." At any rate, it must not do so if the word "intellect" refers here to things that are supposed to occur in the soul, in the traditional sense of the word, rather than in the human body.[23]

In choosing the word "simulation," can I avoid these traps? I am certainly not happy with it, but I prefer it to "representation." For me, the brain is a simulator in the sense of flight simulator and not in the sense of computer simulation. Simulation means the whole of an action being orchestrated in the brain by internal models of physical reality that are not mathematical operators but real neurons whose properties of form, resistance, oscillation, and amplification are part of the physical world, in tune with the external world. A few years ago, we constructed the scheme in Figure 1.5, suggesting that the brain processes movement according to two modes. One, conservative, functions continuously like a servo system; the other, projective, simulates movement by predicting its consequences and choosing the best strategy.

When I talk about simulation, I mean the set of operations carried out by a simulator; that is, a real machine that possesses at least some portion of the properties of physical reality. As recently as twenty years ago, Gurfinkel had the same idea. He rejected computer simulation of all the physical constraints on walking as impossible; a model, however limited, "incarnated" or rather embodied in a robot would do it better. So he constructed a spider with the help of the department of robotics at the University of Moscow. It is in this sense that the word "simulation" is better. The word "representation" is too contaminated by the idea of visual image. "Now that I perceive the thing itself and not a representation, I would only add that this thing is an extension of my gaze and, in general, of my exploring," wrote Merleau-Ponty.[24]

The development of models of neural networks out of silicone components called VLSI (Very Large-Scale Integration)[25] relates to the same idea, its technological objective notwithstanding: biological models must be embodied in physical reality just as the brain is incarnated in the body. This does not necessarily lead to physicalist theories of the nervous system, or at least this drift should be avoided. The notion of simulation is important, to avoid too narrowly identifying the brain with a computer that calculates. It also must be possible to distinguish, as Jeannerod proposes, the semantic aspect from the pragmatic aspect of the control of movement.[26] If the neurons of the colliculus allow a cat to catch a mouse by anticipating its future position, it is because the neurons are *sensitive* to the velocity of movement: they do not calcu-

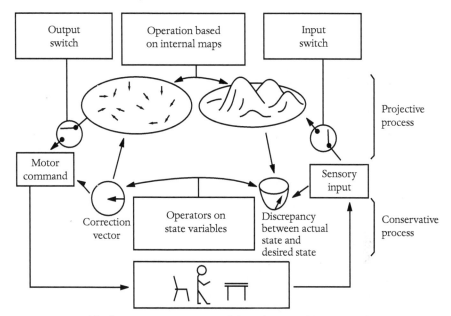

Figure 1.5. The brain activates two parallel modes over the course of an action. The first, which is the oldest, is conservative. It tends to maintain certain variables within limits defined by the intended action. It functions in closed circuits that couple the sensory receptors to motor commands either via reflexes or through central structures that specify preconfigured motor programs in a repertoire of interconnected synergies. It also uses motor primitives—gestures with recognizably distinctive features organized in specialized groups of neurons in the brain (the basal ganglia, the motor and premotor cortex, the cerebellum, and so on). In this mode, the brain functions like a controller to correct for mechanical variables such as velocity and force. The second mode of control evolved more recently. It is projective, using internal maps to simulate movement without carrying it out. These maps are not the same as geographic maps. Although some (like those of the colliculus) are genuine environmental maps, the concept of map here means that the groups of neurons in a particular region of the brain have a structured set of properties analogous to the external world. If the input and output switches (that is, the sensory information and motor activity) cease to operate, this mode can still function as a simulator, as in a dream.

late a velocity. This property allows the cat to survive without having to do calculations. Roboticists also began by making robots that simulated by calculating, but these robots are slow. The new generation of robots is endowed with properties of adaptation and prediction. These robots use synthetic work parameters that simplify calculations and that permit internal simulation of processes for finding new solutions. I will come back to these ideas in Chapter 8.

The fact that perception and simulation of action are embedded in each other has been demonstrated recently in an experiment concerning mental rotation. Imagine that you wish to compare two identical or different but similar objects that are tilted with respect to each other by a certain angle. If you are asked to decide whether they are different or identical, it will take you a certain time to answer (which can extend to seconds). The value of this mental processing time is proportional to the angle between the two objects. This fact has led psychologists to suggest that the brain performs a mental rotation of the objects to compare them more easily. We have shown that if, during the mental time of comparison, the subject is asked to perform a movement with his hand in either the same or a different direction than the mental rotation, a change in the time to respond occurs.[27]

Now that I have provided a general idea of the field, why not tackle the first problem that arises: How do we get from a physiology of sensation to a physiology of perception that acts, judges, and decides on the basis of action and of the future consequences of action?

2

THE SENSE OF MOVEMENT: A SIXTH SENSE?

> Praised be our senses. Magically, they make manifest for us various vibrations as light, color, and heat—not just sound; they capture the forces of chemical attraction as flavors or odors. In short, for our experience of all the enchanting splendor and invigorating freshness of the physical world, we are indebted to the senses and the symbols that serve as their intermediaries, through which we gain news of these things . . . Think of the months, perhaps the years of work that it would cost a physicist to define the color tones of a landscape seen only one time, which our eye apprehends with a single look and is immediately ready to exchange for another image.
>
> —H. von Helmholtz

In the first chapter, I claimed that the senses are not passive receptors. Now I will show that anticipation really is an essential characteristic of their functioning, and that even at this level certain mechanisms allow internal simulation of movement. I will pay particular attention to senses that are not very well known: muscular proprioception and the vestibular system.

Indeed, to the five traditional senses—touch, sight, hearing, taste, smell—we must add the sense of movement, or kinesthesia. Its characteristic feature is that it makes use of many receptors, but remarkably it has been forgotten in the count of the senses. We seem to be stuck at Aristotle's assertion: "In the Psychology we have given a general account of the objects corresponding to the particular sense-organs, to wit colour, sound, smell, flavour, and touch."[1] By what twist did language suppress the sense most important to survival? A plausible explanation is that it is not identified by consciousness, and its receptors are concealed. It seems normal to us to recognize the movement of our

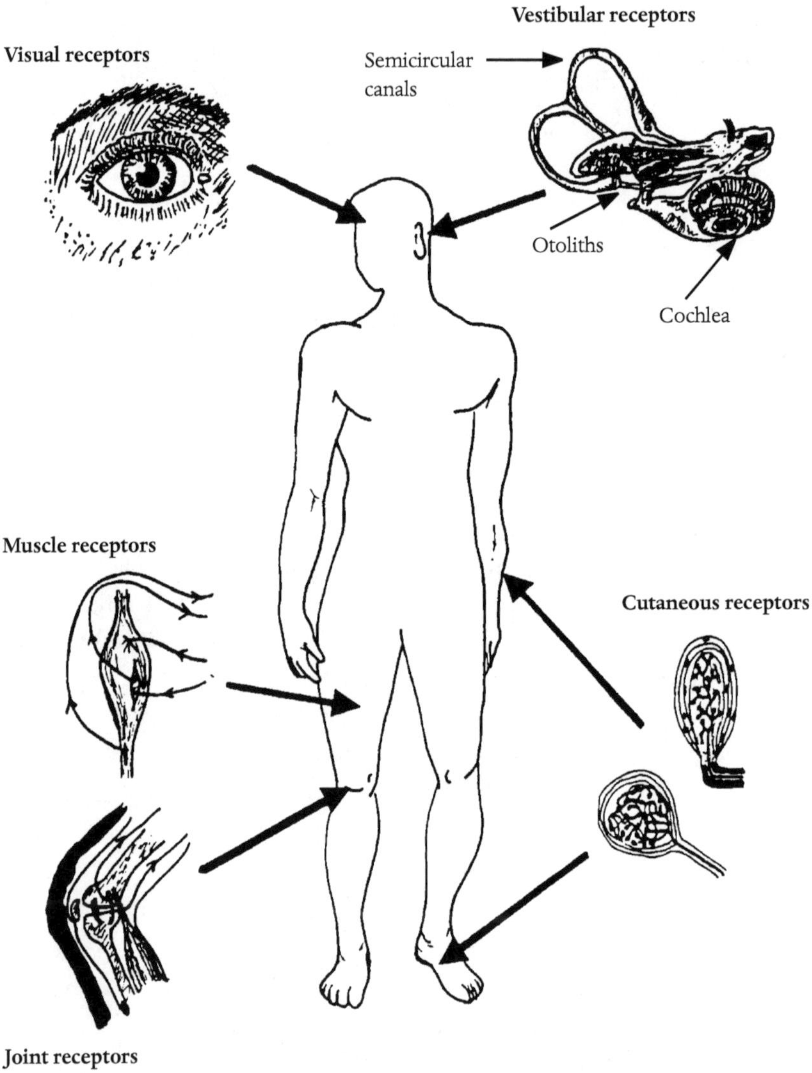

Visual receptors

Vestibular receptors

Semicircular canals

Otoliths

Cochlea

Muscle receptors

Cutaneous receptors

Joint receptors

Figure 2.1. Sensory receptors contribute to the sense of movement. The receptors located in the muscles (neuromuscular spindles, Golgi tendon organs) and the joints (joint receptors) all detect movements of the limbs. This is proprioception. Cutaneous receptors detect skin pressure and friction from contact of the limbs with each other and the external world. Vision detects shifting images of the visual world on the retina, the position of objects in space, shape, color, and so on. Movement of the head in space is detected from a combination of visual and proprioceptive cues, but it is also detected independently by the vestibular receptors. These receptors are situated in the inner ear near the cochlea and comprise three semi-circular canals (horizontal, anterior vertical, and posterior vertical) in three perpendicular planes and the otoliths (the utricle, in the plane of the horizontal canal, and the saccule, in the plane of the anterior vertical canal).

arm or vertical direction, but nothing indicates that we have receptors for stretch and force in our muscles, for rotation in our joints, for pressure and friction in our skin, and that in each inner ear we have five receptors (the utricle, the saccule, and the three semicircular canals) that specifically detect movements of the head (Figure 2.1).

Proprioception

THE SENSE OF POSITION AND OF VELOCITY

In all the muscles of our body, there are very specialized little fibers that run parallel to the muscle fibers themselves. These are the neuromuscular spindles. They are sensory receptors that detect muscle stretch.

Let us conduct an experiment. If a small vibration is applied to a muscle (for example, the biceps), around 50 to 100 Hz, oscillating at an amplitude of only about a few hundreds of microns, we will observe two phenomena.[2] First, if the arm is free to move, we will notice a reflex contraction of the vibrated muscle. Now, if we immobilize the arm on the edge of a table and close our eyes as the biceps is vibrated, we will experience an illusion of uncontrollable displacement of the arm, characterized by two different perceptions: change of *position* in space and *velocity* of displacement. The arm is perceived at the same time to be somewhere else and in motion. This illusion of movement is often accompanied by remarkable muscle activity: the opposing muscle (called the antagonist) to the vibrated muscle is now activated. In other words, the brain activates the muscle, perceived to be in motion, as if it were the perception (and not the sensation triggered by the receptors) that leads to the contraction.[3]

What are the mechanisms of these two phenomena, the reflex contraction and the conscious perception of illusory movement? Sherrington elucidated the first one: the contraction is due to a very short reflex, called myotatic reflex, that connects neuromuscular spindles and agonist muscle.[4] When a muscle is stretched, the spindles and their fibers are stretched as well, and they emit discharges proportional to the length and velocity of muscle stretch in the sensory fibers. The structure of these spindles is extremely refined and enormously diverse: some are more tonic, others more dynamic.[5] The discharge makes its way to the medulla, where it induces the activity of the muscle via very short neural relays. This reflex makes it possible to resist the tension exerted on the limb. The spindles detect not only the length of the muscles but also the first derivative of length, which is velocity. In other words, the spindle discharge is maximal before elongation is complete because velocity increases more rapidly than length. This anticipation allows compensation for the delays between α motor neuron discharge and muscle contraction. A muscle actually contracts very slowly. It attains its maximum force about 80 milliseconds after a neural command. Eighty milliseconds is a very long time if you are trying to get away from a predator. In the language of servo systems, the phase lead of the spindles compensates for the low-pass filter characteristics of the muscles. The result is that when a muscle is stretched, the time it takes to contract exactly compensates the time it takes to stretch owing to a dynamic anticipation made possible by the properties of spindles. Typically, the more a receptor returns a high-order derivative of the mechanical variable it detects, the greater the likelihood the brain will anticipate this variable itself at some future time. Practically all the sensory receptors detect the derivatives of the variables that specifically activate them. Evolution obviously selected receptors capable of predicting the future.

I have explained the first phenomenon, contraction induced by vibration: this contraction excites the sensory endings of the spindles just as natural stretching would; it *simulates* stretching and thus induces a myotatic reflex, which explains the contraction.

What is the mechanism behind the illusion of displacement of the arm? This illusion, too, is due to stimulation of the spindles by the vibration. The cerebral cortex, which receives information about the state of the spindles, interprets this activity as elongation of the vibrated muscle. It thus works out a perception of displacement and activates the muscle that corresponds to this percept. This allows the cerebral cortex to situate the motor response in the global behavioral context of the actions of the subject. When the same vibration is applied to the same muscle in different contexts, for example, when the subject is seated or standing up, leaning on the vibrated limb or free to move, completely different illusions result.[6] The brain assigns a status to sensory information based on its assessment of the general state of the body. We are very far from a simple potentiometer.

The spindles are not passive. The brain can modify their sensitivity. They are accessible to motor intention. Indeed, they are endowed with small muscles at their extremities that are innervated by motor neurons of the medulla called γ motor neurons, themselves under the control of volition and motor intention via corticospinal pathways. Activation of these γ motor neurons enables the brain to *simulate* muscle contraction or stretch. Their discharge or their silence induces activity or rest in the neuromuscular spindles as if the movement had actually taken place, despite the absence of any externally imposed movement of the muscle or movement induced by the α motor neurons that normally spark muscular contraction.

The brain thus uses γ motor neurons as a tool for *modulating* sensory information at its source, to adapt it to the requirements of movement, and to simulate movement. During locomotion, the brain induces modulation of the dynamic sensitivity of the spindles.[7] I studied this property because, being an ergonomist at the time, I had observed that a seated person subjected to low-frequency (4 Hz) mechanical vibrations experienced significantly amplified body oscillations. I wanted to understand the origin of this resonance. Wishing to apply variable forces to human subjects, I had the idea—I think for the first time—of using a motor run on magnetic powder that made it possible to generate variable forces.[8] This technology has since been replaced by torque motors. A person's arm or head, or an animal's paw, was attached to the machine to study the nerve responses to the application of sudden or oscillating forces. This machine allowed me and my colleagues to describe, in accordance with Viviani, the dynamics of the cervical column in humans and to show that tilting movements of the head employ two centers of rotation in different ways according to the velocity of the movement. One day we used the machine to establish the modulation of dynamic γ commands in the course of animal locomotion. We applied sinusoidal stimulation to a muscle tendon during locomotion and recorded the activity of the sensory fibers in the neuromuscular spindles of the same muscle. The goal was not to study the response to the stimulus, but to test, by stretching, the state of the receptors. We noted at the time that spindle discharge in response to stretching entailed a phasic component that varied with each walking cycle, evidence of modulation by dynamic γ motor neurons. Since then, a body of research has confirmed this modulatory sensitivity of the sensory receptors. At each phase of the step, the brain adjusts the sensitivity of its sensors.

The first example of a mechanism of *anticipation* is thus modulation of dynamic properties of the neuromuscular spindles according to the movement required by the task and the context. It is worth noting that this modulation could, in theory, allow the brain to simulate a movement without executing it. Indeed, if the spindles are activated directly by the central nervous system, through contraction of the γ fibers, they induce, as in the case of a vibration, a percept of movement.

THE SENSE OF TOUCH

Among the least studied of the senses and yet one of the most important is tactile sensitivity. Our skin contains numerous receptors that are sensitive to different aspects of contact with the external world. Some—Meissner and Pacinian corpuscles, for example—detect pressure and are endowed with more or less phasic properties: some are sensitive to rapid variation in pressure, others to its prolonged duration, still others, attached to hairs, are sensitive to friction and stroking, and are activated when the hairs bend. There are receptors that detect heat and cold and constitute a class of thermoreceptors. And others transmit painful stimuli, called nociceptors. How rich and how extremely subtle the sense of touch is! In this case as well as in that of the other sensory apparatus, nature has segregated the variables that the different receptors detect and has made some capable of anticipation, since they detect rapid variations (derivatives) in pressure or velocity of an object sliding over the skin.

The distribution of these receptors on the skin is very uneven. They are concentrated on the parts of the body that are the most involved in tactile perception (the mouth, hands, genitals, thighs, soles of the feet, and so on). Their representation in the cerebral cortex is also uneven and reflects these functional differences, which vary according to species. Every region of the skin activates particular neurons of the somatosensory cortex: in other words, the neurons in such regions have receptor fields, as vision does. For a long time these receptor fields were thought to be fixed and unchanging. However, recent research shows that when a finger is accidentally severed, or when a small section of a rat's whiskers is anesthetized, the cortical projections of the tactile sensors are very rapidly reorganized. This reorganization affects not only the cortex. Simultaneous recording of the neurons of the sensory pathways that carry information from the rat's whiskers shows that reorganization affects the thalamus as well as the cortex. It also depends on the use made, for example, of one finger as opposed to another.[9] Moreover, brain imaging has shown that after amputation of a limb, the cortical representations of the parts of the limb are reorganized.[10]

Similarly, the transmission of tactile information to the brain is subject to an intense process of selection. Tactile transmission is facilitated when the rat pays attention to which part of its body is being stroked. So-called presynaptic inhibitory mechanisms can modulate or block tactile information even before it enters the medulla, and other centralized mechanisms perform this filtering at the level of the first neural relays in the medulla or in the brainstem.

Tactile receptors contribute to the perception of force, but they cannot distinguish between the force that I am exerting on the weight I am holding in my hand and the force the weight exerts owing to its mass and movement, since these two forces are equal if the weight does not slip out of my hand.

Other centralized information like the sense of effort or recognition of the motor command I send to the muscles (what is called corollary discharge of the motor command) resolves this ambiguity.

The Sense of Effort: Does the Brain Use Composite Variables to Control Movement?

Some receptors detect the effort exerted by muscle on a joint: these are the Golgi tendon organs, encapsulated receptors located at the junction of muscle and tendon. They fire when force is applied to the extremity of the muscle, even at constant duration. They seem thus to be sensors of force or of variation in force. But does that mean that force is really perceived and detected directly in this way? It has not been proved. What these Golgi receptors transmit to the medulla are signals linked to *changes* in force.[11] The latter is then reconstructed in the medulla by the combination of this information with that concerning the length and velocity of muscle stretch. According to the theory of nonlinear control of robotic movement developed by Slotine, composite variables (a mixture of variables and their successive derivatives) can be used by the nervous system.[12] The advantage of these composite variables is that nonlinear problems of control are made linear and considerably simplified. I and my colleagues have proposed the idea that composite variables are also used by the brain.[13]

It is not yet clear how force is represented in the firing rate of the neurons of the motor cortex. It is doubtful that the so-called pyramidal cortical neurons, which control muscle contractions, each encode the force provided by individual muscles. The neural basis of the sense of effort remains to be discovered. It appears that to detect effort the brain uses the motor command itself. Indeed, a person whose arm is numb (because the sensory receptors have been anesthetized) can still produce an effort of a given magnitude. However, the debate between the two theories—*peripheralist,* which stresses the role of sensory information, and *centralist,* which stresses the role of internal messages—remains unresolved. The respective roles of these two components in perceiving force probably depend on the task and the context in which the movement occurs. A recent experiment elegantly demonstrated the central nature of the sense of effort: A subject was asked to accomplish a movement while walking; then he had to stand still and imagine the movement.[14] It takes the same amount of time to mentally execute a motor task as to actually execute it. This "isochrony" is used by coaches in training champions. If the subject is asked to take the same walk while imagining that he has a weight on his shoulders, the length of the mental journey is modified. Moreover, this virtual weight leads to changes in cardiac rate and several other vegetative indices, as if the effort really were greater.[15] Imagining the effort detected by the receptors thus has the same consequences as its reality.

Do the Golgi receptors, like the neuromuscular spindles, have an efferent control that allows the central nervous system to regulate them through anticipation? So far, it would appear not. On the other hand, as soon as they enter the medulla, their axons are subjected to control by a presynaptic inhibitory mechanism that makes it possible to block transmission of neural messages in the first relays. Via the γ motor neurons, the brain can regulate the stiffness of muscles and thus the way in which they will respond to a force applied to them.

The Vestibulary System: An Inertial Center?

The brain is thus capable of recognizing movements of body segments, selecting and modulating information supplied by the receptors at the source. But proprioceptive receptors can themselves only detect *relative* movements of body masses. They are inadequate for complex movements—locomotion, running, jumping—where the brain must recognize *absolute* movements of the head and body in space. To do this, nature has, since the earliest stages of evolution, used inertial cues, so called because they detect the forces of inertia. These are the vestibular receptors.

THE DISCOVERY: SCARPA, FLOURENS

The Italian anatomist Scarpa deserves the credit for having described, in 1789, the anatomy of the three semicircular canals and the two otoliths located in the inner ear.[16] Flourens subsequently discovered the function of these receptors in 1832.[17] A student of Cuvier at the Museum of Natural History, then professor at the Collège de France, Flourens studied lesions. His approach to the brain ran counter to the most advanced ideas of his era: he disagreed with Gall's localizationalist phrenology and rejected Darwin's theory of evolution. But he was a good experimenter, attentive to the unexpected. One day he introduced a lesion in the inner ear of a pigeon, expecting to get an auditory disturbance; but the animal lost its balance, went around in circles, and fled to the dark corners of the room. Flourens very rightly inferred that the ear does not function only to hear, and he thought, this time incorrectly, that he had discovered a "moderating apparatus of cerebral activity."

Fifty years passed before the laws that govern the directional sensitivity of the semicircular canals were formulated. Around 1910, these receptors attracted the attention of the physicists Mach and Helmholtz, polymaths at a time when that was not so unusual. They understood that the canals detect angular acceleration and the otoliths linear acceleration, and that together they constitute a genuine gravito-inertial receptor system.

The semicircular canals are essentially inertial receptors. They function without a base and, consequently, are as advantageous to the bird flying as to the lion capturing its prey or the monkey jumping from branch to branch. But this ostensibly advantageous characteristic masks a weakness. Indeed, in 1911 Einstein showed that an inertial sensor cannot distinguish between gravity and an acceleration of the same magnitude (9.81 meters per second) due to a movement. The first physicists who became interested in the vestibular receptors discovered the ambiguity of vestibular signals, which confuse head tilt, acceleration, and deceleration. Visual inputs are necessary to resolve this ambiguity, which happens to account for many airplane accidents and otherwise unexplained falls (see Chapter 13).

A Basic Euclidean Frame of Reference

The *semicircular canals* are rings of a sort filled with a viscous liquid—the endolymph. At the end of each canal is a bulge, the cupula, that contains the hair cells that constitute the vestibular receptor. These hair cells are sensitive to the pressure that the endolymph exerts on them. I will not delve into the details of the mechanism by which this pressure is transformed into nerve discharge, but it is important to understand which mechanical variables produce the variations in pressure that activate the receptors: these are angular accelerations of the head—in other words, changes in velocity. Thus, these receptors detect the second derivative of angular displacement; some vestibular receptors are even sensitive to the third derivative of movement, or jerk (expressed in degrees per second per second).

When you turn your head sharply, the endolymph lags behind the head and the membrane of the canal that contains it. This lag is due to the forces of inertia, defined by the basic relationship of dynamics: to put a body into movement takes a force proportional to the product of its mass times acceleration. These forces are connected to the acceleration of the head. The greater the acceleration, the greater the movement of the endolymph in the tube of the canal. The semicircular canal is thus a sensor for angular acceleration of the head.

It also detects braking, which is another kind of acceleration (negative). Try the experiment yourself. Stand up with your eyes closed, and turn around four or five times at a constant speed. Then, stop yourself abruptly by tapping hard with one foot so that the stop is clean. Keep your head still, and you will distinctly perceive an illusion of rotation in the opposite sense. This sensation derives from the fact that the cupula of the horizontal semicircular canal, which is subjected to the forces of inertia during a sudden deceleration, deviates slightly and takes several seconds (around 10) before returning to its position of rest. If you lay your fingers lightly on your closed eyelids, you can even feel the movement of your eyes caused by the vestibular stimulation. This movement is called nystagmus.

The sensitivity of these receptors is exquisite—they detect accelerations of a few tenths of degrees per second per second—and depends on the frequency of movement. Two words about this concept. When we walk, our heads oscillate at a frequency (number of oscillations per second) of around once per second, or a frequency of 1 Hz. When we run, the frequency increases. When we are standing still, waiting for the bus, our heads oscillate very slowly, at around one oscillation every 10 seconds; a frequency of 0.1 Hz. This terminology can also be used in the case of simple movements. For example, if you turn your head once to the left, this single rotation can be regarded as one cycle of a periodic movement. The rate of oscillation of this periodic movement can be determined by doing a Fourier analysis of the movement signal. Low frequencies correspond to slow movements and high frequencies to rapid movements. Each movement contains a certain number of frequencies that determine what is called the band of frequencies. The considerable advantage of this description is that it permits the description of the dynamic properties of movement using only a few numbers, the receptors that detect the properties, and the motor commands.

The geometry of the semicircular canals is adapted to each species.[18] The fluid that transmits the forces of inertia to the sensory cells of the canal—the endolymph—has remained unchanged throughout the millions of years of evolution. Consequently, the diameter of the canal has increased only slightly compared with the size of animals. Why? The geometry of the semicircular canal is adapted to the rapidity of the movements of the species, that is, to the frequency of physical movements (between 0.1 and 1 Hz). Indeed, in going from the mouse to the elephant, it was necessary to attenuate the frequencies to which the receptor is sensitive, since the elephant moves a lot more slowly in walking than the mouse. Geometry solved a mechanical problem over the course of evolution. What extraordinary adaptation!

Thus, the semicircular canals are sensitive to dynamic aspects of head movement: acceleration and its derivative, jerk. The analysis of feedback and control systems, introduced in neurophysiology by Hartline, Terzuolo, and others in the 1960s, provided a quantitative method for evaluating receptor dynamics: a nerve fiber was recorded while the animal was rotated on a turntable that was subjected to sinusoidal movement. Varying the frequency of oscillation (the number of oscillations per second) made it possible to obtain the response of the receptor to movements of differing rapidity. It just required measuring the amplitude of the modulation of neural discharge and the temporal relation between discharge and movement (phase).

Consider the case of rotations in the horizontal plane, a movement you make hundreds of times each day in turning your head. When an animal is subjected to sinusoidal horizontal rotations alternating clockwise and counterclockwise and whose frequency of repetition goes from one cycle every 100 seconds (0.01 Hz) to five per second (5 Hz), the neurons of the vestibular receptors fire with a frequency that also varies in a sinusoidal manner. The amplitude of their modulation is weak for the slowest movements (0.01 Hz), then increases with frequency, remains stable between 0.1 and 1 Hz (the frequency of everyday movements), then increases again beyond that. The delay with which the neurons fire in relation to the displacement of the head varies with the frequency of oscillation. They are in phase with acceleration for slow movements, which is normal since the semicircular canals are receptors for angular acceleration; then they are in phase with the velocity of the head for the frequency of natural movements (between 0.1 and 1 Hz) and with very rapid movements, like those made when falling down; and they come back into phase with acceleration, which makes possible the rapid reactions needed for getting back up.

Generally, the signals from these receptors enable anticipation of the future position of the head owing to their sensitivity to derivatives (jerk, acceleration, and velocity) of movements of the head. These receptors not only transmit signals to the brain, they are also subjected to centrifugal control from the central nervous system that modulates their properties. As a matter of fact, efferent neurons project from the brainstem to the receptors of the sensory cell. The exact mechanism of centrifugal control is not yet completely understood. Nor is it the only one: interactions between receptor cells enable retrocontrol of sensory messages; that is, rapid and subtle adjustment of the sensitivity of the cell to the origin of the sensory signal.[19] The sensory message is thus processed by the brain at its source. The same two major principles already described for the muscle receptors also apply to the canals: detection of derivatives of movement that enable anticipation, and modulation of the message at the source by the central nervous system. One of the basic functions of the vestibular system is thus to detect movements of the head in a Euclidean reference system.

Very early in evolution the canals appeared in certain animals: the lamprey, one of the oldest among the fishes, has two sets. Mammals have three of them on each side of their heads. The canals are situated in three approximately perpendicular planes, though not the ones you might guess merely by observing animals from outside. One of the planes corresponds to the horizontal plane of the head. To locate it in humans, just look at a person from the side and plot a line from the meatus of the ear to the external edge of the eye. The horizontal canal is in a plane that overhangs this line by 20 degrees (see Figure 2.4).

The two remaining planes of the canals are at 45 degrees with respect to the frontal and saggital planes of the body.[20] Why is this anatomical detail important? Because the three planes so constituted form a basic egocentric frame of reference with respect to which, as I will discuss later, our entire perception of movement in space is organized. The geometry of the canals dictates the organization of the cerebral analysis of visual movement and perhaps also other movements. It might yet prove to be the basis of Euclidean geometry.

Poincaré was interested in this question. In his book *La valeur de la science* (The Value of Science), he commented on the theories that pitted Cyon, on the one hand, against Mach and Delage, on the other. According to Cyon, "The three pairs of canals would have as sole function to tell us that space has three dimensions. Japanese mice have only two pairs of canals; they believe, it would seem, that space has only two dimensions . . . The lampreys, having only one pair of canals, believe that space has only one dimension, but their manifestations are less turbulent."[21] Poincaré objected to this view and maintained, along with Mach and Delage, that the semicircular canals "contribute, therefore, to inform us of the movements that we have executed, and that on the same ground as the muscular sensations."[22] For Poincaré, "the sense-organs are designed to tell us of *changes* which happen in the exterior world. We could not understand why the Creator should have given us organs destined to cry without cease: 'Remember that space has three dimensions,' since the number of these three dimensions is not subject to change."[23]

Elsewhere Poincaré asks an important question: Why are Euclidean movements perceived as actual movements? He says, "Our mind adapted itself to the conditions of the external world through natural selection; it adopted the geometry most advantageous or, to put it another way, most convenient to the species. Geometry is not true; it is advantageous."[24] "One geometry cannot be more true than another; it can only be *more convenient*. Now, Euclidean geometry is, and will remain, the most convenient: 1st, because it is the simplest, and it is not so only because of our mental habits or because of the kind of direct intuition that we have of Euclidean space; . . . 2nd, because it sufficiently agrees with the properties of natural solids."[25]

What are the relationships between geometric space and representative space? Are their properties similar? "It is often said," Poincaré writes, "that the images we form of external objects are localised in space, and even that they can only be formed on this condition. It is also said that this space, which thus serves as a kind of *framework* ready prepared for our sensations and representations, is identical with the space of the geometers, having all the properties of that space."[26] However, he argues, geometric space is continuous, infinite, three-dimensional, homogeneous (all its points are identical), isotropic (all the straight lines that pass through the same point are identical). In contrast, representative space (which is primarily visual space) is two-dimensional (retinal space); it becomes three-dimensional owing to convergence and accommodation; it is not homogeneous, because the part of the retina that is most sensitive to shape, the fovea, is nonhomogeneous with the periphery. It is thus not isotropic.

Poincaré concludes from this that representative space is only an image of geometrical space distorted by the rules that govern how our perceptive apparatus works. "Thus we do not *represent* to ourselves external bodies in geometrical space, but we *reason* about these bodies as if they were situated in geometrical space."[27] Modern psychophysics confirms this intuition.

Poincaré introduces a fundamental notion that is essential to my own theory: "To localize an object simply means to represent to oneself the movements that would be necessary to reach it. I will explain myself. It is not a question of representing the movements themselves in space, but solely of representing to oneself the muscular sensations which accompany these movements and which do not presuppose the preexistence of the notion of space."[28] But then, if we cannot represent geometrical space for ourselves, and if our representation of space is in fact only what I call a simulation, that is, the movements that we need to make to travel it, the question becomes one of knowing how the idea of geometric space could even have come about. We must, says Poincaré, look at the way we evaluate how change affects an object. An object can undergo a change of state or of position. In either case, the change is signaled *"by a modification in an aggregate of impressions."*[29] By this Poincaré means something very close to what I call a configuration of states of sensory receptors. Inasmuch as position can be corrected, the brain distinguishes change of state and change of position: "We could restore the primitive aggregate of impressions by making movements which would confront us with the movable object in the same *relative* situation . . . [A] motionless being," he adds, "could never have acquired [the idea of space], because, not being able to correct by his movements the effects of the change of position of external objects, he would have had no reason to distinguish them from changes of state."[30] Sight and touch would not be able to give us a sense of space without the "muscular sense."

These processes require that objects be *rigid*. If there were no solid bodies in nature, there would be no geometry. For Poincaré, geometry is thus the description of phenomena called displacements, that is, external changes that can be compensated for by changes in our own bodies detectable by the vestibular receptors.

The Distinction between Tilt and Acceleration

The second type of receptor in the vestibular system—the *otolith*—is also an accelerometer that detects linear translations (it is even found in the spiny lobster, where it is called a statocyst). In addition, it detects the direction of gravity, and thus tilting of the head. Gravity is a particular acceleration, indiscernible, as Einstein demonstrated in 1911, from a linear acceleration (Figure 2.2). When the head tilts, a component of gravity acts on the receptor and thus supplies an indirect measure of tilt of the head with respect to earth's field of gravity. Animals obviously discovered early in the course of evolution that gravity is a very useful frame of reference, present at any point on the earth and constant during any movement. Even before crawling out of the water and having to construct mechanisms to compensate for the mechanical effects of gravity—"this sticky force," in the words of an astronaut returning from a space trip and a little sad at finding himself glued to Earth—marine animals used gravity as a reference.

In insects and batrachians, the otoliths also pick up mechanical vibrations, which permits the animal to detect the presence of a predator and to initiate flight reactions. So in these animals, the otoliths are a sort of ear that is sensitive to very low-frequency vibrations. In fish, birds, and mammals, the otoliths have reached a high degree of refinement and simultaneously detect both linear acceleration and static tilt of the head. They are composed of two cavities—the utricle and the saccule—that contain crystals floating in the same fluid (the endolymph) as that of the semicircular canals (Figure 2.3; see also Figure 2.2). When the head is displaced in translation, force is exerted on the crystals owing to the linear acceleration of the head. This force induces a tiny

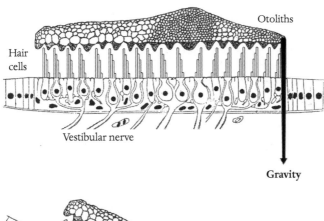

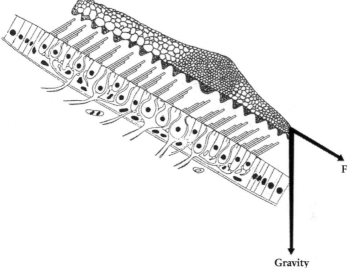

Figure 2.2. The otolithic receptors. The upper part of the figure shows schematically the mass of crystals floating in the endolymph, which together constitute the otoliths. The endolymph is not shown. The hairs of the sensory neurons are situated beneath the crystal mass. The lower part of each drawing depicts the base of the receptor (called the macula). Ionic changes triggered by movement of the hairs result in discharge of the sensory cells that travels along the vestibular nerve to the brain, which is thereby informed about the magnitude of acceleration of the head. In this case, gravity has no effect on the receptor, because it is perpendicular to it. If the receptor is tilted, the gravitational component is in the plane of the macula. This receptor indicates both acceleration and tilt of the head. The receptor shown here is the utricle, which detects horizontal accelerations of the head; another receptor, called the saccule, is located perpendicular to it and detects accelerations and gravitational force in the vertical plane.

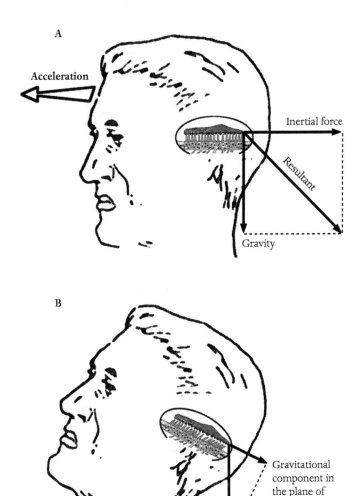

A

Acceleration

Inertial force

Resultant

Gravity

B

Gravitational
component in
the plane of
the otoliths

Gravity

Figure 2.3. Ambiguity results when the head is simultaneously accelerated and tilted. *(A):* If the head accelerates forward, a backward inertial force is exerted on the crystals of the utricle. The combination of this force and the downward gravitational force produces a resultant tilting force. The brain has the illusion that this resultant is gravity. And because it uses gravity as a fixed reference with respect to the earth, the brain deduces that the head is tilted. *(B):* When the head is motionless but tilted backward, a gravitational component in the plane of the receptor gives the impression of a forward acceleration. (The size of the receptor has been considerably enlarged.)

displacement of the crystals, which leads to activation of the vestibular nerves. The same activation is produced by tilting the head. In this last case, tilting induces a modification of the magnitude of the component of gravity acting on the crystals. They are thus both accelerometers and "tiltometers." One could simply say that the saccule (the sensor of vibrations in insects and batrachia), whose plane is close to that of the vertical anterior semicircular canal, detects vertical linear accelerations of the head (when you jump up and down with your feet together, for example), and that the utricle, whose plane is close to that of the horizontal semicircular canal, detects horizontal linear accelerations of the head (when your car brakes or accelerates, for example).

These two kinds of vestibular receptors, the canals and the otoliths, detect movement of the head in space without needing a reference point. They constitute a genuine inertial system like those mounted on board airplanes. Moreover, they use gravity as a reference to supply the brain with information about the static tilt of the head. In microgravity—in space stations—these receptors can thus always detect angular or linear accelerations of the head, but can no longer rely on gravity for detecting angle of tilt. The brain has to reinterpret the information transmitted by these receptors to be able to compare them with sensations from other proprioceptive and visual receptors.

Finally, let me reiterate a point that is often poorly understood: vestibular receptors are not only sensors of movement. They also signal immobility. Indeed, these sensors have a basic discharge whose lack of variation is interpreted by the brain as immobility. They are essential for assessing what is called the subjective vertical (see Chapter 4).

Although they make it possible to recognize head movements, vestibular receptors alone are not enough, because of the ambiguous information they provide. For example, they cannot distinguish between acceleration of the head in one direction and braking in the other direction. Tactile or visual information helps to resolve this ambiguity. This clearly shows how perception requires cooperation among the senses.

Another ambiguity that has already been mentioned is the distinction between the angle of tilt of the head and linear acceleration. This is known as the *problem of gravito-inertial differentiation,* and I will refer to it again in Chapter 13. To this problem nature found a solution, called frequential filtering. It consists in separating the two components of movement at the level of the first sensory relays, in the so-called vestibular nuclei and perhaps even at the level of the fibers that connect the receptors to these first relays. In fact, in the vestibular nuclei there are two kinds of neurons. Some respond especially to slow variations in acceleration, but are indifferent to rapid movements. The solution is an elegant one, because detecting the angle of tilt of the head involves slow neurons, and detecting acceleration involves rapid neurons. Others are highly activated by very rapid movements of the head and only slightly by slow movements. Differentiation of gravito-inertial forces is resolved by segregation at the very level of capture of sensory information.

This very early capacity for segregation is a basic feature of sensory analyzers. I will discuss it further in Chapter 3.

A FUNDAMENTAL DISSOCIATION: ROTATION AND TRANSLATION

One of the most puzzling questions raised by the study of the properties of the vestibular system is that of dissociation between detection of *rotations* by the semicircular canals and detection of *translations* by the otoliths. Does the geometrical and kinematic distinction between the concepts of rotation and translation have a biological basis? It is tempting to think so.

More generally, the nervous system, of which the senses are an extension, actually begins to segregate information from the initial inputs. Detection of angle, length, and force is dissociated at the level of the muscle and joint receptors. In the case of the vestibular system, rotation and translation are dissociated at the level of the receptors. Next, the first central sensory relays begin filtering out inertial movements and static tilt of the head, as I discussed earlier with respect to the function of otoliths and their perceptual ambiguities. Segregation also happens very early in the visual system (see Figure 3.1). For example, already at the level of these first relays—the lateral geniculate nucleus—a definite segregation occurs between two pathways. The first is the so-called magnocellular pathway, whose neurons are insensitive to color and sensitive to movement, with high sensibility to contrast and a weak capacity for distinguishing visual elements that are close together (low spatial resolution)—in other words, favorable for transmitting information about movement. The second pathway is the so-called parvocellular layer, whose neurons are sensitive to color, slow to respond, and have weak sensitivity to contrast and a high spatial resolution. This segregation is enhanced at the next level of the primary visual cortex. Visual information arriving via the deep layers, which are different for movement and color, is transmitted to the superficial layers where aggregates of neurons sensitive to shape and color are found, whereas information about movement is transmitted to the next cortical relay in the mediotemporal (MT) lobe, also called V5. This segregation is maintained in different forms in the visual areas V2, V3, and V4, the last of which, for example, is specialized for processing color. Later on I will show that information about visual movement borrows yet another pathway called the accessory optic pathway, where geometric segregation happens early on.

It is as if the senses were specialized organs for detecting variables pertinent to the survival of each species, nature having put in place very peripheral mechanisms for selecting these variables and dissociating them. Once the brain has these bits of information, it reconstructs them into more complex messages and encodes them for adaptation to various central operations needed for perception and action. Gibson stressed the idea that the senses are analyzers that do not merely detect the physical variables likely to stimulate their endings, but that in a way they have integrated into their functioning the elements of nature important for the animal's repertoire of actions. The early segregation of sensory information by the senses and their first analytical neural networks supports this idea and does seem to suggest that the transduction of physical variables (light, sound, pressure, etc.) responds to preexisting questions that the nervous system asks the world.

The Functions of the Vestibular System

POSTURAL STABILIZATION

The vestibular system ensures postural stability. If we trip over a root, our head accelerates, and the canals and the otoliths are activated, exciting the so-called vestibulospinal reflex pathways that initiate the reactions needed to readjust posture: lifting the head and the rest of the body by means of proprioceptive reflexes whose receptors are located in the neck. I will not describe this function, known since the work of Magnus[31] and Rademaker[32] at the beginning of the century and studied by the Italian school of neurophysiology during the 1950s, the Japanese school of Fukuda, and by Roberts at Edinburgh. What I will stress is the anticipatory properties of these reflexes. The anticipatory power of the vestibular receptors comes from the fact that they detect acceleration. Now, acceleration is maximal from the beginning of a fall or loss of balance. The reflexes for getting back up are thus very rapid since they are initiated as soon as the perturbation begins. The conjunction of the discharge of the tactile receptors of the spindles sensitive to the velocity of stretching—for they are sensitive to variations in pressure—and the vestibular receptors sensitive to acceleration supplies the nervous system with preliminary information about the nature of loss of balance. The muscle contractions that lift the body are the result of this anticipation, and the swiftness of the postural adjustment is further facilitated by virtue of the brain's repertoire of ready-made postural reactions. These stereotypical reactions are triggered by specially configured sensory signals that set in motion what are called synergies (see Chapter 7). The vestibular receptors are initiators, not just detectors. The information they convey is a sign, not just a signal.

Gaze Stabilization

Perceptual stabilization is the second major vestibular function. Indeed, if the world appears stable when we move, it is first and foremost owing to reflexes of vestibular origin that stabilize the image of the world on the retina. These reflexes connect the vestibular receptors to the muscles of the eye. The anatomy of neural connections is such that moving the head in one direction leads to moving the eye in the other, which has the effect of suppressing, or diminishing, the shifting of images on the retina (retinal slip).

You can very easily see for yourself. Just look at a point on the wall in front of you. Close your eyes, turn your head while imagining the point on the wall, and open your eyes. You will notice that your gaze has remained on the point that you memorized. Your eye made a movement in a direction contrary to that of your head, and of the same amplitude. Its origin is vestibular: it is produced, in the horizontal plane, by the semicircular canals. But how do I know, you say, that the muscle receptors in the neck have not detected the movement of my head? To answer your question, you have only to sit in a rolling chair and ask someone to repeat the experiment by turning you to one side or the other. This time, as your eyes are closed during rotation, only the semicircular canals detect the rotation (the tactile receptors alone are not sufficient to detect a rotation angle). You will see that your gaze has remained fixed on the same point on the wall. Thus the semicircular canals are able to do the job by themselves. (However, the receptors in the neck do help during natural rotations of the head, cooperating with the vestibular receptors.)

But, you say now, how do I know that it is a reflex? Since I am introducing the memory of a target on a wall, it is very likely that I am also involving my cerebral cortex and the structures that are implicated in spatial memory. You are correct. In fact, the vestibulo-ocular reflex is nothing like a simple reflex; its amplitude is regulated by the cortex, and I will show in Chapter 3 that vestibular information is transmitted to the cortex and that the zones that receive it project in their turn to the vestibular nuclei.

Around 1930, Lorente de Nó discovered a network of neurons in the brainstem that connects the semicircular canals and the muscles of the eyes.[33] He suggested that the reflex is subtended by an arc of three neurons (Fig-

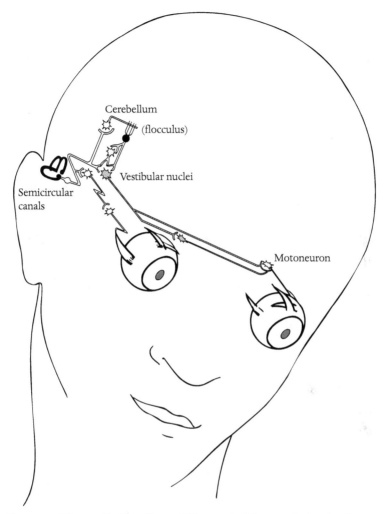

Figure 2.4. The vestibular reflex stabilizes retinal images during head movements. What is called the vestibulo-ocular reflex is the product of the network of neurons that join the vestibular receptors to the muscles of the eyes. Vestibular information is relayed to the cerebellum via the Purkinje neurons, inhibitory neurons that coordinate and modulate the reflex (black circle). The part of the cerebellum involved in rotations is the flocculus.

ure 2.4): a peripheral neuron that connects the semicircular canals to the vestibular nuclei of the brainstem; an intermediary neuron, called the second-order vestibular neuron, whose cellular body is in the vestibular nucleus and which projects to the motor nucleus called abducens (motor nucleus VI); and finally a neuron that causes the muscles of the eyes to contract (motor neuron). A second reflex pathway includes a mechanism for generating the rapid phase in the form of a cascade of neurons situated in the reticular formation. Around 1965 Szentagothai established the exquisite correspondence between each canal and each of the three pairs of eye muscles.[34] The basic anatomy of a reflex was thus pieced back together.

It was Eccles's perfection of the methods of intracellular recording that furthered the understanding of the synaptic organization of the reflex.[35] This organization turns out to be both very simple and very subtle. A network of excitatory and inhibitory synapses underlies it and ensures a very rigid kind of "push-pull" functioning. Moreover, indirect pathways alter signals of movement and produce vestibular nystagmus. Finally, circuits that traverse the cerebellum exert feedforward control on the reflex. The discovery that I made, with Baker, of the role of the *prepositus hypoglossi nucleus* as a supplementary reticular relay for control pathways further complicates the organization of this reflex, which is composed of a series of parallel pathways that interact one with the other.[36] But this is not the place to describe this remarkable mechanism.

Just when the complexity of neural organization could have trapped electrophysiologists in a profusion of details, systems theory emerged with several welcome simplifying concepts.

The Concept of the Neural Integrator

The way the vestibulo-ocular reflex operates involves a paradox that is worth analyzing to understand how the nervous system manages to handle speed and precision at the same time. As noted earlier, the vestibular receptors are accelerometers (measuring degrees per second per second) that detect movements very rapidly, since certain sensory cells are even sensitive to jerk, which is a derivative of acceleration. But the command that displaces the eye in its socket necessarily produces a rotation, specified as degrees of angular measure; so although the brain gains from an early, dynamic piece of information, it must transform it by delaying it to control the position of the eye with respect to the environment: it must be subjected to low-pass filtering, or integration. An initial filtering, or first integration (going from acceleration to velocity), is carried out at the level of the receptors themselves. In fact, they are endowed with viscoelastic properties that delay the signals in a variable manner according to the rapidity of the movement. But a second integration must take place in the brain to go from velocity to position.

A part of integration is certainly carried out at the level of the vestibular nuclei, for the response to rotation is delayed at this level. If, instead of being subjected to sinusoidal stimulation, the head is briefly rotated, the fibers of the vestibular receptors respond very quickly, and their response diminishes spontaneously after about 5 to 12 seconds once the head has stopped moving. The neurons of the vestibular nuclei do not stop firing until about 20 to 25 seconds afterward. You can feel this effect by closing your eyes and turning your head very rapidly, and concentrating on the impression of rotation that follows: you will continue to feel it for about 20 seconds. But a supplementary integration intervenes between the vestibular nuclei and the movement of the eyes.

The research on this central neuronal integrator has greatly fascinated physiologists. To avoid confusion, let me make clear that I am not talking about the concept of integration as Sherrington used it, in the sense of a complex combination of multiple signals to "integrate their messages," but rather in the literal mathematical sense of progression from the derivative of a variable of movement to its integral; that is, of acceleration to velocity or of velocity to position. The theory of servo systems, which was developed mainly during World War II to control radar and which flourished in the 1950s, was very useful for expressing these dynamic transformations in quantitative terms. Engineers use concepts borrowed from the analysis of servo systems to describe these dynamic transformations.

At least four mechanisms have been proposed to explain this integration. Lorente de Nó was the first to suggest that simply returning discharge from a neuron to itself in a so-called re-excitation or recurring excitation loop sufficed to maintain activity and produce integration. The same effect could be produced by an extended long loop action involving other groups of neurons. A third hypothesis was the intervention of inhibitory mechanisms that could produce the same effect (lateral inhibition). Finally, a cascade of successive partial integrations in a nucleus close to the vestibular nucleus, the prepositus hypoglossi nucleus, could lead to this progression from velocity to a signal of position. The question has not been resolved, although Baker and his colleagues have identified precise zones of the brainstem in the fish where these networks of neurons that enable integration are probably located. It is also possible that integration is due to so-called intrinsic properties of the same neurons. For example, certain neurons can respond to a burst of activity with a continuous discharge. This mechanism involves ion channels in the neuronal membrane that depend on the concentration of calcium ions. Settling on the right mechanism is a matter of time. It will require new methods that are only now being developed and that will permit manipulation of the properties of membranes in the intact brain.

How does the brain handle problems of geometry? Control of a reflex as simple as the vestibulo-ocular reflex does not consist merely in resolving relational problems among acceleration, velocity, and position—in other words, problems of dynamics. The six muscles of the eye and the thirty muscles of the head, situated in very varied orientations, also must be controlled by the three semicircular canals and the otoliths, which are located in three planes. Restating the problem will better indicate its complexity. Imagine that you are sitting in a boat and a wave comes and rocks the boat; your head will move in rotation with the boat. If you want to stay upright, you will have to make a movement of straightening your body, equal to and in a contrary sense from that of the boat. Your semicircular canals (ignoring the otoliths for now) detect three projections of this rotation in three perpendicular planes. Thus you have three values that represent the projection of the true rotation in a system of special coordinates whose axes are at right angles. Put simply, the rotation is encoded as covariant coordinates. The numerical values of these coordinates constitute a vector that can be represented by an arrow whose magnitude and direction are measures of the intensity of the sensation detected by the receptor.

A system of coordinates encodes the movement that the eyes, head, and even the trunk must make to compensate for this rotation due to the motion of the boat. These systems of coordinates are the directions of tension of the muscles of these different parts of the body. The brain thus has to transform the sensory information, encoded in covariant coordinates, into motor commands expressed in the systems of coordinates of the muscles and their planes of action.

Pellionisz and Llinás maintained that certain portions of the brain transform sensory coordinates into motor coordinates.[37] They observed that as there are in general more dimensions in muscular frames of reference than in sensory frames of reference, in principle the problem of transformation has many solutions. One of their solutions was the idea that perhaps the brain functions like a tensor. A tensor is a group of mathematical operators called matrices that carry out transformations between vectors. They have properties that are irrelevant to the discussion here. P. Churchland's book *Neurophilosophy* contains a summary of this theory.

Pellionisz and Llinás applied their theory to two examples: the vestibulo-ocular reflex, which stabilizes gaze, and the vestibulo-nuchal reflex, which lifts the head when the boat rocks.[38] In the case of the vestibulo-ocular reflex, they assumed that rotations of the head are encoded covariantly by the vestibular receptors and that several successive transformations precede the motor command that turns the eye in the coordinates contravariant to the six muscles of the eyes. In another model, they suggested that the cerebellum was the structure that carries out the most important transformations.[39] I will try to explain broadly the basic idea of these models. When the motor command is sent to the motor neurons of the neck muscles, it is distorted by the geometrical properties of the components of the forces (contravariant projection) and by the mechanical properties of the head and neck. To obtain a movement of the head that is in perfect opposition to that of the boat such that the vestibular receptors can detect it, one solution, invented by roboticists, is that the command signal itself be already distorted in an inverse way before it is sent to the muscles. The simplest image I have been able to come up with to represent this process is that of a casting mold, which is in a way the inverse of the object it is intended to shape. In fact, we use this strategy in many movements; for example, if we need to lift an object that is hanging at the end of a rubber band, we make a much bigger movement, because we are anticipating the elasticity of the rubber band. The advantage of the internal model that accomplishes the inverse is precisely that we do not have to produce a different motor command: it is automatically transformed. This inverse transformation is assumed to take place in the cerebellum.[40]

To sum up: The movement of the boat is detected by the vestibular receptors in three planes. It is thus represented by three vectors. This information is sent to a nerve center (in this case, the cerebellum), which transforms these signals to supply them with properties that will, in advance, compensate for the distortions the signal will undergo in the movements coordinated by the muscles, as a function of the mechanics of the limbs. The signal is then sent to the muscles; it undergoes contravariant transformations of the coordinates, and the head is stabilized.

Even if the brain is not a tensor, and if today the subscribers to this theory are not legion, the fact remains that the theory of tensors has forced a generation of physiologists to take very seriously the question of geometry and the solutions that nature appears to have found to simplify neurocomputation.[41] In addition, this theory contains an important idea, that of the internal model, which can be found today in the most progressive theories. I will introduce other examples in Chapter 8.

I would like to make a general observation, central to the premise of this book. The examples taken from the theory of servo systems and of tensors have in common that they consider reflexes and sensorimotor systems to be continuous chains of transformations that proceed from sensation to motor command. Although this approach incorporates the idea of an internal model and, thus, an important anticipatory mechanism, it gives short shrift to the influence of action on sensory processing. It is thus very far from a projective conceptualization of the nervous system.

Seeing Movement

The study of the role of vision in perception keeps an enormous number of physiologists, psychologists, and mathematicians busy doing research on cerebral sensory mechanisms. I will not attempt to review this work here. A hegemony of research and ideas concerning cortical vision has become established in the neurosciences that has overshadowed the importance of subcortical vision, which, however, preceded cortical vision by several millions of years with refinements whose subtlety we are nowhere near to understanding. This hegemony of research on cortical vision is particularly troubling because it leads to isolation of its functioning from that of other senses. Moreover, assumptions about how the visual cortex is organized—for example, its architecture in columns—have totally obscured the importance of the transversal organization of cortical functions for reasons that have to do as much with sociology and science as with the rigor of the data.

This dictatorship of vision had several causes: First, the idea that vision is the most highly developed sense in primates and in humans and that, together with language, it is what makes humans distinctive. Furthermore, the manipulation of visual stimuli is still very simple in comparison with manipulating stimuli to the other senses; so much so that it has easily allowed implementation of the dominant stimulus–response paradigm in neurobiology. In addition, vision is a fascinating model of sensory activity, because psychophysical and neural research can be pursued in parallel.

Finally, cortical vision induces a conscious perception that is obviously connected to the organ, whereas it is impossible to introspectively connect conscious perceptions or actions resulting from the subcortical visual mechanisms of the vestibular system and proprioception to the receptors that produce them. For example, if I am able to catch a tennis ball in midair practically without having to think about it, it is thanks to subcortical mechanisms of movement detection and visual prediction of the trajectory, which have nothing conscious about them.

This ignorance is at the source of the surprise provoked by the discovery of what was called blind vision. It took several years of intellectual battle before Weiskrantz could get people to accept the existence of the so-called residual vision that persists after destruction of the visual cerebral cortex.[42] One unhappy consequence of this resistance is the extremely conservative attitude of educational institutions toward visually impaired children. I describe below an extraordinary experiment in teaching speed sports to visually impaired children taken out of school and for whom sports are normally prohibited.

VISUALLY IMPAIRED CHILDREN PLAYING BASKETBALL

One day in the 1980s, I was visited by Messrs. Chaumiène, a professor of physical education from a high school for visually impaired children in Montgeron, and Leguern, a researcher at the French national institute of sports. They asked me to take a few minutes to look at a videocassette of a basketball game between children, and then asked me the following question: "One of the teams is made up of children who see normally. The other is made up of amblyopic children who demonstrate less than one-tenth of normal vision in classical ophthalmological tests, read in Braille, and do not attend normal educational institutions. For safety reasons, administrative regulations forbid these children to participate in sports. Which is the handicapped team?"

It was impossible for me to tell! Using an enriched environment, these two instructors had managed to teach visually impaired children to play ball, to fence, and so on. The faster the movement, the more the children seemed to succeed. We worked together for several years to try to understand this remarkable achievement. For me, it was an opportunity to discover many aspects of this disability. For instance, the ophthalmological files for most of these children were incomplete. We had to organize bus trips to the hospital in Créteil to get tests done. Then, thanks to the sympathetic understanding of the school administrators, whose help was indispensable, we began studies of how these children performed visuomotor tasks that we hoped would also interest other researchers in the field.

This fascinating experience convinced me that ophthalmological tests examine only a tiny fraction of visual function and, in any case, totally ignore perception of movement. This is still true today. Although millions of people wear progressive lenses that modify the apparent velocity of visual motion in a nonlinear manner on the surface of the lens, no test worthy of the name examines dynamic vision, which is the most important for coordinating gestures and perceiving three-dimensional shapes, as research on the contribution of movement to perception of curvature shows. We are working on this question in collaboration with glass manufacturers.

Perception of self-motion is among the major functions of vision. Remarkably, this function of vision was completely ignored for close to fifty years despite the pioneering work of eminent scientists like the physicist Mach.[43] Moreover, anatomists labeled it an accessory system, and it was not until 1975 that someone worked out its properties. Areas that process information about optic flow and that appeared later in evolution (like MT and MST), have only recently been studied in the monkey, following the pioneering work of Wurtz.

We have all felt the compelling illusion of moving forward when, seated on a stationary train, we observe the train on the next track start to pull away; or when, looking from a bridge at the river below, we get the feeling that the bridge is moving forward. Mach called this illusion of bodily self-motion, induced by a visual displacement, vection. At the beginning of this century, he constructed several machines intended to elicit it.

For example, he rotated rugs on a revolving circular platform to study the illusion of rotation of the body induced in subjects exposed to this stimulation, which today is called optokinetic. He deduced from his findings that in fact what was involved was a basic mechanism of detection of bodily self-motion. But the problem languished until 1970, when a psychologist, Lee, attacked it with renewed interest and opened a period fertile with discovery. Lee's original experiment consisted in hanging from the ceiling of an auditorium of his university a wooden box about 3 meters on a side.[44] One side of the box and the bottom were left open. Lee caused this hanging chamber to oscillate slowly. A subject standing still within the box had the undeniable illusion that the box was stationary and that the University of Edinburgh was rocking! Again, the situation was one of vection such as Mach had described.

Lee did not stop there. A student of Gibson, he attributed this effect to what he called the "proprioceptive function" of vision. He described it qualitatively and showed, in particular, that during the illusion, the subject's body began to oscillate and thus that the perception was accompanied by active postural readjustments. He also showed that two-year-old children could be thrown off balance by the effect and fall flat on their back.

Between 1970 and 1975, three groups generated the first quantitative descriptions of vection and the motor effects associated with it. First, in Jung's laboratory in Fribourg in Switzerland, Dichgans and Brandt studied circular vection in a horizontal plane (obtained by rotating a cylinder around the subject);[45] then Young's group in Cambridge, Massachusetts, in collaboration with Dichgans, studied vection induced by the rotation of scenes in the frontal plane, and linear vection induced by flight simulators; finally, in my laboratory in Paris, we investigated the properties of linear vection that we manipulated at the same time as vestibular perception by placing subjects on a trolley, shown in Figure 2.5.[46]

It turned out that the intensity of vection is proportional to several variables of movement in the visual world. The first is surface: the greater the surface of the visual field in motion, the stronger the intensity of vection, though a very small surface field is also likely to induce vection, as will sometimes even a simple point of light.

This idea is in line with my suggestion that perception is a function not so much of the intensity of a stimulus as of the agreement between the stimulus and an assumption the brain makes. This can be tested on an airplane before takeoff. Shortly before the plane starts to move, a few minutes go by during which everyone is anticipating a slight acceleration. It only takes a luggage trolley or other vehicle passing by in front of a window, setting in motion a tiny part of the visual field, for the movement of the vehicle to be interpreted by the brain as that of the airplane. Anticipation is crucial to the perception of self-motion. A similar kind of anticipation, which involves not vision but the otolithic system, can happen in an elevator when, thinking you have reached the floor you want, you perceive an illusory deceleration.

Another important variable for determining the intensity of vection is the velocity of scenic motion. Below a certain threshold, vection is imperceptible; above that threshold, the intensity of sensation increases with velocity up to a maximum beyond which vection disappears abruptly if velocity increases. In front of a scene that moves by very quickly, we have the impression that we are motionless. This perceptual inversion is absolutely spectacular on the highway. It is very familiar to drivers of fast cars and auto-racing champions. Above 200 kilometers per hour, instead of feeling that they are gaining on the cars in front of them, they suddenly have the extraordinary conviction that the cars in front are approaching them. I don't recommend that you try this experiment! These drivers have lost vection, that is, the sensation of self-motion, and have the illusion that they are motionless before a world that is hurtling toward them.

This perceptual inversion is very dramatic. But why does it happen when the visual stimulus is so marked? Perhaps the velocity of the visual scene is too great compared with the range of velocities of bodily self-motion. The brain can no longer interpret the retinal slip as the result of a natural displacement. The question is one of a genuine perceptual decision, not disappearance of perception.

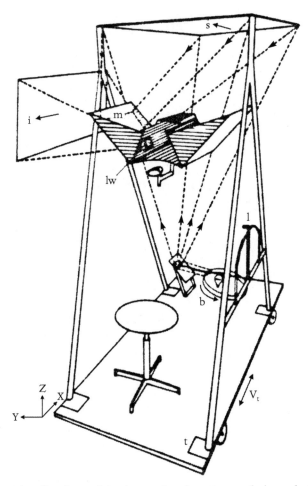

Figure 2.5. A trolley for studying interactions between vestibular and visual detection of translational movements. The subject is seated on a stool placed on the trolley *(t)*, which is driven forward or backward by translational movements that are controlled by a computer. A projector (represented by a light bulb) displays a reel of transparent film *(r)* with various shapes (squares, points, and so on). Theses images are projected by a set of mirrors *(m)* onto a screen *(s)* located above the subject. The mirrors create virtual images of the screen on each side of the subject, who sees them through lateral windows *(lw)* in such a way that he has the illusion of moving in an optical tunnel. This visual movement makes him feel that he is advancing or moving backward in a direction contrary to the movement of the visual scene (called vection). The vestibular system can be stimulated by movement of the trolley, so that the interaction between visual and vestibular information in the perception of movement may be studied.

The spatial frequency of the scene, that is, the number of elements per unit of surface, also has an influence on the intensity of vection, which is greater when you go through a village that has many visual elements per unit of surface than when you cross the desert. Road accidents are probably caused by disparities between the information supplied from both sides of the road; night, especially, gives a driver erroneous impressions of changing directions due to the difference in vection induced by each retina.

The final element is the distance of the scene in motion and, particularly, its position as backdrop or forestage. The most distant part of the environment is what determines the intensity and direction of vection. It is yet further proof of the extraordinary assessment of the perceptual context carried out by the brain.

If a subject is exposed to visual movement for a long time, vection diminishes in intensity. In the example of the highway, driving for several hours induces underestimation of vection and, thus, underestimation of the sensation of velocity. When drivers leave the highway, they have a tendency to drive too fast, which is why speed ramps exist. Adaptation to stimulus is a familiar phenomenon in psychophysics. Disappearance of vection can be interpreted as the suppression of a repetitive signal that has nothing more to contribute to the ongoing action. If that were true, the question would appear to be one of an interesting mechanism that would be fairly easy to test. But it seems to me that, on the contrary, one must assume that an active neural mechanism opposes the stimulus, the brain creating an internal movement that blocks the stimulus supplied by the external movement. When the stimulus is suppressed, all that is left is this active neural construction, which was counterbalancing it. At issue here is not a simple mental object, but a *mental movement*. An illusion that probably involves similar mechanisms is called "the waterfall." If you look at a waterfall for a long time and then cast your eye on the surrounding landscape, you perceive apparent movement resulting from the adaptation of your brain to the visual scene in motion. I will return to the meaning of these illusions in Chapter 13.

Another especially clear example of this type of mechanism is provided by what is called consecutive optokinetic nystagmus. This odd-sounding expression describes a very simple phenomenon. Suppose that you are on a bus, looking out the window. Your eye follows the visual scene, then turns back. That is optokinetic nystagmus. Now suppose that the journey lasts several minutes and that you are suddenly plunged into darkness. Instead of stopping, your eye will continue to move. This consecutive optokinetic nystagmus is the result of a still mysterious mechanism: it is as if a neural memory of movement had "charged" (as you charge a capacitor in electronics) a signal connected to the velocity of the eye, which runs down in the dark. It is likely that this perceptual memory is of the same nature as that involved in a long sea voyage. Indeed, sailors know that several days after their return, they will still feel as though they are moving, even though they are on terra firma. Here, again, the brain has constructed a dynamic internal response to the motion of the sea, probably using visual and proprioceptive vestibular information. During the voyage, this internal neural activity—which predicts the disruptive effects of the oscillations caused by the waves—opposes the sensory stimuli induced by pitching and rolling. Hence sailors are able to live at sea without feeling that they are moving and are able to coordinate their own movements.

The decision to interpret a movement in the visual world as a movement of the body is thus the result of multiple central neural processes, and yet it seems so obvious. How does the brain succeed in giving perceptual unity to all these variables and all their values? It is thanks to special mechanisms for building coherence, like those that make it possible to perceive a single visual object after it is analyzed in the first relays of visual pathways. I will return to these mechanisms later on.

3

BUILDING COHERENCE

Sensory receptor function has a predictive quality. Receptors can detect the derivatives (that is, velocity, acceleration, changes in force and pressure) of the physical variables that stimulate them. Detecting changes in a variable allows the receptors to predict the value of that variable at a future time. In addition, modulation of receptor sensitivity preselects receptors that specify future movement by endowing them with properties that anticipate the nature of the movement, tonic or dynamic. Such modulation allows simulation of the movement without having to carry it out. In fact, simple activation of γ fibers simulates a displacement of the arm by triggering spindle activity, and this spindle activity produces the perceptual illusion of movement, as vibration experiments demonstrate. Finally, filtering of messages by mechanisms such as presynaptic inhibition increases the efficiency of preselection even more.

In this chapter, I will show how the combination of signals from the different senses—the essentially multisensory character of perception—and the brain's use of endogenous (that is, produced by the subject) signals further increases the predictive power of the brain.[1] Indeed, a true physiology of perception must abandon the idea of splitting up sensory functions and instead approach them by way of their multisensory character. Aristotle himself wondered about this question:

> We perceive things as a whole, rather than what some may refer to as a continuity of their parts. Yet we can say that things do not always appear to us as they are; and that is why the size of the sun as we see it is not its true size. But let us return to our earlier question, whether we can perceive several things simultaneously, that is, in a single part of the

soul, in an indivisible moment. It has been proved that the soul perceives all sensations with one and the same faculty, which collects the information from all the senses. Yet this faculty, though numerically one, differs in its accounts: it is the same soul, but differently disposed.[2]

The significance of this message was neglected until very recently. For fifty years, the entire field of the physiology of perception had isolated the senses. Consequently, laboratories studying vision proliferated, even though as early as 1876 Ferrier was writing: "Without these labyrinthine impressions, optic and tactile impressions are of themselves unable to excite the harmonious activity of the centres of equilibration."[3] The same is true for hearing, since vision can modify auditory perception in a decisive way.[4] Ventriloquists provide the best example of this phenomenon: they create the illusion that their words are coming from their dummy. The mere fact of not moving their lips and working those of the dummy displaces the perception of the source of the sound. I will describe the neurophysiological discoveries that may explain this effect. Similarly, the art of entertaining demonstrates that the flavor of a dish depends no less on the way it looks than how it smells. The same agreeable smell at the beginning of a meal may become disagreeable by the end, owing to a property of the brain called allesthesia—taste is a source of pleasure when a person is hungry and of repulsion when the person is sated.

The importance of combining sensations does not appear to have interested physiologists before 1960, although a hundred years ago Sechenov tied it to how the image of one's own body is constructed during active exploration: "The child often sees a toy in the hands of its mother and in its own hands. The first of these sensations is simple; the second one is complex, including also tactile and muscular elements. This happens many thousand times. The two sensations become separated from each other, and the child's own hand appears in its consciousness with a tinge of self-consciousness."[5]

In work with animals, pioneers such as Held and Hein showed the importance of activity in the development of visual functions and the coordination of sensory systems.[6] In the 1960s, Gibson suggested the basis for a new experimental test of perception. "We shall have to conceive the external senses in a new way, as active rather than passive, as systems rather than channels, and as interrelated rather than mutually exclusive. If they function to pick up information, not simply to arouse sensations, this function should be denoted by a different term. They will here be called *perceptual systems*."[7] One compelling explanation for the multisensory nature of perception is that every animal is only interested in certain aspects of reality. It selects configurations of features that are useful to it as well as the relationships between the properties of objects. For example, as Turvey says, the height of the steps in a staircase expressed in centimeters is less important than the relationship between the height of the steps and the height to which we are able to lift our feet.[8] This explains why the steps in the staircase of a French château are not very high, since horses had to climb them. Similarly, the height of the stairs at the opera has to be adjusted to accommodate the step of women in evening gowns.

Gibson called these relationships that define the constants useful to a particular species "affordances."[9] They are a measure of the feasibility of carrying out actions important for the species in question, taking into account its sensorimotor repertoire. Sometimes opposite properties are interesting to different animals. Thus the degree of firmness (which can be measured as the relationship between pressure and displacement) of the ground allows humans to walk, whereas its friability permits the earthworm to move about. A single object such as a slipper can serve as a warm refuge for a person's foot, oral stimulation for a puppy, and a scrumptious meal for a larva. The geometrical constants essential for humans have ceased to be essential for other species.

What are the neural mechanisms underlying the multisensorial and action-related character of perception? When microelectrodes were invented in the 1950s, it became possible to stimulate different sensory receptors with electrical impulses and to record neuronal activity in anesthetized animals. It turned out that multisensory convergences occur at all levels of the nervous system.

They occur first of all in the medulla, where the most important interneurons—that is, neurons that link sensory and motor neurons—(like interneuron Ia) are the site of convergences from muscular spindle receptors, tactile receptors and the Golgi tendon organs. In the cerebellum, also the center of intense multisensory interactions, each Purkinje cell in the cerebellar vermis receives visual and proprioceptive information, just as the neurons of the flocculus receive vestibular information, and so on. The superior colliculus is an important structure for orienting movements. It receives visual and proprioceptive information and is another focus of convergence between auditory and visual signals. Convergences also occur in the so-called sensory thalamus and in many cortical areas. Multisensory convergence is thus the rule. What is not known is what fundamental role it plays. It was not until the 1970s that recordings of neurons in alert animals shed some light on this question. The last five years, however, have yielded an exceptional harvest of data on this subject. An example from our own work will show how vision and the vestibular system work together to reconstruct movements of the head in space.

How Vision Detects Movement

An engineer with two sensors at his disposal for detecting movements of the head—the retina and the semicircular canals—runs into a problem of geometry. Indeed, the semicircular canals detect movement of the head in their three perpendicular planes, so information is encoded in three dimensions; that is, in a Euclidean frame of reference. The retina encodes retinal slip on a spherical, or two-dimensional, surface. It receives information of great complexity. To better understand the geometrical problem posed by the fusion of information about movements of visual and vestibular origin requires a brief discussion of the concept of optic flow.

When people move about in the real world, the image of the environment on their retinas is displaced and becomes distorted in a very complex way. This distortion of the image on the retina during displacement is called optic flow. It can be represented by a velocity vector at any point on the retina that represents the direction and velocity from the point of the visual environment that is projected onto the retina.

Gibson, who worked for the American army, looked into the problem of airplane landings and described the form of these fields of vectors, which depend both on the geometry of the environment and the movement of the airplane. He observed that optic flow is very straightforward during a rotation because all the points are animated by an identical angular velocity. In contrast, during a forward translation—for example, moving forward while walking, or landing an airplane—optic flow takes the form of a complex expansion of the field of vectors, which appear to start at a vanishing point at the edge of this perspective: the "focus of expansion." During lateral translations, as when we are looking out the window of a train, optic flow takes on a more subtle complexity: all along the track the visual world moves in a sense contrary to the movement of the train, whereas the distant landscape appears to move in the same direction as the train.

How does the brain achieve this fusion of three-dimensional messages supplied by the canals and by optic flow? The last twenty years have provided new data on the neural mechanisms of the contribution of vision to the perception of self-motion (vection), how moving objects are sensed, and the "blindsight" of movement. I will summarize these findings here.

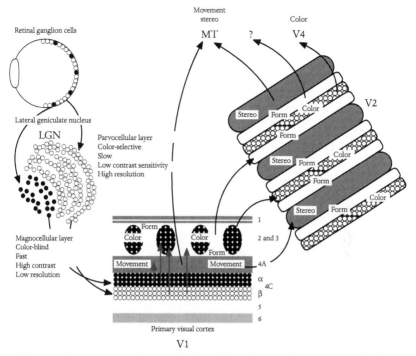

Figure 3.1. Segregation of inputs for identifying shape, color, motion, and depth in the first relays of the visual system.

Movement is detected mainly in the retinal periphery. Three kinds of ganglionic cells can be distinguished in the retina. In primates, two of these are sensitive to movement: first, the cells that activate both movement and strong contrasts between light and darkness and flickering; second, neurons sensitive to slow movements. Segregation of information about shape, color, depth, and movement, which is carried out on the retina, is maintained in the first central relay in the thalamus, the lateral geniculate nucleus. This nucleus is divided into two parts, the magnocellular zone and the parvocellular zone (Figure 3.1). From there the information is directed either toward the superior colliculus (a very old structure, called the optic tectum in birds) or toward the primary areas of the visual cortex—V1, V2, and V3 (area V4 mostly receives information about color). Segregation of featural elements of color, shape, and depth still occurs at this level.

Two areas of the cerebral cortex are involved more particularly in processing information about visual movement. They are called MT (middle temporal area) and MST (medial superior temporal area).[10] The MT receives impulses from the visual cortex, and its neurons are directly influenced by movement on the retina. The activity of the MT is transmitted to the MST, where the visual data are combined with signals of movement from the eyes and vestibular signals; in other words, extraretinal signals. The neurons in this area also solve an important problem for the visual perception of movement: they are able to "decide" whether an object is moving in the environment or whether it is stationary. Shadlen and Newsome recently investigated this "perceptual decision."[11] Moreover, positional signals of the eyes and head compensate for alterations in optic flow that occur when you turn your head while walking; they also keep the perception of the heading direction constant.

Modern brain-imaging techniques bring daily findings of the existence of other zones of the cortex that are involved in processing visual data on movement. For instance, it has just been discovered that the sensitivity of the neurons of the visual cortex is modified by the behavioral context, that is, the action in which the subject is involved.[12]

Ocular Pursuit Predicts the Movement of Targets

Say you are on a bus and you are looking out the window. Your eye follows the landscape. The movement of your eye is the result of cooperation between two mechanisms. The first is ocular pursuit. This continuous tracking of a moving target is possible only for animals that possess a fovea, the monkey and man; it does not exist in fish, rabbits, or even cats, whose eyes can only skip from one visual space to another in motions called saccades. Ocular pursuit is not very rapid. Move your finger in front of you with a back and forth motion. You will observe that when your movement goes faster than once per second (1 Hz, or about 60 degrees per second), you can no longer follow your finger. But the brain of primates and man has compensated for this slowness by endowing pursuit with great predictive capability: after a few back and forths, the movement of your eye anticipates the movement of your finger. Similarly, if you suddenly stop moving your finger, your eye will continue to follow your absent finger for an instant, which suggests an anticipatory neural activity. The pursuit also continues when the light is briefly turned off or when the target disappears behind an obstacle. This ability to follow a target behind an obstacle appears in infants during the first months of life.

Ocular pursuit is produced along a pathway (Figure 3.2) that transmits the movements of the target to the MT and MST areas, then to the frontal cortex in an area called the frontal eye field (FEF). In the MT, neurons are assigned a particular direction of the movement of the target. Each neuron fires when the target moves in a given direction. In the MST the convergence of sensory signals and internal messages signaling movements of the eyes or head (extraretinal) endows the neurons of this area with a special property: they can fire

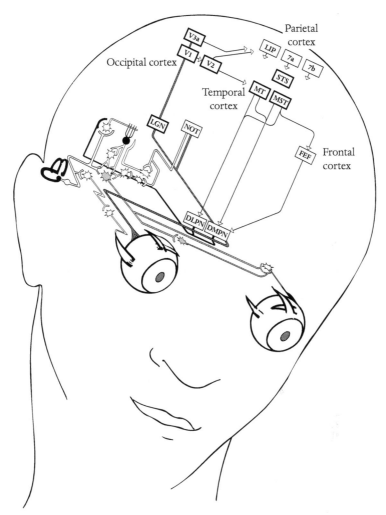

Figure 3.2. The main centers involved in processing information about visual movement. *(1):* The accessory optic system. In this system, visual movement is broken down into three planes (one horizontal and two vertical) that are aligned with those of the semicircular canals. Shown here is the nucleus of the optic tract (NOT), which encodes horizontal movement of the visual world, that is, shifting images on the retina that are in the plane of the horizontal semicircular canal (when the eye is centered in its socket). *(2):* Pathways for visual pursuit of objects in motion. In the monkey and in humans, the image of the object in motion is transmitted to the lateral geniculate body (LGB) by retinal neurons sensitive to movement, and from there to the visual cortex (cortical areas V1, V2, and V3). Messages about movement are next transmitted to the parietal cortical areas (7a, 7b, LIP, and so on), the areas of the mediotemporal cortex (MT and MTS), and the frontal oculomotor field (FOF). From the cerebral cortex, control and pursuit signals are sent to the dorsolateral and dorsomedial pontine nuclei (DLPN and DMPN).

during pursuit of a target even if the target disappears for a brief moment. Messages about eye movement, which are transmitted from the centers of the brainstem and control eye muscle contractions, activate these neurons for a brief instant after the light disappears and thus ensure that a person's gaze will be in the right spot when the target reappears. This convergence of external and internal signals also enables us to make tracking movements toward targets that are actually motionless but that we perceive to be moving, as when a row of targets is illuminated one after another and the brain creates a percept of movement.[13]

Tracking signals next activate the centers of the brainstem (the pontine nuclei) and of the cerebellum, linking the vestibular nuclei and other premotor nuclei that project to the motor neurons of the eye muscles.

Visual Movement and Vestibular Receptors

The second mechanism that enables us to follow the landscape from the bus is a very old reflex action, the optokinetic reflex, present even in the fish and the frog, who have no fovea. Actually, all sighted animals possess this mechanism, which allows them to follow the movements of the visual environment. It also contributes to maintenance of posture. If you tilt the aquarium in which it is swimming, a fish will tilt its body to align it with whichever surface of the aquarium is lit.

The nerve pathway that processes these signals thus also influences the centers that control posture. Anatomists discovered it a long time ago and called it the accessory optic pathway for want of understanding its function. What is remarkable is that all along this pathway neurons respond to visual movements in preferred directions aligned along the planes of the semicircular canals of the vestibular system.[14]

This alignment is also characteristic of the neurons of the retinal ganglia, which fire preferentially when the visual movement is in the plane of the semicircular canals. This geometrical segregation becomes more pronounced in the second relay. The neurons sensitive to movement in the plane of the horizontal canal are united in the so-called nucleus of the optic tract (NOT) following a very precise orientation, the temporonasal direction. Another nucleus, called the medial terminal nucleus (MTN), contains the neurons sensitive to downward visual movement, and yet another, the dorsal terminal nucleus (DTN), contains those sensitive to upward visual movement. Other groups of cells belonging to the same accessory optic system, situated in a relay portion of the ventral tegmental area, encode combinations of rotational and linear components of optic flow.

Consequently, just as information is broken down into its main components (color, movement, contrast, texture, and so on) in the primary cortical visual projections (V1–V4), the geometric properties of movement represented in optic flow are also segregated. This early segregation in the sequence of neural processing is further refined in the next relay, the dorsal motor nucleus of the inferior olive, a brainstem structure that projects to the cerebellum, where some of the visuovestibular convergence takes place. This dorsal motor nucleus contains three groups of neurons that correspond exactly to the three planes of the canals. The next relay, the cerebellum, also contains neurons, located in the flocculus, whose preferential planes of activation are in those of the canals.

The existence of separate components organized according to the planes of the semicircular canals clearly facilitates the convergence, that is, the fusion, of these inputs, whose origin is visual, and those from the vestibular receptors, with which the visual messages converge.

The geometrical correspondence between the visual system and the semicircular canals extends to the motor system. Indeed, despite the migration of the eyes from a lateral to a frontal position, the plane of action of the three pairs of extraocular muscles, which enable rotation of the eyeball in all species, has remained approximately parallel to the planes of the semicircular canals.[15]

The demonstration of an organization in three planes was made using tensor analysis and the theories of Pellionisz and Llinás. They proposed that the muscles be represented by their eigenvectors, that is, their own virtual vectors that represent the direction and amplitude of the force exerted by each muscle. These vectors are classical mathematical quantities, but are very easy for nonmathematicians to grasp. They have the advantage of being accessible to experimentation: in animals or in humans, all that is needed is to establish points of attachment of the muscles and to reconstruct the skeletal anatomy of this motor apparatus, and then to record the activity of these muscles during activation by tilting the head, which initiates the so-called vestibulo-nuchal reflex. The eigenvectors of the neck muscles of a cat are situated in the plane of the canals. The geometry of the semicircular canals is thus basic to the perception and control of movement.

When a monkey jumps from branch to branch, when a diver leaps from his diving board, or a gymnast grabs the high bar or jumps on a trampoline

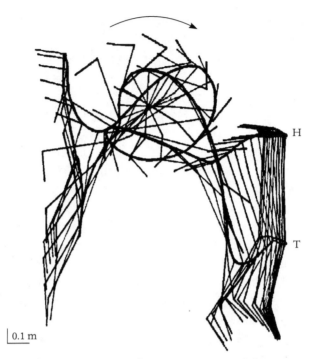

Figure 3.3. A video camera connected to a computer recorded movements of an athlete's body during trampoline jumping. The computer traced the lines of motion of markers placed on his head *(H)*, trunk *(T)*, legs, and feet. Jumping was broken down into three main phases: First *(left)*, the body goes up, the head is stabilized in rotation, and the brain is able to use information from vision and the vestibular system. Second, the brain launches a very rapid rotational movement (greater than 600 degrees per second) during which vision is useless because images on the retina are moving too quickly. The brain relies only on vestibular cues. Finally, the rotation ends, and the body lands vertically. During this phase, the brain uses vision, the vestibular sensors, and proprioception—as well as internal models of the effects of gravity—to plan a soft landing.

(Figure 3.3), the brain has to reconstruct the movement of the head in space very rapidly and with the most exquisite precision. How does it do it using visual and vestibular information, and proprioceptive information from the neck?

From the first central synapse, the velocity of the head in space is, in fact, detected by combining the information supplied by the vestibular accelerometers and the visual tachometer. This discovery was made by recording the neurons of the vestibular nuclei in the fish and, later, in the monkey.[16] These nuclei are located in the brainstem and constitute the first central relay of vestibular information; that is, the fibers of the semicircular canals project directly to them. They are called vestibular nuclei (see Figure 2.4) for purely anatomical reasons.

The famous experiment of Dichgans and his co-workers is a good illustration of the way vision and the vestibular receptors cooperate in detecting movement of the head. Dichgans and his group put their fish on a turntable and subjected them to three separate experiments. First, a rotation in the dark, broken down into three stages: acceleration, then constant velocity, finally deceleration. We can replicate this stimulus by making a few turns with our eyes closed. The form of the stimulus is a trapezium of velocity: one or two intervals of acceleration separated by constant velocity. The vestibular receptors are activated only by the acceleration and do not fire until the animal begins to turn or when it brakes. The neurons of the vestibular nuclei respond to such stimulation by increasing their discharge followed by a slow diminution after 10 to 20 seconds. Consequently, the brain receives no information about the rotation of the head during the period of rotation at constant velocity. It is the same for us if we twirl around for some time: when the effect of the transitory activation of the canals wears off, we can no longer perceive any movement if all we have is the vestibular information.

In the second experiment, the animal was immobilized and the visual world rotated around it, using a rotating cylinder. The researchers observed that these neurons, called vestibular, are also activated by visual movement. This activation is due to transmission along the pathways of the accessory optic system. The activity increases very slowly at first, then holds steady during the rotation, and finally diminishes slowly, the animal immobile the whole time.

The third experiment involved rotating the animal in light. Combining visual and vestibular stimuli by imitating a natural rotation, the neurons of the vestibular nuclei, like the velocity of rotation of the head, fire in a trapezoidal pattern. In other words, the angular velocity of the head is perfectly reproduced from the first neural relays by means of the combination of dynamic properties of the visual and vestibular receptors. This discovery was repeated in several species (frog, rabbit, cat, monkey): it is thus very likely true for humans as well.

We are aware of the angular velocity of the head thanks to the sensitivity to rotation of both the visual and vestibular systems. Vision alone is not sufficient—it is too slow. The contribution of the vestibular receptors is to inform the brain about the movement under way by virtue of their having detected derivatives of displacement (velocity and acceleration). The term "dynamic complementarity" is certainly apropos.

Moreover, the neurons of the vestibular nuclei do not only receive visual and vestibular information: they are also sensitive to activation of the muscle receptors in the neck. Very short synaptic pathways connect them to the neuromuscular spindles (numerous in the muscles of the neck), which detect the length and velocity of muscle stretch. When the head turns, the brain is informed of the rotation by this combination of visual, vestibular, and proprioceptive information. From the very beginning—from the first synapse of the central nervous system—the detection of movement is multisensory.

THE DIRECTION OF GAZE MODIFIES THE VESTIBULAR PERCEPTION OF MOVEMENT

Among the various neurons contained in the vestibular nuclei, the neuron called the second-order vestibular neuron is very interesting. Actually, it is a sensory neuron, because it receives visual, vestibular, and proprioceptive information. But it is also a motor command neuron, because it projects directly to the ocular motor neurons, and in some cases also to the neck. Reflex theories explain it as a simple relay of sensory information in a sensorimotor loop. We recorded the activity of its axons during spontaneous ocular movements and were very surprised to record activity connected to the discharge of the ocular motor neurons.[17] What we believed to be a sensory neuron turned out to be activated by a motor discharge! A similar phenomenon is observable in the first visual relays.[18]

Some time later, I invented a technique that makes it possible to record the activity of a neuron under physiological conditions, to analyze its entire structure, and to determine the target structures of the axon.[19] We were able to penetrate the neuronal axon (whose diameter is around 5 microns) thanks to a very fine glass microelectrode whose tip measured 1 micron in diameter. We presented targets to the animal to induce spontaneous movements of its eyes (saccades), and we rotated the animal in the dark to stimulate the semicircular canals and induce a nystagmus. The power of this method derives from the injection of a tracer, horseradish peroxidase, which makes it possible to reconstruct the connectivity of the neuron.

We observed that these vestibular neurons, which normally respond to rotation of the head, and thus to its angular velocity, by increasing their firing rate, exhibit a basic activity that changes completely as a function of the direction where the animal is looking. A corollary signal, "a copy of the motor command" (Figures 3.4 and 3.5), is thus present as early as the first sensory relay.[20]

What is the role of this positional signal of the eye in vestibular neurons?

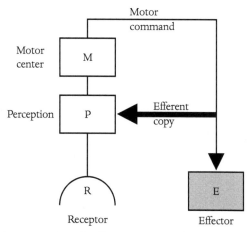

Figure 3.4. The principle of efferent copy according to von Holst and Mittelstaedt. Sensory signals are sent to the neural centers where perception is derived. A motor command travels in a loop from these centers to the effectors (the muscles). But a copy of the motor command is also sent to the perceptual centers, where it alters sensory information according to the ongoing action. The brain is thus able to anticipate the results of movement without having to wait for information from the sensors.

The question has not yet been answered. For some, it is a simple change of co-ordinates; for others, it has to do with the coordination of the eyes and the head. In all these cases, the data show that the senses are not isolated. Action influences perception at its source.

Am I in my Bed or Hanging from the Ceiling?

I have described the nerve pathways that transmit the movement detected by vision to the cerebral cortex. I have also analyzed the vestibular detection of head movement and the mechanisms by which vision is enhanced by the vestibulo-ocular reflex. But vestibular information does not remain at the level of the first sensory relays. It is also transmitted to the cerebral cortex, where it contributes to many functions: conscious perception of orientation, gaze movements, control of posture, coordination of gestures. It also plays a role in constructing a coherent perception of the relationships between the body and space. This vestibular contribution to the highest cognitive functions was neglected until recently. Yet, a few pioneers such as Penfield had considered it.

Penfield's subjects reported that they felt vertigo or rotating sensations, characteristic of the vestibular system, during electrical stimulation of a tem-

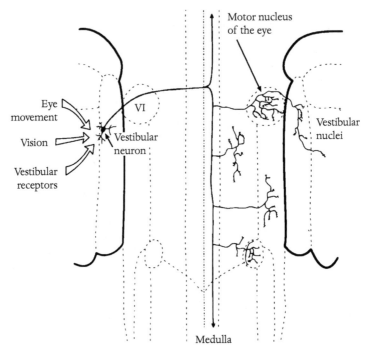

Figure 3.5. Second-order vestibular neurons are influenced by the direction of gaze. A second-order vestibular neuron transmits movements of the head detected by the semicircular canals to the muscles of the eyes. One can see the axonal branching to various centers of the brain that process this information. This neuron is the first sensory relay for vestibular information, but it receives signals from vision and receptors in the muscles and joints as well. It also receives copies of the motor commands for eye movements.

poroparietal region.[21] It also has been known for a long time that illusions of dissociation of the body are induced during the aura—that is, alteration of consciousness—that often precedes seizures in patients suffering from temporal lobe epilepsy.[22] During an epileptic fit, these patients sometimes have the feeling that they are simultaneously lying in their bed and floating at the level of the ceiling, as if their body were split. The area of the brain responsible for these illusions is located deep within the lateral sulcus, around the superior temporal gyrus, at the caudal portion of the insula and along the retroinsular region, near the auditory area.

Messages of vestibular origin contribute to the most cognitive aspects of the body schema and of spatial representation.[23] In humans, the existence of this vestibular cortical area was recently confirmed by brain-imaging methods.[24] Patients with lesions of the parieto-insular zone present with deviations of the vertical subjective: the world appears to them to be leaning to the side opposite the lesion. The effect is so strong that they take skewed photographs.[25]

The Neural Basis of Vestibular Influences on Cortical Function

The parieto-insular vestibular cortex (PIVC) was first identified in the monkey.[26] It is the principal center for relaying vestibular information to the rest of the cerebral cortex, and is located in the temporal portion of the lateral sulcus. Like the vestibular nuclei, this vestibular cortical area is multisensory. There is no known direct route for visual inputs to the PIVC, but there are several candidate pathways. Its role in the multisensory representation of movement of the head in space was confirmed by recordings of the activity of neurons in the cortex of a monkey during vestibular stimulation in the dark.[27] Two-thirds of the neurons of this area respond to vestibular stimulation. The other third respond mainly to somatosensory stimulation of the neck and shoulders. Almost all of the neurons that respond to vestibular stimulation are also activated by movements of the visual environment and by somatosensory stimulation.

The characteristic feature of PIVC neurons is their sensitivity to angular rotations in the various planes and to visual movements in the direction opposite to that of the head. In other words, they respond in a way that is perfectly adapted to detecting movements of the head in space, since when we make a rotation while the head is turning to the right, the visual world is turning to the left. These experimental findings suggest that the PIVC neurons encode movement of the head in space based on vestibular and somatosensory visual information, the latter mostly likely transmitted by the receptors in the neck and the torso.

One might speculate that one of the processes accomplished in this part of the cortex is encoding of movement of the head in planes more numerous than those of the vestibular nuclei. Optokinetic and vestibular coding of the movement of the head (through the accessory optic system) is organized according to the geometry of the planes of the semicircular canals. This geometric selection is probably very useful in maintaining a connection between these two signals and combining them at the level of the brainstem. However, at the level of the cerebral cortex, the representation of movement of the head is in all likelihood combined with signals from temporal visual areas like the MT and MST, which mainly process information about visual movement. Strict separation in three planes is thus insufficient. What is required is a multidimensional representation of the movement of the head.

According to Grüsser, the cortical areas involved in processing form an inner vestibular circuit that processes information about movements of the head in space and maintains continuously activated neural representations useful for simulating movements internally and predicting their consequences.[28]

Neurophysiology has confirmed the functional importance of vestibular influences on the cortical areas.[29] I will describe next two examples of the influence of these vestibular messages on vision.

Why Does the Obelisk Look Vertical Even Though my Head Is Tilted?

The neurons of the primary visual areas are activated by the border between light and dark, the orientation of the border, and the direction of its movement. They can be activated, for example, by showing an animal bars of light. At birth, most cats and monkeys respond to orientations of these bars of light in all directions, whereas only a few show a preference for horizontal and vertical orientation. Consequently, selection based on geometry occurs here also from the first sensory relays.

This directional selectivity is determined very early during a critical period that in the cat extends from 3 to 10 weeks.[30] The active movement of the eyes detected by the receptors of the ocular muscles influences the neurons of the visual cortex and leads to a preference in these cells for given lines of orientation. In this way, projection of the proprioceptors of the eye (and perhaps of the neck) on the neurons of the visual cortex contributes to their directionality. It suggests as well that the muscles of the eye constitute a frame of reference with the same capacity as the semicircular canals. Because the planes of action of the eye muscles are close to those of the canals, it is difficult to favor one or the other of these two hypotheses. Perhaps embryology will be able to answer this question.

Proprioception is not the only factor that influences the directionality of the neurons of the visual cortex. Indeed, a recent discovery shows a vestibular influence on supposedly purely visual areas, like area V2: 40 percent of the neurons of this area (which is one of the primary visual areas) are sensitive to the contours of visual shapes, but their activity is influenced by the tilt of the head.[31] The preferential direction of the contour to which each neuron is sensitive goes in a direction opposite to that of the head with respect to the vertical. This vestibular influence makes it possible to keep the perceived orientation of the contour constant. When I tilt my head and the Obelisk at the Place de la Concorde still looks vertical to me, it is perhaps thanks to this mechanism. It is not a reflex that makes my eye turn (moreover, the torsional vestibulo-ocular reflex is almost inactive), but a mechanism of mental rotation of visual representations.

The Vestibular System and Face Recognition

Everyone knows that it is extremely difficult to recognize a face tilted at 90 degrees or, especially, upside down, even though the image of the face on the retina remains quite clear. The reader also knows that it is absolutely impossible to look at a slightly crooked picture. So gravity may exercise a constraint on the perception of visual forms.

The role of gravity in the perception of visual forms was studied in our laboratory in the particular case of the perception of symmetries aboard space flights that put humans in a situation of microgravity. On earth, when a subject is asked to say whether a shape is symmetrical or not, he does it more rapidly if the axis of symmetry is horizontal or vertical. When the axis is diagonal, the subject takes a little more time. In particular, the symmetry is especially easy to identify when its axis is vertical. But when the astronauts of the space station MIR were subjected in space to a recognition test of symmetrical shapes (polygons, for example), they no longer showed any preference for the vertical. This same lack of preference for the vertical obtains when you are lying on your back and the gravitational vertical is no longer aligned with that of the axis of symmetry of the visual object.

Otolithic detection of gravity thus plays an important role in our visual perception of the environment.

A Neural Compass?

Sit in a revolving chair. Look at an object in front of you, close your eyes, then make one or two turns. Even with your eyes closed, you will find you can still point to the object. A neural mechanism of vestibulary origin assesses the direction of the head in the horizontal plane, memorizes it, and updates it during the movement. One of the most important discoveries of recent years was that of the existence of neurons whose discharge is connected to the orientation of the head in the horizontal plane. These neurons have been identified in the thalamus of the rat.[32] The thalamus transmits sensory information to the cortex and the postsubiculum, the nucleus of the hippocampal formation, which, as I will explain later in detail, plays a role in spatial memory.

These neurons signal the direction of the head no matter where an animal is in a room, a property that is called local invariance. Thus they constitute an internal compass. It is a sophisticated compass, because it indicates a wide range of directions. For example, one of these neurons, silent when the rat moves its head in many directions, suddenly starts to fire when the animal turns toward the north, no matter where the animal is in the room. These neurons are, however, very dependent on the local visual environment. The animal has to be familiar with the room it is in for the directions preferred by each neuron to be well stabilized. When the animal's location in the room is changed, its directional neurons are reorganized. But what is really remarkable is that when the animal is familiar with a location, the neurons do not lose their sensitivity to a given direction in the dark. It follows that these neurons participate in spatial memory.

Vestibular information about head orientation is transmitted from the medial vestibular nucleus to the dorsolateral thalamus and from there to the hippocampal formation. It is also known that the hippocampus contains another category of neurons, the place cells. These cells, independent of the direction of the head, discharge when the animal is in a given location. It is thus possible that memory of the object, in the experiment above, is due to vestibular information, which can update representations of space in the hippocampus. I will come back to this topic in Chapter 5.

I will come back to this topic in Chapter 5.

NEGLECT

Constructing a coherent perception of the relationships between the body and space thus depends on both hierarchical and parallel mechanisms that combine sensory cues and signals connected to action. So it is not surprising that lesions or deficits in certain circuits induce a breakup of this coherence. One example, which I will mention very briefly, is spatial neglect.

Patients who have a lesion of the right parietal cortex often present with impaired space perception. Although they are looking at space in its entirety, they can only perceive the right side, and they neglect the left side. Hence the name of this illness: hemispatial neglect. If you ask these patients to draw a flower, they will only draw the petals on the right side; if you ask them to draw a clock, they will only draw the hours from 12 to 6; if you ask them to paint objects, they will only paint one half of each object; finally, they will only eat what is on one side of a dish of strawberries, and so on. These patients do not have all these symptoms at once. In fact, it is possible to distinguish several kinds of neglect: for example, personal, a person who neglects his left arm; extrapersonal, a person who neglects to draw objects on the left side of the room; or representational, a person who cannot picture the left side of his living room, of his apartment, or of his city. It is also possible to distinguish perceptual from motor neglect. Motor neglect consists in not using a limb even though its motor function is intact. Several authors have made the remarkable observation that the simple injection of cold water into the left ear—that is, introduction of a vestibular caloric stimulus—causes some of the symptoms of neglect to disappear temporarily.[33] This effect is not due to the deviation of the eye induced by the stimulus.

Patients with spatial neglect sometimes have associated neurological deficits, such as hemianopia (loss of vision for one half of the visual field) and hemianesthesia (anesthesia of one half of the body).[34] These deficits, apparently primary—affecting primary structures in the sensory areas of the cortex—also sometimes disappear after injection of cold water; in other words, following a vestibular stimulus.[35] Almost complete—albeit temporary—remission of hemianesthesia following damage to the right cerebral cortex is also obtained as well as improvement of other impairments of the perception of the body.[36] For example, certain patients no longer feel it when someone touches their hand. Bottini and co-workers found that vestibular stimulation restores perception of touch in a transitory manner.[37] Brain activity in such a patient was recorded with the aid of positron emission tomography. In normal subjects, tactile and vestibular signals share projections to the neurons of the putamen, insula, somatosensory areas of the cortex, premotor cortex, and the supramarginal gyrus. Some of these areas that were intact in this patient were activated by the vestibular stimulus at the same time that sensation reappeared. Thus, "the somatosensory primary cortex has no priority for tactile perceptual consciousness; rather, multiple sensory representations of the body exist in the brain and any of these can contribute to perception."[38]

Vestibular stimulation has a spectacular effect on self-awareness of body parts. In particular, it offers relief from somatoparaphrenia. Patients with this illness deny possession of their own limbs.[39] For example, one patient denied that her left arm was her own.[40] She said: "This is my mother's arm." We injected water into her ear (caloric stimulation of the vestibular receptors), and this woman reclaimed possession of her arm. Unfortunately, the effect was only temporary.

These observations suggest that spatial neglect has a single cause connected to the representation of the body schema, even though it has multiple manifestations. The efficacy of the vestibular stimulation could be due to a basic imbalance induced by the lesion and compensated for by an imbalance in the opposite direction produced by the stimulus through the ascending vestibular pathways described above, as well as through those that correspond to the "directional neurons of the head." If this assumption is correct, the fact that the vibrations of the neck, the transcutaneous stimulation of the tactile receptors, optokinetic stimulus, and wearing prism goggles also produce a remission of spatial neglect is not surprising, given the multimodal convergences toward all the centers of the vestibular pathways.[41] I emphasized above that several structures formerly called vestibular are in fact multimodal and are involved in reconstructing orientation and movement of the head in space.

These considerations are important because they provide a biological basis for so-called out-of-body parapsychic phenomena. Quacks who may profit from such bizarre perceptions to exploit the credulity and pocketbooks of people by selling these illusions as supernatural phenomena should be denounced.

DISTANCE PERCEPTION

The perception of distance is another example of the multisensory nature of perception. But, here again, the combining messages originate in part from copies of the motor commands. In fact, the perception of distance is the result of visual messages and signals from the convergence of the two eyes. Action is used as information about the world.

Lorenz suggested that fish evaluate the distance of objects from the bottom of the sea using a combination of visual information about disparity (that is, the difference between the two images of the object supplied by the two eyes) and to convergence of the two eyes on the object.[42] The idea that the perception of distance is due to a multisensory interaction was also formulated by Poincaré: "However, sight enables us to appreciate distance, and therefore to perceive a third dimension. But everyone knows that this perception of the third dimension reduces to a sense of the effort of accommodation which must be made, and to a sense of the convergence of the two eyes, that must take place in order to perceive an object distinctly."[43]

Perception of distance is actually due to a combination of visual and motor inputs.[44] In the primary visual area V1, certain neurons are activated by the disparate images of the same object on the two eyes, with some connection to the distance of the object. By presenting the animal with two patterned planar images called "Julesz stereograms"—duplicating the two perspectives the two eyes will have of the object—the apparent distance of the patterns can be made to vary by varying only the disparity (or gap). Three kinds of neurons from area V1 are activated respectively when the image is located behind or in front of the point on which the eye is focused.

This sensitivity to retinal disparity can be modified by a nonvisual (extraretinal) signal. In fact, if the subject puts on prism goggles that bring the target closer without changing the disparity, the two eyes converge. A neuron activated by a visual target 20 centimeters from the eye is activated again when the target is 80 centimeters from the eye because of the prisms. This experiment shows that a signal connected to the convergence of the two eyes modulates the response of these neurons. The distance can also be ascertained by measuring how much the eyes cross! The exact origin of these signals of convergence is unknown. Two possibilities are the ocular muscle receptors or an efferent copy of the motor command for convergence; in other words, a motor signal or a combination of the two. Many structures of the visual cortex—V1, V2, MST, LIP (lateral intraparietal), PO (parieto-occipital), and so on—receive information about the position of the eyes in this way.

The idea that variables such as distance are constructed by unconscious inferences, as Helmholtz suggested, is not accepted by all the philosophers. Even Merleau-Ponty, whose ideas have been validated by experimentation, was confused on this point: "How is it possible," he writes, "to assert that perception of distance is a construct of the apparent size of objects, of the disparity between the retinal images, of the accommodation of the lens, of the convergence of the eyes, and that the perception of depth is a construct of the difference between the image supplied by the right eye and that supplied by the left, when, if we hold to phenomena, none of these 'signs' is clearly presented to our consciousness? How can we reason in the absence of premises?"[45]

The Coherence between Seeing and Hearing

What of the cooperation between two other sensory modalities, sight and hearing? Everyone knows that you can catch an object more easily if, at the same time you see it, you hear it coming. This cooperation brings into play a special structure of the brain, the superior colliculus. Essentially a biological machine that recognizes moving objects and identifies their distinctive features using multisensory cues, the superior colliculus holds the key to how the brain directs multisensory fusion and extraction of pertinent signals. It controls orientation and avoidance reactions, among other behaviors, and it is a splendid example of a sensory and motor structure that directs the execution and correction of movements carried out by multiple effectors like the eyes, the head, the trunk, and the limbs.

It is impossible to catch moving prey without focusing in front of where the prey is and very rapidly correcting the trajectory in the event of error. The colliculus is thus involved in anticipation and motor prediction. Its presence also explains how the brain assembles motor commands in a manner sufficiently general for the movement to be accomplished by a variety of effectors (tongue, hand or paw, foot, and so on) with differing dynamic properties.

The colliculus (from the Latin *collis,* hill), called the optic tectum in non-mammalian vertebrates, is composed of two symmetrical parts (hence the expression "the colliculi"). There is a superior colliculus and an inferior colliculus. I will consider the superior colliculus here. Lesions of this organ cause impaired reactions of orientation and of catching a target. A damaged animal can catch a mouse, but less rapidly, which indicates its ecological importance. The superior colliculus is composed of seven layers of nerve fibers and neurons, and its complexity is formidable. There are, however, three principal layers, called, prosaically, superficial, intermediary, and deep. These neurons receive information from more than twenty cerebral structures. The afferents of the superior colliculus are mainly visuosensory, proprioceptive (also called somatic), and auditory. But it receives many cortical projections as well as projections from structures such as the basal ganglia, the cerebellum, and so on. Other afferents, whose functions are poorly understood, come from structures such as the raphe nucleus and the locus ceruleus. As Stein and Meredith sum up, "Some of these latter inputs may play a role in varying the likelihood of a given movement depending on the prior experience and immediate needs of the organism."[46]

First I will address the visual responses. If you move a point of light in front of an animal, the neurons of the colliculus, like those of the visual cortex, will respond with a discharge of action potentials. The response to the visual stimulus is selective for a particular region of visual space (this is the idea of a receptor field). The receptor fields of these neurons are expansive. They are bordered by an inhibitory region: if a target is present simultaneously in the center of the receptor field and at the periphery, the response from the center is reduced. The size of the effective stimulus is, in general, less than that of the receptor field. Some neurons are particularly sensitive to movement in certain directions. Finally, repeated exposure induces a habituation that results in extinction of the discharge. The colliculus participates in detecting the unusual. If the animal turns its attention to the visual target, the sensory response is increased. This amplification of the response to a stimulus adds weight to arguments in favor of the active character of perception. Amplification also takes place just before an ocular saccade toward the target; further proof that intended movement modulates the sensitivity of the first sensory relays.

In the deep layers of the colliculus are neurons that project to the brainstem and the medulla where they induce ocular movements toward the target (saccades) (Figure 3.6). A crossed projection is involved in orienting movements of the head, the eyes, the trunk, and perhaps the limbs. Another, un-

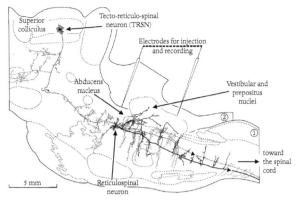

Figure 3.6. The neurons of the superior colliculus project to the motor centers of the eyes and the head. The colliculus and the motor centers of the brainstem interact to control movements and posture during orientation of the head toward an object. The tecto-reticulo-spinal neuron (TRSN) connects the colliculus to the brainstem. It is active when an animal directs its gaze to a visual target. It projects to neurons of the brainstem that activate the motor centers of the eyes, head, and perhaps the arms and trunk.

crossed, initiates recoiling, or flight when a predator is approaching. Other neurons project to the thalamus and probably inform the cortex, via the thalamus, of ongoing movements. Still others, projecting to the contralateral colliculus, undoubtedly ensure the coordination of the two sides and play a role in reactions to approaching objects. These neurons are arranged in such a way that their location in the deeper layers of the colliculus is in the same register as the retinal representation of the higher layers.

In other words, if a target (a bird, for example), appears in one's visual field, the population of neurons of the colliculus that will be activated in the superficial layers is situated exactly above the neurons of the deep layers that have to be activated to make a combined movement of the eyes and head toward the bird. To put it simply, the colliculus contains two sorts of maps: sensory maps (in the superficial layers) and motor maps (in the intermediary and deep layers) in the same conical section.

Many of the neurons in the deep and superficial layers of the colliculus respond to multisensory stimuli. There are, for instance, three sensory maps that share the same neurons: a visual map (retinotopic), a map of auditory space (audiotopic), and a map representing the parts of the body (somatotopic). These maps are particular to each species. For example, in the rat, whiskers play a fundamental role in exploring space and in trapping; they are represented extensively in the colliculus. In primates that have a fovea, the central region of the retina is represented in numerous neurons. The receptor fields of each of these modalities are different, and it is important to beware of taking too simplistic a view of these three neuronal representations of sensory space. The essential point is that these three domains of perception share a common sensoritopy: retinal space.[47]

This representational coherence is very important; it goes a long way toward simplifying the coordination of receptor space to enable a quick, single orientation, directed toward a goal. At the same time, I will show that these maps are the center for multiple phenomena of selection and of dynamic readjustment that, provided they are investigated, will one day help to explain the predictive character of orienting movements.

The colliculus is an interesting structure for understanding how the brain handles the problem of spatial and temporal coherence of messages supplied by the different senses. It is obvious, in fact, that if you suddenly hear the sound of a door lightly closing on your right and you turn your head quickly to direct your gaze toward the door, you have to coordinate at least three spaces: vision, that is, the retinal space on which the image of the room you are in is projected; hearing, that is, the auditory space detected by your two ears; and the space of your body, which has to turn to look in the direction of the door that has shut. These adjustments of retinal, auditory, and somatic space are accomplished in the superior colliculus. Let us first look at an example of adjustment between auditory space and vision.

Although auditory signals are received by another structure, the inferior colliculus, the neurons of the superior colliculus are also activated by sound. These neurons do not have exactly the same properties as those of the primary auditory pathways. The latter are sensitive to narrow bands of frequency and constitute what is called a tonotopic map; that is, they gradually allow the animal to recognize sounds emitted by others of its kind, as is the case with color recognition via visual pathways.

At any rate, no representation of auditory space in the classical hearing pathways can be found; it is encoded in the superior colliculus. Its neurons are relatively insensitive to pure sounds. They prefer complex sounds like those produced by clapping the hands or whistling—in short, natural noises like those prey or predators might make. They are very sensitive to novel sounds. Most of them are binaural: they respond to stimulation of both ears. They construct large receptor fields based on this binaural detection, taking into account the (very short!) time between the arrival of sounds in both ears.

Remember that reciprocal reinforcement exists between visual and auditory inputs at the level of each neuron where these two modalities converge. How can this cooperation be demonstrated? A neuron that responds with a strong discharge to a bird singing cannot respond to either of the two aspects of the stimulus—the image of the bird or the song—presented alone.

A fundamental property of multisensory integration in the colliculus is this: the receptor field is the pertinent reference for multisensory integration, not the external space. Merging of receptor signals occurs within the space of the receptor fields and not by centrally reconstructing the external Cartesian space. Here there are rules that will be encountered again later with respect to the neurons in the putamen and area 6 of the cerebral cortex that are activated by both visual and tactile stimuli.[48]

The auditory map constructed in this way can be shifted by movements of the eyes or ears or by wearing prism goggles that shift the visual space with respect to that of the head. What explains this flexibility and this influence of motor signals on neural maps?

Say that a bird appears at the periphery of your visual field. Because the position of the bird in space is given on the retina (in retinal coordinates), its position on the retina will depend on the direction of your gaze. Try this experiment. Look in front of you, eyes straight ahead. If the bird appears at 45 degrees to the right and begins to sing or to make a little chirp of fright, the two maps—visual and auditory—will coincide, and the same population of neurons of the colliculus will be activated and will produce the same orienting movement of the eyes and the head. Now, if you keep your head facing front but shift your eyes, the two maps may not coincide. The influence of the direction of gaze on the colliculus recalibrates the two maps without the brain having to make complicated calculations.

Other interactions of a spatial nature modify the neurons of the colliculus. The response to two stimuli from the same place is favorable to the selection of objects of interest by the configurations of stimuli that characterize them. Another mechanism that allows selection of interesting targets is the depression that accompanies the simultaneous presentation of two stimuli that are not located in the same place. In fact, each multisensory neuron of the colliculus has a zone at the periphery of its receptor field that inhibits response. This property is very widespread throughout the nervous system and in the visual cortex: it is believed to function in spatial selection.

Another problem the brain has had to resolve to enable the fusion of multisensory information is that of time shift. Different lengths of time are actually needed for sensory signals to reach the colliculus. For example, a sound introduced to the ear takes around 13 milliseconds to get there; a touch around 25 milliseconds; a visual stimulus around 40 to 150 milliseconds, according to the alertness of the animal.

But it gets even more complicated. If a bird sings in a tree 50 meters away from a cat, the sound transmitted through the air at 330 kilometers per second will take about a tenth of a second to arrive at the colliculus of the feline, whereas light, which moves at 300,000 kilometers per second, will arrive at the retina almost instantaneously.

The solution adopted by the nervous system is one called "temporal windows." It is extremely simple and extremely flexible. Electrophysiological recordings show that a light stimulus induces a discharge that can be maintained for more than 100 milliseconds; if a sound arrives after several hundred milliseconds, the amplification of the response can still occur. The neuronal network of the colliculus thus develops a memory that maintains the sensitivity of the multimodal neurons during a certain time, hence the name temporal windows.

Temporal windows would explain in large part the success of predators. Stein and Meredith note:

> The presence of these features hardly seems coincidental. Quite the contrary, they are likely to have developed and been maintained during evolution because: (1) the presence of wide temporal windows gives critical flexibility in detecting and responding to minimal, albeit important stimuli at different distances from the animal . . . and (2) the appearance of maximal depression when spatially disparate stimuli are in close temporal proximity is a means of focusing attention on the strongest, and presumably the most important, stimulus in the presence of potential distractors.[49]

But the nervous system has other mechanisms for resolving the problem of time shift. The cerebellum receives sensory information from a wide range of sources, in particular the proprioceptors of the limbs. It is important for this structure to be able to compare the messages from the receptors of the feet, arms, and neck, for example, because it is involved in coordinating movements, especially via the precise temporal organization of gestures and relationships between posture and movement. The solution used by the nervous system may be to vary the velocity of conduction of the nerve fibers so that all the signals arrive at the same time. The signals from the receptors of the feet, located more than a meter from the cerebellum, are transmitted more rapidly than those of the receptors of the neck located a few dozen centimeters away! But the problem of time shifts is so important that the brain has probably found numerous different solutions that should interest roboticists.

Many philosophers have emphasized the cooperation between vision and the sense of touch. Merleau-Ponty wrote:

> We must get used to the idea that everything visible emerges from what is tangible; that every tactile being is destined in some way to be seen; and that there is encroachment and overlap not just between touched and touching but also between the tangible and the visible embedded in it, just as, conversely, the tangible itself is not devoid of visibility, that is, not without visual existence. Since the same body sees and touches, visible and tangible belong to the same world. A wonder too seldom noticed is that any movement of my eyes—better yet, any displacement of my body—has its place in the same visible universe, that through these movements I examine and explore, just as, conversely, seeing occurs somewhere in tactile space.[50]

Sartre, too, stressed the reciprocal analogies between vision and touch: "1. A series of kinaesthetic (or tactile) impressions can function as analogue for a series of visual impressions . . . 2. A movement (given as a kinaesthetic series) can function as analogue for the trajectory that the moving body describes or is assumed to describe, which means that a kinaesthetic series can function as analogical substitute of a visual form."[51]

The neuropsychologist De Ajuriaguerra, known for his rigorous analysis of the relationships between mothers and infants, focused on the complex nature of perception in his published work and in his courses at the Collège de France. "Perception can be divorced neither from ocular motor function," he wrote,

> nor from cognition, nor from the affective life of the subject. Perception confirms the world of objects: there can be no cognizance of things and beings without exploration: tactile, for the blind; tactile and visual, for the sighted. Although it has been proposed that the two domains (optic and haptic) are controlled by different laws, studies of blind subjects who have recovered their sight show that they do experience some kind of transfer between the two modalities.[52]

SEEING WITH THE SKIN

The Braille alphabet gives congenitally blind people access to reading. To the extent that it allows them to construct a tactile representation of a page of text, it is an example of sensory substitution and equivalence between touch and vision.

This functional equivalence between the visible and the tangible was brilliantly demonstrated by the experiments of Bach-y-Rita on the use of vibrations as a visual substitute for the blind.[53] He perfected a technique of stimulating the skin with a matrix of small vibrators activated by an image on a television camera. The boundaries of tactile perception had been known since Mountcastle's work on vibrational sensitivity, comparing the thresholds of activation of cutaneous receptors in humans and in monkeys. The curve of sensitivity to cutaneous vibrations has a bell shape with a maximum around 80 to 100 Hz. Actually, the bell shape conceals two curves that correspond to the two types of cutaneous receptors activated by the vibrations: the Meissner corpuscles, which have a maximum sensitivity around 40 Hz, and the Pacinian corpuscles, which are especially sensitive around 80 to 100 Hz. These delicate vibrations applied to the skin were of very weak intensity; they acted mainly on the cutaneous receptors and induced no muscular reflexes or illusions of displacement of the vibrated limb.

In this way, a tactile vibratory image was created on various parts of the body (hand, abdomen, back, and so on). Remarkably, the perceptions induced by these tactile images all had the normal properties of visual perception such as parallax, perspective, enlargement (as when an object approaches the body, for example), and depth perception. Tactile and visual perception were shown to be equivalent. If a ball was thrown in front of the subject, even if the image of the ball was projected by the vibrators attached to the subject's back, he still perceived it as being thrown in front of him. Thanks to this tactile matrix device, a blind subject could play a simplified version of table tennis and thus predict the trajectories of moving targets in space.

When a person using this television camera trains with a matrix placed on the abdomen, the training persists when the camera is placed on the forehead or on the hand. The reason for this equivalence is that the image is not perceived to be skin-deep but in the external space, as one perceives the ground at the end of the walking stick and not in the palm of the hand that holds the handle and where the force feedback is felt. The image is projected into space by the brain. What a wonderful illustration of what I mean by the need to construct a projective theory of perception!

This was the technique used to help congenitally blind people to read: a little vibrator placed at the end of the finger projected letters onto the skin. However, the method ultimately had some limitations and did not replace the Braille tactile alphabet. There is probably something in the active character of

Braille that cannot be replaced by techniques that passively transmit sensory information. Indeed, although the subject can manipulate the television camera to explore the visual space, he probably does not produce the combination of active movements and sensations made possible by the Braille method. I will return to the importance of active exploration in perception.

These data suggest that tactile inputs have access to the centers of the brain that process visual information. Is there presently proof of this? Recordings of cerebral activity in early blind people have shown that the first central sensory relays that process visual information in the cerebral cortex are activated by tactile stimulation when these subjects are trying to imagine an object.[54]

I have discussed how the visible and the tangible are combined in perceiving shapes and movement. These two modalities are also involved in the coherent perception of one's own body and in establishing multiple frames of reference connected to different parts of the body. The neurons in several zones of the brain—the putamen, the frontal cortical area 6, and the parietal area 7b—respond when a target appears in certain parts of the visual field and are also sensitive to touch. By this I mean that they fire when certain places on the skin are stroked or brushed; they are bimodal.[55] These bimodal neurons have fascinating properties. For example, if a neuron fires when a monkey's cheek is touched, the same neuron will also fire when the hand is brought close to the cheek without touching it. The visual spatial receptor field is anchored to the tactile receptor field (Figure 3.7). Thus the activity of the neuron can be triggered by placing the hand or an object within a space of about 10 centimeters surrounding the cheek. This discovery has several important implications.

First, it confirms the equivalence between visual and tactile stimulation. It also explains why, when I bring my hand to my cheek, I feel my hand on my cheek even before it actually touches me. This convergence, this bimodality, actually entails something more profound than simple equivalence. In this case, the visual perception is not only the analog of tactile perception, as in the experiments of Bach-y-Rita—it anticipates it. Mere proximity becomes contact through anticipation of the area of the body that will be touched.

Gross described a remarkable cell that has a tactile receptor field extending from the eyebrows and a visual receptor field within a cone whose summit is the eyebrow and that projects in front of the subject, in the direction of his gaze. This neuron is activated by any object that approaches the eye; in a way, it extends the skin's zone of receptivity. Vision is indeed, as Merleau-Ponty said, "the brain's way of touching."

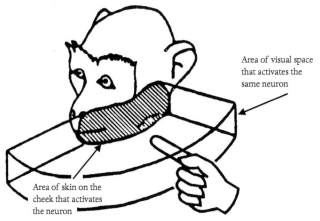

Area of visual space
that activates the
same neuron

Area of skin on the
cheek that activates
the neuron

Figure 3.7. Vision and touch: regions of the skin and visual space for which a bimodal neuron of the putamen is activated in the monkey. When the finger advances in the visual receptor field of the neuron, the neuron discharges, and discharges again when the finger touches the skin in the tactile receptor field. When the finger gets close to the skin but does not touch it, the monkey will have the impression of contact even before contact is made.

But Gross's discovery had an even more profound impact on our understanding of multisensory functioning and mechanisms of interaction. Take the properties of the neurons of the arm that are sensitive to stroking. In the putamen, bimodal neurons have a tactile receptor field that is anchored to the visual receptor field: if a monkey's arm is placed on a table before it, the tactile receptor field being, for example, on the wrist, the visual receptor field will be within a cone connecting the gaze to the wrist or in a space surrounding the wrist. If then the monkey's arm is moved, its visual receptor field will follow to stay in the same spatial register as its arm. This spatial updating is probably a response to proprioceptive signals; it can be induced by copies of motor commands.

To Be Touched or to Touch? That Is the Question

I have repeatedly emphasized that the multisensory character of perception includes signals that derive not from the senses but from the intention to move. The active character of perception is evidenced by this profound influence of the intentional character of gesture. Here is an example.

Close your eyes, and touch your left hand with your right hand. You will feel two very different sensations that either overlap or cancel each other out. Merleau-Ponty was absolutely fascinated by this observation:[56]

Husserl[57] refers to an experiment of touching. When I touch my left hand with my right hand, my touching hand grasps my touched hand like an object. But suddenly I notice that my left hand is beginning to feel. The relationships are reversed. We experience an overlap between the contribution of the left hand and that of the right hand, and a reversal of their function. As a physical object the hand always remains what it is, and yet *it is different according to whether it is touched or touching.*

Merleau-Ponty often comes back to this example, which for him remains one of the great mysteries of perception. I have put the final part of the last sentence in italics, because it suggests very clearly how the idea of perception is different in these two cases, one where the limb is touched passively, and the other where it is touched by a motion the subject makes himself. This long philosophical introduction is just to say that evidence now exists of a difference in activity of the neurons of the brain between these two cases.

Tactile perception is not only connected with vision, it is also influenced by the active character of visual attention. In one zone of the superior temporal sulcus (STS) in the monkey, neurons are activated when the hand of the monkey is lightly stroked with a stick; but if the monkey grabs the stick himself, the same neurons stop firing. The active movement suppresses the transmission of tactile information. Perrett attributes this disappearance of response to the predictive character of the contact, and observes that the neurons of this area are very sensitive to unpredictable tactile stimuli.[58] In contrast, when the movement of the animal induces the contact, a still mysterious mechanism of inhibition suppresses the activity. This observation corresponds very well with the subjective impression that we have of a very reduced sensation of contact when we touch ourselves.

> When the monkey's active exploration leads it to encounter repeatedly an object of particular texture and compliance at a particular location, the object's tactile properties can be said to become "expected." As shown above, tactile feedback, which results from the monkey's reaching and contacting such expected objects, fails to activate somatosensory cells in the STS. If, however, a novel object of different texture/compliance is introduced at the same location, tactile feedback does result in activation of STS cells.[59]

This selectivity also holds true for vision: the cells of the STS sensitive to the entry of visual objects into the visual field of the monkey lose their sensitivity when the arm of the animal comes into the field. These effects are not due to the simple fact of attention. They are actually very selective inhibitions of an aggregate of multisensory properties (texture, visual aspect, space) of a part of the body of a monkey or a known object. Perception is selection and anticipation. Time will reveal their mechanisms.

This selection by anticipation is not an isolated case: there are many accounts of reduced effects of sensory stimuli due to self-activity of the animal. For example, the Purkinje cells in the cerebellar cortex—an organ important for the coordination of gestures—respond much less to cutaneous stimuli when animals make endogenous movements than when experimenters touch them. Generally, the effect of a sensory stimulus is reduced during endogenous movements. However, when the task is new, or when a stimulus is unexpected, sensitivity is restored. In a recent review, Prochazska concluded: "We have seen many examples of task and context-dependent sensory transmission in widely ranging species and widely ranging motor behaviours. From all this it seems safe to conclude that anticipatory gain control of sensory transmission is indeed a fundamental strategy of motor systems."[60]

Force Feedback and Visuomotor Functions

Cooperation between vision and the different senses extends to other proprioceptive functions—to the sense of effort, for example. To study such cooperation, in this case the contribution of haptic information to the visual perception of the properties of objects, my team and I built a force-feedback joystick. But I am getting ahead of myself. If I lay a pair of glasses or a pen on my desk, I can, thanks to my vision, glean quite a bit of information about the volume and the geometry of the object lying there. I can even infer certain dynamic properties; for instance, the number of degrees of freedom, the object's weight, its elasticity. Now, if I close my eyes and simply take the object between two fingers (try it with a pair of glasses or any complex object nearby), the perception of the force the object is exerting on my fingers given me by my haptic sense—that is, the combination of proprioceptive information from receptors in the skin, muscles, tendons, and joints—allows me to confirm many of the dynamic properties of the object: its mass, its inertia, the number of degrees of freedom, its viscosity, and so on. I use the term "force feedback" to describe the nervous system's use of force as information.

Here is an example that explains which mechanisms underlie this multisensory cooperation. Suppose that I want to follow a rigid surface with my finger, so as to position a peg in a hole. My brain can adopt any of three strategies for following this surface.[61] First, it can ignore any forces and carry out a simple visual tracking of the contour of the surface. Second, it can apply a continuous force to the surface and detect that my finger loses contact when the force becomes nil. To keep my finger in contact with the surface, I just have to control the intensity of the force detected by touch, as with a servo-control. Finally, my brain can adopt a third strategy: it can form an internal model of the shape of the object to follow (here a surface) and program a movement to follow a trajectory that passes *slightly inside* the surface. At the same time, the muscles are controlled so that my fingertip acts like a supple spring (in robotics terms, I might say that the strategy is compliant). The difference between the programmed position of my fingertip (inside the surface) and the actual position of my fingertip (on the rigid surface) will cause the virtual spring to compress, resulting in a slight pressure of my fingertip on the surface. In addition, the spring will expand or contract to absorb moderate differences between the actual surface and the internal model, thus maintaining contact with the surface throughout the movement.

These conditions make clear the respective roles of vision and force feedback in each of these strategies: in the first, vision is essential; in the second, it is scarcely involved; in the third, it serves to construct an internal model of the trajectory that makes it possible to anticipate *and control* the consequences of the movement.

I want to stress the essentially predictive character of the third strategy. The fact of having constructed an internal model of the trajectory allows my brain not to have to wait until my finger has run the entire course to correct errors but, on the contrary, to anticipate certain properties of the trajectory and to make assumptions. Perception thus becomes a comparison between an expected state and an actual state. This is the heart of my thesis.

How can this type of process be studied? Suppose, first, that this surface appears on a computer screen and that the object can be made to move with the aid of a joystick. I can ask the subject to follow the object with the joystick. This task is similar to that of playing video games at a bar. But I can also introduce a motor between the joystick and the apparatus that measures its movement. Each time the object touches the surface on the screen, the motor exerts a force on the joystick to give the impression of resistance. This impression will be all the stronger the more the subject leans on the joystick. In this way I can simulate the force feedback that the real surface exerts on a subject's hand when he actually carries out the task. This device shows that the brain uses a mixture of the second and third strategies. Consequently, vision and the haptic sense cooperate in a task of contour-following by constructing an internal model of the surface.

The Problem of the Coherence and Unity of Perception

Models of multisensory interaction are interesting because they enable us to make predictions about the way sensory receptors combine their signals. But they do not do anything to solve the fundamental problem, which is that of the unity of perception, or coherence. How the perceptual unity of a visual object can be achieved by temporal synchronization of the discharge of neurons that respond to different properties of the object was discussed in Chapter 1. But constructing coherence means solving many problems, because sensory inputs have properties that separate them and make their fusion difficult.

Sensory inputs are ambiguous. The problem is one of the difference between acceleration in one direction and braking in the other in the canals and otoliths, of Coriolis accelerations in the canals, and of speed in vection. Sensory inputs do not occur in the same systems of coordinates. The canals are in a Euclidean frame of reference connected to the head; vision occurs in the retinal space; proprioception of the eyes, neck, and limbs occurs in the space of the muscles, so sensory space is, in this case, utterly linked to motor space; hearing takes place in the space of frequencies. I have offered one example of a solution: the similarity of the planes of detection for visual and vestibular information.

Sensory inputs are staggered in time. Try to move your finger and your foot in synchrony. Wait! Don't start right away. You will notice that your first attempt will not be synchronous. It will take you several tries to synchronize your movements. In fact, signals from the muscle receptors of the foot arrive at the cerebellum with a delay over those of the fingers that can be considerable when you think that nerve impulses circulate in the medulla at velocities of 10 to 80 meters per second. Thus in humans it takes 10 to 100 milliseconds for contact of the foot on the ground to reach the cerebellum and two times less long for signals from the finger. Ten times less for signals from the tongue! It is thought that the brain anticipates the movement of the foot or retards the information from the finger. I have shown several solutions to this in neural operations in the colliculus.

Sensory inputs do not cover the same range of velocities. The vestibular receptors are very fast and detect accelerations; vision is slower. The receptors of the muscular spindles have widely varying response times, which is also true for the tactile receptors, whose bandwidths are different from those of vision, whereas very tight perceptual relationships exist between sight and touch.

Finally, sensory inputs are often fuzzy. They produce innumerable random fluctuations that are unrelated to the magnitude detected. This noise is due to the chemical and mechanical properties, among others, of transduction, but also to neural noise introduced into central relays.

So the problem of coherence is not only a problem of geometry or of dynamics. It assumes active central mechanisms that permit resolution of ambiguities, catching up with or anticipating differential delays between receptors, unifying space via clever biological mechanisms that are more than just changes of coordinates, and so on. A genuine theory of coherence has yet to be constructed.

I will start from the supposition that constructing coherence is not only an effect of convergence, but also the product of a central activity that depends on the knowledge of a priori mechanisms that each species possesses. "I realised," wrote Marcel Proust,

> that it is not only the physical world that differs from the aspect in which we see it; that all reality is perhaps equally dissimilar from what we believe ourselves to be directly perceiving and which we compose with the aid of ideas that do not reveal themselves but are none the less efficacious, just as the trees, the sun and the sky would not be the same as what we see if they were apprehended by creatures having eyes differently constituted from ours, or else endowed for that purpose with organs other than eyes which would furnish equivalents of trees and sky and sun, though not visual ones.[62]

Perception is an interpretation; its coherence is a construction whose rules depend on endogenous factors and on the actions that we plan. The difficulty in building a theory of coherence is that there is most likely not one single coherent theory for all of perception. There are often several ways of arranging sensory data to construct a coherence. This range of possibilities is probably a key to the way illusions are manufactured.

Many mathematical models have been proposed to account for multisensory interactions, but those that emphasize the problem of coherence are rare. In general, these models merely execute a combination of sensory signals with more or less predictive operators.

Droulez and Darlot proposed a model of multisensory fusion that directly addresses the problem of coherence.[63] Their central idea is that the brain is not interested in the physical variables of classical mechanics, or which stimuli are specific to which receptors, in and of themselves. The brain seeks to reconstruct variables relevant to the behavior and action of the organism. To this end, it uses the fact that messages about movement, supplied by the array of sensory receptors, are in general redundant. In other words, the same movement variable can be calculated or estimated by several combinations of receptors, in addition to any specialized receptors.

The angular velocity of the head in space during movement provides a good example. It is an important variable because it is a derivative of displacement. One can thus integrate (in the mathematical sense) and obtain a displacement that the brain will use either to activate stabilizing reflexes or to estimate trajectories or even to coordinate the movements of the eyes, head, hands, and so on.

According to the basic assumption of the model of Droulez and Darlot, the brain estimates this velocity of the head in at least two ways. The first, direct, is detection by a specialized receptor, the semicircular canals, which detect angular acceleration of the head in its plane, called the horizontal plane. But this specialized receptor transmits a velocity to the brain, because its viscoelastic properties delay the signals. This is precisely what an integrating filter does. The second, indirect, way that the brain estimates the head's velocity is a multisensory convergence of other variables—for example, the velocity of retinal slip and that of the eye in its socket. Adding these all together gives an estimation of the angular velocity of the head.

The main measure of this model is thus the discrepancy between a specific sensory input and a prediction of this variable from combinations of other signals. It was first suggested in 1989 to account for the fusion of the sensory messages supplied by the semicircular canals, the otoliths, and vision. Droulez and Cornilleau-Pérès wrote:

> The fact that individual sensors provide reliable information only in a limited (and sometimes variable) working range, in both spatial and temporal domains, and the fact that the degradation of the measure by noise can depend on the context (see for instance the analysis of optic flow) and the existence of multiple interpretations suggest that sensory signals cannot be processed as a set of measures but as a series of constraints upon the internal estimates. We suggest that sensory signals are not used to *directly estimate* the relevant variables but to *estimate the mismatch* between internal estimates and measures.[64]

For example, it really does seem that in vision the brain employs at least two constraints: continuity and rigidity. Indeed, the objects that surround us are in general locally continuous (that is, they are not divided). The painters called pointillists provided a vision of nature that only makes sense because it is filtered by the continuity of perception. Moreover, objects are generally rigid. A body of evidence shows that the brain makes these assumptions of continuity and rigidity, which helps in constructing coherence. The brain also favors symmetry and thus tends to impose symmetry on the world it perceives.

But the neural basis of coherence is still not well understood. In all likelihood, coherence is due, in part, to the simultaneity of neural activities in several parts of the brain.[65] Many recent theories and experimental results in fact suggest that establishing relations among features of perception that have been analyzed separately in several areas of the brain might also be achieved by temporal synchronization; that is, simultaneous activation of several groups of neurons. This temporal encoding most likely takes a variety of forms that have yet to be discovered.

Autism: The Disintegration of Coherence?

It seems to me that the disintegration—the fragmentation—of coherence is the source of numerous symptoms of several major illnesses that have as much to do with psychiatry as with neurology. A few reflections on autism will illustrate this thesis. Autism is a developmental disorder that strikes very young children. It is not known what causes it. The main symptoms of autism are, first, a qualitative alteration in social interaction: autistic children do not look at the people speaking to them; when they are put in a group, their gaze does not follow the activities of the members of the group. Their behavior isolates them. Next, they begin to exhibit impaired communication: they hardly speak, do not return a smile or respond to facial expressions of emotion, and so on. Finally, they engage in stereotypical behaviors that they repeat endlessly, totally divorced from any social context or even any connection with the activity in which they are engaged. They suffer from a perseveration syndrome. Autistic children show disturbed initiation of action and, more generally, disturbed intention. Like patients with Parkinson's, they freeze in initiating certain actions. By "freeze" I mean that they are perfectly capable of making a move, but that they decide not to begin it.

They display what appear to be nonmotor disturbances of anticipation of the hand, since if they suddenly commit their hands to action, muscle tone returns. They exhibit visual and auditory disturbances. Some of these children can be classed in subgroups that present particular symptoms to a greater or lesser degree.

It is possible to associate these symptoms with several types of dysfunction: initiation of walking to disturbances of the basal ganglia, the absence of posture and of anticipatory synergy to disturbances similar to those described by Babinski in patients with lesions of the cerebellum, disturbances of motor intention, and perseveration with prefrontal and frontal deficits. But it is also possible that these disturbances have a common origin. These children are incapable of relocating local aspects of their action in a global context.

In *The Enigma of Autism*, Uta Frith gives several examples of the disintegration of coherence that are helpful in explaining what I mean by coherence. She cites the example of little Jerry:

> According to Jerry, his childhood experience could be summarized as consisting of two predominant experiential states: confusion and terror. The recurrent theme that ran through all of Jerry's recollections was that of living in a frightening world presenting painful stimuli that could not be mastered . . . Nothing seemed constant; everything was unpredictable and strange. Animate beings were a particular problem . . . I would like to suggest that the repetitions and sensations that were such poignant experiences in Jerry's life (and also in the lives of many other autistic people) are two sides of the same coin: part of the same underlying problem. This problem is what I have already labelled lack of central coherence. Let us consider what it would be like if perception reflected a fragmented rather than a coherent world.[66]

Autistic children develop what Uta Frith calls a "local coherence." She cites another case; that of Elly, a child who had constructed a system of telling time based on the shadow of her body cast by the sun.

> (For instance, Elly paid close attention to shadows because they were for some reason important and meaningful to her, and very relevant to her moods. When she travelled to a different time zone, she was alarmed that her shadow at 6 PM was not where it would have been at home. She could not relax until her mother had explained to her that 6 PM on her watch meant it was only 5 PM at the new place.) This example suggests that Elly had a limited but strongly coherent scheme about the position of the sun at a certain time and the length of her shadow. It really mattered to her when there were unexpected discrepancies. The scheme had to be kept coherent. This insistence on sameness is a type of local coherence. It is not at all like *central* coherence.[67]

I think this example is very interesting because Elly had devised a method but had great difficulty placing it in a broader context—in thinking about changes in time zones, for example.

Interestingly, studies in cognitive psychology on the problem of decision making show that our cognitive apparatus makes many mistakes, due in large part to our inability to depart from stereotyped schemas characterized by Uta Frith's local coherence. We have trouble revising according to context. The perseveration of autistic children, their immutability, is perhaps only an extreme version of many widespread behaviors.

But the most interesting aspect of Uta Frith's theory is that she goes even further in the analysis of the importance of coherence: coherence, she says, is necessary not only to construct a perception of the body or of its relationship with the environment, but also to elaborate what is called a theory of mind. This notion was invented by psychologists to refer to the fact that we attribute thoughts to others, that we have an idea, a theory of what they have in their minds, of their intentions, and so on. Premack showed that this ability to attribute thoughts to others shows up in nonhuman primates, and he gives several examples of it. And actually anybody who has a dog will have noticed this faculty in the animal. Uta Frith says with respect to the autistic child: "We have hypothesized that they [autistic children] do not have the basic propensity to pull together vast amounts of information from events, objects, people and behaviour. Even if they had the cognitive prerequisites that enable them to mentalize, they would only form 'small' theories about mental states, but not a comprehensive theory of mind. Autistic children are behaviourists. They do not *expect* people to be kind or to be cruel. They take behaviour as it is."[68]

She is right in saying that if autistic children have no coherent representation of the world, they cannot construct a theory of mind for others and thus communicate. It is certainly not possible to make any internal assumptions about the intentions of others if one has not succeeded in giving coherence to one's perception of relationships between oneself and the environment, and with all the information it contains.

Uta Frith completed her analysis by observing in autistic children the impairment of what she calls "planning functions" (working memory, control of impulses), which are normally a function of the frontal lobe of the cerebral cortex. There is a paradox here, for although the impairments of autistic children imply a deficit of the frontal lobe, patients presenting with this deficit are not autistic. It is possible that the question is one of differences connected to stage of development.

In any case, how is it possible to imagine that children can coherently evaluate the people they see if they cannot evaluate relationships between their own bodies and the environment? How is it thinkable that children who perceive only a fragmented universe would have any desire to talk? How can beings whose brains are the center of multiple incongruities have even the slightest desire to communicate with a world with which they cannot identify?

Finally, I would like to suggest a reason why autistic children have difficulty in perceiving the "global" features of the world and are confined to "local" events. My hypothesis is that they may not be able to mentally manipulate "allocentric" relations in their perception of space and remain stuck at "egocentric" coding of space, as discussed in Chapter 5.

4

The ethologist Golani's description of the exchange of looks between two dogs about to chew out each other's throats will serve as an excellent introduction to the problem of frames of reference. Golani used the system of notation invented by the Israeli choreographer Eshkol to plot the movements of dancers in three different but coexisting frames of reference: a so-called body frame of reference that I will call egocentric; an environmental frame of reference, also egocentric, but accounting for movement with respect to the external framework of the space where the dance is taking place; and an interactive frame of reference based on the connection of two partners that I will call relative because it is defined by the relationship between the two dancers.[1]

Golani based his account of the extraordinarily rigid connection arising from the fixed gaze of two hostile dogs on the third frame of reference. He showed that a line of attention as unmalleable as a steel rod is established between the eyes of the two animals; the dominant dog has only to tilt his head as a first step, before ever touching, to fell the other. Any deviation means death for one of them. In fact, the distance between their two heads is fixed, and their relative position has only to vary several centimeters to facilitate a fatal aggression. This finding would have been impossible without recognizing that the frame of reference on which both dogs base their movements must be relative; they ignore their environment just as the moon and the earth maintain a connection only through their mutual attraction.

This chapter explores one question: Which frames of reference does the brain use to organize perception and action?

The concept of a frame of reference is tied to that of space. Our actions un-fold in a space organized, according to Grüsser, into "personal" space, "extra-personal" space, and "far" space.[2] Each of these spaces is itself organized into several subspaces explorable by different mechanisms that provide distinct frames of reference.

The first experimental proof that the distinction between personal and extrapersonal space had a neural basis was supplied by Hyvarinen[3] and con-firmed by Mountcastle and his students.[4] Neurons situated in the parietal cor-tex of the monkey fire each time a person enters its grasping space with a grape. The sight of the grape (monkeys are fond of them) outside of this space is not enough to trigger the activity of these neurons.

Personal space consists of self-space (egocentered). It is perceived by the internal senses and is in principle located within the limits of a person's own body. But there are some important distinctions. First, the body can be per-ceived as an external object, especially by vision. The hand that I see at the end of my arm is not necessarily my own (in Chapter 2 we described the case of a patient who denied possession of her arm). Thus, attributing an element perceived by the body to itself requires a perceptual decision. The dissocia-tions and denials of ownership of elements of the body that manifest in cer-tain illnesses are all too familiar. Moreover, a perception of the body can occur even in the absence of parts of the body. The case of phantom limbs, for in-stance, reveals the existence of mental representations of the body—that is, in-ternal models of the body independent of its presence. Patients whose arms have been severed feel their arms perfectly and even feel the movement of their hands. Sometimes the presence of the whole body is perceived as ambig-uous, and the subject may experience a phenomenon called autoscopy.

Grasping space is itself divisible into intraoral and perioral space—espe-cially important during the first six months of life—as well as into other local spaces connected to specific activities related to grasping. It is marked out by visual, tactile, and olfactory cues. Indeed, the perception of the body can be extended by a tool. For example, you feel the tip of your hand as if it were at the end of the pencil you are holding: the body extends into the tool. Contact with the ground is felt at the end of stilts, the driver of a car feels the wheels on the ground, and it is well known that pilots feel the wheels of the airplane on landing as if they were their own feet. Haptic sensations play a decisive role in these perceptions. This property of the body—integrating with the physical elements it has grasped—is very important, for it often determines the frame of reference. What is remarkable about this extension is that the object is perceived where it is supposed to be in the extrapersonal space and not at the point of contact of the instrument with the body. The brain is able to construct a spatially correct extension.

This ability to extend the body and to localize the point of contact is probably acquired very early, during multiple movements the baby makes in carrying things to its mouth. If the object is long, the baby's brain learns to correlate the tactile sensations of the hand holding the object with the sensations of the mouth. The essential element of this connection seems, to me, to be the rigidity of the object, which introduces covariations of pressure between the two zones (hand and mouth) costimulated by the object. The assumed rigidity of objects is an important property that turns up again in the visual perception of their curvature and shape. In this case, it is said that the brain formulates a rigidity hypothesis. The brain makes assumptions about the world based on which it constructs internal models of reality. In the case of extension through a tool, over the course of development a simple correlation between two parts of the body could become the simulation of a correlation between a part of the body (the one that is holding the tool) and a point in space.

Egocentric and Allocentric Frames of Reference

The concept of reference frames excited the imagination of many researchers, who devised different reference systems, just as geographers devised several systems of projection to make maps of the world. Which frames of reference does the brain use, and does it use them in the mathematical sense of the concept?

It is possible to represent the position of objects in several ways. Suppose that you wish to describe the relationships between two objects in a room (a chair and a table) and your body. One way of encoding these relationships is to relate everything to yourself, to estimate the distance and the angle of each of these objects in relation to your body. This polar type of encoding is typically egocentric. The position of the two objects can also be encoded in Cartesian coordinates, using the two axes formed by the frontal and sagittal planes of your body, say, or the perpendicular planes of the vertical semicircular canals, and so on. The frame of reference is still egocentric.

A second means of encoding spatial relationships is to use the relationships between the objects themselves or relate them to a frame of reference external to your body. Take as a reference point the door of the room and this time evaluate the position of the chair and the table with respect to the door.

There is no reference to your body. This encoding is called allocentric, centered elsewhere than on your body. The distinctive feature of this encoding is that it moves the basic point of reference away from the center into the external space. Often the term "exocentric" is used in place of "allocentric"; these two terms have similar meanings.

The difference between these two types of frames of reference is important. It seems that most animals are capable of egocentric encoding, but only primates and humans are genuinely capable of allocentric encoding. The power of the latter is that it enables mental manipulation as well as manipulation of the relationships between objects without having to continuously relate them to the body. The ability to think about the distance between the village bell tower and the police station, between the synagogue and the pastry shop, to wonder whether one is larger than the other is what is distinctive about the human brain and leads to speculation about geometric relationships. Moreover, allocentric encoding is constant with respect to a person's own movement; thus it is well suited to internal mental simulation of displacements. It appears rather late in children, who first relate space to their own bodies.

One particular example of an egocentric frame of reference is our perception of the midsagittal planes of our bodies: that is, we perceive very easily whether things are to our left or to our right. This perception is impaired in patients with right parietal lobe lesions who exhibit a syndrome called "spatial neglect." The midsagittal plane shifts at the same time that they ignore the left part of space, or the left part of their plate when they eat, or the left part of their body. Using functional brain imaging in normal healthy people, my colleagues and I have recently shown that the perception of this midsagittal plane involves activity in a pair of areas in the parietal and the frontal cortex; the same areas involved in the neglect syndrome when they are lesioned.[5] Other data from Galati and his colleagues, soon to be published in *Experimental Brain Research*, suggest that the allocentric frame of reference comprises in addition an occipitotemporal system involving the parahippocampus, whose role in the processing of spatial information has also recently been demonstrated.

Natural Frames of Reference

GRAVITY

The semicircular canals of the vestibular organ define a basic Euclidean frame of reference that may be at the root of our geometric perception of space. By its very structure, it provides a reference frame only for movements of the body. It is an egocentric system.

Nature bequeathed us another frame of reference connected to external space: gravity. From our perspective, this omnipresent force has several very important features. First, it does not vary in magnitude or direction with respect to a plane tangential to the ideal surface of the earth—it is a constant of terrestrial space. Second, it can be detected by specialized receptors, the otoliths. Finally, it constitutes a reference point external to the body and consequently, as Paillard says, "an external plumb line" related to bodily movements in a frame of reference he calls "geocentric."[6]

To demonstrate the importance of this second frame of reference, I will examine the bodily motion of a person who is jumping in a complicated way, or running, or doing moguls down a ski slope; or the way storks fly, lions run, and so on. My examination will reveal that during what André Thomas calls "surplus equilibrium" (l'équilibre de luxe) with regard to movements in sports or in dance, the head is stabilized in rotation.

I discovered that this property also applies to humans during daily locomotor movements such as running. I took some photographs that Muybridge had made of people in motion,[7] and I drew on the photographs the line that connects the external canthus (the corner) of the eye to the meatus of the ear (Figure 4.1). This line is approximately parallel to one of the planes of the semicircular canals. Making a montage of the photographs superimposing the meatus of the ear on all of them revealed the head to be remarkably stable during rotation. Using video cameras connected to a computer, we showed that during walking, rotation of the head is stabilized about positions determined by the direction of gaze. In other words, we are not unlike birds when they fly or gazelles and ostriches when they run. If you look at them on film, you will see that their heads remain perfectly stable in relation to the vertical. Most likely, the otoliths estimate how much the head is tilted with respect to gravity.[8] The direction of gaze determines the plane of stabilization.

It is as if the brain creates a stabilized platform to coordinate movements of the limbs. In these complex movements, the feet rarely touch the ground, so much so that the earth's surface cannot serve as a reference point. The brain uses the gravity detected by the vestibular system to stabilize the head and create a mobile platform as a frame of reference. The advantage of this solution is that because they do not undergo rotation, the visual and vestibular receptors are freed from the problem of gravito-inertial differentiation that I discussed in Chapter 2. They can cooperate better to detect translations based on the information supplied by optic flow. Engineers who must control the

Figure 4.1. In these drawings of a man running, the head is stabilized to control posture and coordination of movements. Photographs by Muybridge were superimposed at a single point of the head, the auditory meatus. While running, the head is stabilized in rotation at an angle that depends on the direction of gaze. This stabilization relies on the vestibular system, which detects the angle of the head in relation to gravity. The straight line connects the canthus of the eye with the meatus of the ear. It indicates roughly the plane of the horizontal semicircular canal.

movement of satellites in space adopt the same principle. They attach a small platform to the body of the satellite and stabilize its position with reference to the stars.

Gravity intervenes in the organization of movements at specific times in development. When its influence on young rats is altered, locomotion is significantly delayed.[9] Thus there is a period critical to motor function, around ten days after birth, during which the nervous system needs gravity as a reference for organizing coordination of movements. This experiment supplies additional evidence for the existence of critical periods such as that already demonstrated in the development of the visual system.[10] In this latter case, proprioceptive information is important for specifying the directional sensitivity of the neurons of the visual cortex.

THE VERTICAL SUBJECTIVE AS A MULTISENSORY CONSTRUCTION

One of the clearest examples of the essentially multisensory character of perception is that of perception of the vertical. All of us can close our eyes and accurately indicate the direction of the terrestrial vertical; that is, the direction of gravity. On the other hand, if we are in a lit room tilted in relation to the true vertical, we might make a misjudgment due to our perception of the vertical based on the visual cues in the room—what we call the visual vertical. Thus vision contributes to our perception of the terrestrial vertical and may override vestibular cues.

Several laboratories have demonstrated the influence of vision on perception of the vertical by constructing tilted rooms where the subject not only perceives the room's vertical as the true vertical and thus neglects in part the information from the otoliths, but also unconsciously tilts his body to align it with the visual vertical. If a subject is placed in a darkened room and if the only thing illuminated is a rod tilted toward him, he will perceive the vertical as intermediate between the gravitational vertical and the visual vertical (the Aubert effect).[11] Paradoxically, subjects are able to perceive the vertical tilt more than the tilt of the rod (the Müller effect)[12] when the angles are significant. In consequence, perception of the vertical is the result of a multisensory compromise. The brain receives messages from the visual vestibular sensors of vision and proprioception and compares them with the intrinsic references of the axial direction of the body.

Patients who have suffered vestibular lesions use static information from the visual environment to work out a vertical. Astronauts also use visual cues from the space station to compensate for the absence of gravity. The following experiment, carried out by our laboratory in conjunction with the French national center for space research over the course of several space flights, established the role of vision in the reorganization of sensory control of posture in the absence of gravity.[13] We asked an astronaut to maintain an upright posture. On earth, he could do this in an instant, but trying to do it aboard a space vessel makes no sense because the vertical is no longer signaled by gravity. So we asked him to hold himself erect, perpendicular to the floor of the station. We used an experimental trick, a sort of shoebox lit from within and attached to his head, covering his eyes. He thus had no visual information about the position of his body in space. Moreover, given that his body had mass but no weight, the forces exerted on his feet had changed, such that he could no longer rely on tactile pressure to evaluate the position of his body in relation to the floor.

The first day of the flight, the subject leaned forward markedly but responded that he was perfectly upright and perpendicular with respect to the space station. At the same time, the tonicity of the muscles that maintain the upright position—the extensors and flexors of the ankles—was redistributed. Instead of the tonic activation of the extensors that on earth keep us from falling forward, the flexors were tonically activated. The combined information from all the proprioceptors was consequently insufficient to detect the angle of the astronaut's body: this angle was not calibrated, the body schema was distorted, and tonic motor activity was asymmetrical. The subject remained unaware of the distortion. Since he could maintain a perpendicular posture without difficulty when the box was removed and his view of the station was restored, vision was essential to this recalibration of proprioception. A few days later, the same subject, placed in the same conditions, was able to hold his posture perpendicular to the station. His brain had come up with new mechanisms for calibrating proprioception.

The brain seeks to work out a reference; it extracts from the physical world a relevant variable that simplifies the neural processing of sensory information and that guides action. Action is tied to a frame of reference.

RECEPTORS FOR GRAVITY IN THE STOMACH

I have described how the brain uses a basic egocentric frame of reference, the vestibular system, that enables it to refer to the vertical plumb line. But it seems that it also makes use of another more mysterious reference point called the idiotropic vector. This is an innate perception of the longitudinal axis of the body in humans.

A subject is placed in an armchair that tilts laterally in relation to the terrestrial vertical. With his eyes closed, he is asked to indicate what he perceives to be the direction of the terrestrial vertical. He is observed to be slightly tilted with respect to the true direction of gravity. The direction he indicates is intermediate between the axis of his body and that of the true vertical, showing that the perceived vertical (often called the subjective vertical) is the resultant of two verticals, that of gravity detected by the otoliths, and of the axis of the body or idiotropic vector.

Mittelstaedt, who studied the mechanisms underlying the perception of the gravitational vertical, argued for adding another sense to the existing list.[14] Indeed, he discovered that receptors located in the stomach also detect gravity.[15]

Ferrier had already noted this possibility:

> We have thus far considered the influence of tactile, optic and labyrinthine impressions on the functions of equilibration and co-ordination, and it has been shown that the influence of each is capable of experimental demonstration. Though these are apparently the main factors in the general synaesthesis, the possible participation of other afferent fac-

tors in the general result is not absolutely excluded . . . But there appear to me grounds for attributing some influence to visceral impressions. It is well known that cats and other members of the family Felidae, including animals which possess in a marked degree the faculty of equilibration, have in their mesentery relatively large numbers of Pacinian corpuscles, which are specially adapted for transmitting pressure stimuli to the sensory or afferent centres . . . It would seem, therefore, not improbable that the viscera are in relation with the centres of equilibration, and that they mutually affect each other. This is supported by the phenomenon of a distressing form of dyspepsia, characterised by sudden attacks of giddiness, described by Trousseau under the name of *vertigo a stomacho laeso*.[16]

Mittelstaedt recently concluded that this physiological system for somatic detection of gravity might be located in the kidneys or in the vascular system. Only additional experimentation can resolve this issue.

Next I will consider more generally the multisensory character of perception. The contribution of so many sensory receptors to the perception of the vertical raises a theoretical question. There are two conflicting ways of explaining a single percept such as the subjective vertical. The first regards it as the resultant of all the verticals signaled by the various sensory systems. This is the accepted approach. Another holds that there is one or several brain structures organized in such a way as to constitute an internal model of the bodily vertical, or rather the axis of the body, and that this internal organization, this "body schema" according to Gurfinkel's idea,[17] is modulated or altered according to the configuration of receptors. Although you might conclude the two ideas are similar, they do not at all imply the same neural organization. One proceeds from the periphery to the center (bottom-up organization), the other from the center to the periphery (top-down organization).

THE ROTISSERIE EXPERIMENT

Vision is not the only sense that cooperates with the vestibular receptors in determining the orientation and movement of the body in space. Touch plays a fundamental role that is worth examining, for it reveals the richness of multisensory interactions.

To begin with, skin plays a role in detecting asymmetry in the control of posture. Rademaker established this function of cutaneous receptors. He showed that when a dog—or a horse—is lying on its side, reflexes of vestibular origin (righting reflexes) automatically trigger the coordination of motor activity required for the animal to get back up. He also discovered that simply pressing a board held parallel to the surface of the ground to the exposed flank of the prone animal inhibited the righting reflex.[18]

Interpreting this observation is straightforward. Righting reflexes are triggered by asymmetric impulses of vestibular origin and by the visual detection of body tilt in relation to the visual vertical. But when a board is pressed to the flank of the animal, the tactile information supplied by the pressure of the ground on one side and of the board on the other gives the brain symmetrical information. Apparently, the brain's confidence in this tactile information (engineers would call it gain, neurophysiologists synaptic weight) is sufficient to block the righting reflex. This old experiment argues for my hypothesis that the brain uses configurations of receptors to work out a perception and initiate action.

Experiments in space have contributed further proof.[19] On the ground, a subject was positioned before a disk rotating in a vertical plane, which caused a circular vection.[20] A very slight tactile stimulation of his shoulder, which triggered a tactile asymmetry, markedly increased the discrepancy between the true vertical and the perceived vertical. In space, aboard the American space shuttle, floating and holding on only to a bite bar with his teeth, the astronaut felt a considerably stronger vection and had the impression that the vertical he perceived was turning along with him. The absence of gravity and the lack of contradiction between visual and tactile information increased the credibility, for the brain, of bodily rotation.

Here, now, is the rotisserie experiment. A subject lying on a bed can discern which way is up without difficulty. Lackner had the idea of placing a subject in a machine developed during the 1950s by American air force researchers interested in how the otolithic organs of the vestibular system detect the vertical.[21] This machine looks like a rotisserie. Subjects are laid in it horizontally, as if they were brochettes or chickens, and turned at a constant velocity around a rotational axis perpendicular to terrestrial gravity.

When they are in the light, with their eyes open, subjects have no trouble perceiving the direction of their rotation. If the light is switched off, they have only vestibular and tactile information to rely on for detecting the axis of their rotation. The semicircular canals are stimulated by the initial acceleration and the otoliths by the sweep of the components of gravity in the planes of the respective maculae of the utricle and the saccule. Then, because the velocity of the rotation is constant, and the angular acceleration therefore nil, the response of the semicircular canals subsides after about 20 seconds. Only detection by the otoliths of the angle in relation to gravity persists. Consequently, during rotation at constant velocity, only the otoliths and the skin contribute to the perception of the direction of the body with respect to the vertical.

In general, the subject correctly perceives the axis of rotation in the dark. But Lackner demonstrated that just manipulating tactile cues is sufficient to completely alter the perception of orientation of the body. If the subject's feet are suddenly pressed, he has the impression of tipping up and rotating in a vertical position. If his buttocks are pressed, he feels as though he is sitting and spinning in a chair. If his head is pressed, he experiences conical motion about his head at the apex. In other words, despite the sweep of pressure on the side of the body, the brain treats the *local* point of pressure as a point of reference that determines the perceived center of rotation. In this case, the tactile cues determine the frame of reference in which the rotation occurs.

Sometimes the postural context affects how the brain interprets tactile cues. An experiment of Gurfinkel's is a nice illustration of this property. Lay your hand flat, palm up, on a table and ask someone to draw a "p" on your palm with a pencil. Now put your hand behind your back and ask the person to repeat the drawing. You will note that in the second case you perceive the letter as a "b" and not as a "p." In other words, the same sensation yields a different perception according to the orientation of the hand in relation to the body.

The extremely discontinuous character of these alterations in perception shows that the brain carries out genuine *perceptual decisions* based on the state of several receptors, which we call a configuration of receptors.

Multiple Frames of Reference Connected to the Limbs?

Is it really necessary to have a single reference point for planning movement? The points of reference used by the brain are flexible, and depend on movement. When you lift your arm and point your finger at a target, the brain simplifies the command in a remarkable way by monitoring only two kinematic variables: elevation and azimuth.[22] The other aspects of movement are controlled by rules of phase relations among the parts of the limbs. When you move an object on a table, the angle made by your upper arm and forearm, and by your arm and trunk covary: a single command suffices to execute the two movements. This way of controlling movement assumes that the brain uses a frame of reference. It might use, for example, the spatial vertical in relation to which the polar coordinates of elevation and azimuth are calculated.

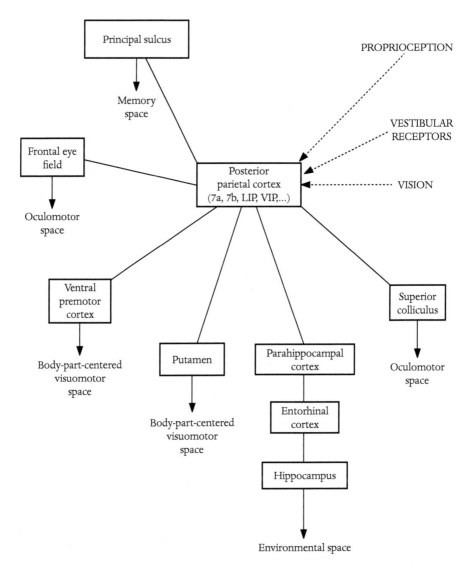

Figure 4.2. The brain uses multiple frames of reference. The information supplied by the sensory receptors (proprioception, vestibular receptors, vision) converges in the parietal cortex, where it is integrated with many other signals about movements and planned actions. Actions are subsequently encoded in varied frames of reference that correspond to multiple spaces relative to the body or the environment, even to internal memory space.

But during hand-pointing tasks the brain also uses a point of rotation located on the shoulder. In this way it constructs a local reference point to simplify neurocomputation. The brain chooses a frame of reference connected to the limb that is executing the movement. Yet again, the advantage is that it reduces the number of variables that need to be monitored (engineers say degrees of freedom), limiting the need for neurocomputation to the single part of the body involved in the movement.

The idea that the brain can choose multiple frames of reference depending on the task and the context is shown by the neurons discovered by Graziano and Gross that I described in Chapter 3. Recall that they found bimodal neurons in the putamen and parietal cortex whose visual and tactile receptor fields remain anchored in space during the course of a movement. One interpretation of these findings is that the spatial encoding of movement and the position of the limbs is not experienced with respect to a single reference frame, as the egocentric frame of reference might suggest. Instead, there are many frames of reference, each related to a part of the body (eyes, limbs, trunk, and so on) (Figure 4.2). Action appears to be organized in modules based on a repertoire of specific types of actions (gaze, reaching, locomotion), which are themselves organized by specific basal ganglia-thalamo-cortical loops.

The convergence of sensory information is the result of integrating data from the senses important for each limb. For example, neurons encode the position of the head based on convergence of visual cues and other cues from the muscles of the neck that detect movement of the head on the trunk, as well as cues from the vestibular receptors, which also detect movement of the head.

Selecting Frames of Reference

All of these findings suggest that the brain uses not one but multiple frames of reference according to the task at hand together with essential or available sensory cues.

Perrett and co-workers showed that in the monkey, the neurons of the temporal cortex are involved in recognizing faces.[23] The retinal image of the face is first broken down into fragments corresponding to the principal channels of the primary visual pathways that dissociate color, shape, movement, and so on; then other neurons reconstruct the facial features. Certain of these neurons are activated by the eyes, others by the hair, still others by the nose, and so on. These features then converge on neurons that respond to faces, but which are also sensitive to behavior, such as the direction of gaze. This ability to detect the direction of gaze is all-important to the monkey; it probably helps to identify the intention of another of its kind: is it friend or foe?

In this chain of processing, the meaning of a facial image is gradually reconstructed.[24] In fact, there are neurons that detect whether the face is turned sideways or facing front, and so on. At the next level, the face is situated in the context of the body, and neurons fire when the face is associated with a body facing front or turned sideways, and so on. Finally, other neurons fire when the face is familiar: a cognitive level is reached that brings memory into play. Up to that point, even in context, the face is processed in a frame of reference centered on the observer. The multiplicity of neurons that respond to the same face, whether it is viewed from different sides or from a different angle, enable the brain to configure the activity of neurons that will respond to a given face whatever the angle from which it is seen. This flexibility implies a level of abstraction that is independent of the frame of reference in which the visage is perceived.

It is also possible to imagine that frames of reference are constructed in connection with action—for example, linked to an object, a goal, and so on. The multiplicity of possible representations enables the brain to construct ad hoc frames of reference. This hypothesis has yet to be verified, but will be extraordinarily fruitful if it turns out to be correct. It would explain in part our ability to do geometry, which requires us to mentally change our perspective as we view tangible objects and the environment. For Kosslyn and his collaborators, categorical spatial relations (front, back, above, and so on), which maintain constancy of certain relationships between objects or parts of the body, are encoded by different mechanisms of metric spatial relations that specify distances.[25]

RELATIVE FRAMES OF REFERENCE

Frames of reference are not only useful for establishing relationships between the body and space. Arbib formulated the idea of "opposition space" between the thumb and the other fingers in grasping tasks (Figure 4.3).[26] He rejects the idea of a single basic frame of reference to which all other frames of reference refer. He writes: "The task of perception is *not to reconstruct* a Euclidean model of the surroundings. Perception is *action-oriented,* combining current stimuli and stored knowledge to determine a course of action appropriate to the task at hand."[27] Building on the findings of Jeannerod, who demonstrated that the hand preshapes to the size and orientation of the object it is about to grasp at

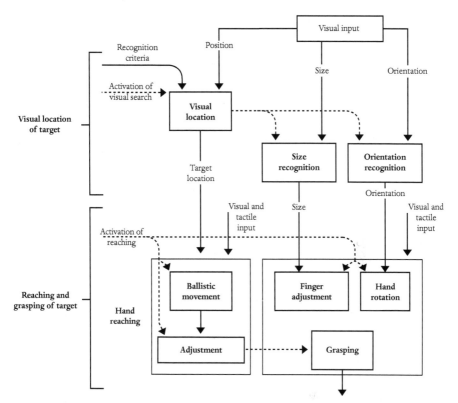

Figure 4.3. How to grasp an object. The brain analyzes visual inputs with respect to three properties: position, size, and orientation. Similarly, neural processes analyze movement of the hand close to an object, adjustment of the fingers to the size of the object, and a rotation of the pincer motion depending on the orientation of the object.

the instant the movement is begun,[28] Arbib suggests that the frame of reference used by the brain is relative, particularly the relationships between the fingers (opposition space) that are about to grasp the object: "There is no one absolute space represented in one place in the brain, only a coupling of sensory and motor spaces in such a way as to yield movement to achieve some goal."[29]

A point of reference can thus be constructed on the basis of relationships between parts of the body. It can also be constructed actively. For example, I said earlier that the head is used as an inertial platform that is stabilized during complex movements and especially when there is no reference to the ground. In this case, the brain uses gravity to stabilize the head, which can then serve as a point of reference for coordinating the limbs. I have emphasized that a movement's point of reference depends on the task.[30] If I am holding a full glass of champagne while walking or bending over, I necessarily have to adjust its movement to the direction of gravity so I do not spill it. But if I am reading a book while walking, the objective of my movements is to minimize the relative movement between the page and my retina, and thus I adjust the movement of my hand to that of my head.

It also must be considered that the brain constructs a series of frames of reference for each phase of the same movement.[31] This is illustrated by the trampoline jumping shown in Figure 3.3. A new physiology is needed; one that will investigate these rapid changes, these swings from one frame of reference to another. This flexibility is so basic that it must be programmed into the very structure of the nervous system.

Explicit and Implicit Frames of Reference: The Theory of the Equilibrium Point

Little is known about the way the nervous system processes frames of reference, but the ideas derived from Bernstein's so-called equilibrium point model deserve mention. Feldman,[32] in Moscow, recently took up this theory again and gave it a formal expression that was in turn elaborated by several Western groups, including Bizzi's at MIT. This theory is interesting because it is at the heart of the question of anticipation and of the way the brain implicitly or explicitly simulates external reality.

Feldman's model emerged from investigations of the functioning of motor neurons (the neurons that control muscle contraction). It is well known that the discharge of a motor neuron is connected to its membrane potential. Normally, this membrane is hyperpolarized (negative potential inside with respect to the outside). A central command or the influence of sensory receptors can depolarize the neuron and bring it to its firing threshold. Thus, the motor neuron produces action potentials that lead to muscle contraction.

Classical models of control of movement (called "α models") assume that a central command directly regulates discharge of motor neurons. But Feldman reasons that the brain regulates an indirect variable, *the firing threshold of the motor neuron*. This could be defined as the value of the membrane potential at which the neuron fires. Why is this interesting? Because it also enables us to control the position of the arm. How? If my arm is in a given position, normally the neuromuscular spindles fire and activate the muscle, which contracts reflexively. But if I change the firing threshold so that, for this length of muscle, the motor neuron—even if it is activated by the sensory receptors—is no longer activated, my arm does not move. The λ variable that controls the threshold is thus equivalent to a position, an angle of the arm. Feldman says that it has a spatial dimension, and that the position is reached by regulating the λ variables of the two antagonist muscles. According to this theory, there is no need for space to be represented explicitly, for it is implicitly contained in the regulation of the threshold. This fits the theories of Viviani and Flash on the morphogenetic and creative properties of geometric trajectories and the law of minimum jerk (Chapter 6).

Feldman shows that although his hypothesis is based on neuronal functioning, it has an impact on the general problem of the frames of reference according to which movements are organized.

> These theoretical results can be understood in terms of physics. Body movement is defined as a change in position in relation to another object, frame of reference, or system of coordinates. But the Galilean principle of the relativity of movement is implicit in the concept of frame of reference: movement can also be induced by a displacement of the frame of reference. The λ threshold can be considered as the point of origin of a frame of reference for the mobilization of motor neurons. In displacing λ, the level of control specifies a new reference point and produces movement. In this way, displacements of the positional frame of reference underlie the control of movement even though activation of motor neurons and production of forces are a consequence of this process.[33]

The scientific community is actually divided on the validity of the λ model. However, the model demonstrates the plausibility of theories about the control of movement that reject the idea that space is explicitly represented in the brain, even if it is consciously perceived that way.

Ideas about frames of reference have evolved from a monolithic conceptualization of a single point of reference to the notion of flexible frames of reference connected to the task at hand. An illustration of this evolution can be found in the remarkable fountain sculpted by Bernini—the personification of the four rivers—in the Piazza Navona in Rome. The sculptor shifted the fountain completely off center in relation to its central column by creating extraordinarily varied frames of reference for the action of his figures. The frames of reference refer to a universe that is relative for each of the characters.

So the problem is more general and very basic: it is the nature of the ego. Indeed, if there is no single egocentric frame of reference but instead multiple representations of the body, it becomes necessary to construct a theory that explains the unity of perception of the body. An explanation of how these diverse local neuronal subsystems are integrated in a body schema is needed.

The situation of this science is a bit like the global political situation of the 1990s. In the wake of ideologies aimed at imposing a single model, there has been a veritable explosion of nationalisms. There is no single frame of reference, but rather a fascination for the multiplicity of cultures. How will we manage to put together from this a world that can account for both individual differences and what might be termed the unity of man? This is the same mammoth challenge facing the neurosciences.

5

A MEMORY FOR PREDICTING

> Perceived in its nascent state, a movement is always movement that is going somewhere, which is absurd for the physicist, who defines the movement not by its objective, but rather by what brought it about. The movement as it is perceived, however, proceeds from its point of arrival to its point of departure.
>
> —M. Merleau-Ponty

Perception is essentially multisensory: it uses multiple, labile frames of reference adapted to the task at hand. It is predictive; receptors detect derivatives, and the brain contains a library of prototype shapes of faces, objects, and perhaps movements and synergies. Nature has devised simplifying laws among the geometric, kinematic, and dynamic properties of natural movements. But the predictive nature of perception is also—perhaps especially—due to memory. For memory is used primarily to predict the consequences of future action by recalling those of past action.[1]

What are the recent findings on memory from psychology, neurophysiology, and neuropsychology that will best identify memory's role in the relationship between perception and action? Modern neuropsychology distinguishes several types of memory that are implemented in different parts of the brain (Figure 5.1). Memory can be declarative, implicit, working, episodic, procedural, short-term, long-term, iconic, topographic, spatial, semantic, lexical, motor, and so on. Neuropsychologists and neurophysiologists have made remarkable strides in understanding the neural organization of multiple facets of memory. Memory is a basic property of the brain that is expressed at all levels of cerebral functioning. Even muscular fibers have a memory; some have so-called catch properties that allow them to remain contracted for sev-

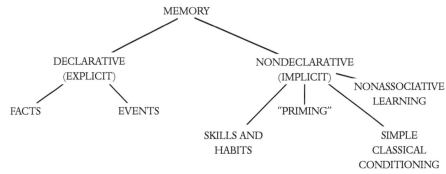

Figure 5.1. Classification of the components of memory. Declarative (explicit) memory is memory of facts and events, and it is a function of the temporal lobe of the cerebral cortex. Nondeclarative (implicit) memory involves a repertoire of skills independent of the temporal lobe. Nonassociative learning concerns memory that suppresses sensitivity to a stimulus when it is repeated (habituation) or, on the contrary, is maintained and increased. "Priming" refers to the change in response observed when a subject is supplied a clue (which he holds in memory for a brief moment) to what the stimulus will be just before it is applied. This schematic diagram does not distinguish between short-term and long-term memory, nor does it include the concept of working memory developed by Baddeley.

eral moments in certain cases. The motor neurons of the medulla possess what are called plateau potentials that maintain the depolarization of their membranes. Other mechanisms at the cellular level, such as long-term potentiation, which is most likely responsible for properties of memorization in the hippocampus and the cortex, or long-term depression, discovered in the cerebellum by Ito and probably involved in plasticity, or even the slow potentials due to calcium ions, are the elementary manifestations of memory, controlled by molecular mechanisms among which the NMDA (*N*-methyl-D-aspartate) synaptic receptor plays a major role. The phenomenon of the neural integrator, which we discussed earlier with respect to the vestibulo-ocular reflex and which transforms velocity signals into positional signals, is a specific case of memory arising from the intrinsic properties of neurons or resulting from reverberating circuits in which a collateral signal is returned (positive feedback). Other mechanisms call into play activities supported by oscillations of internal neural circuits.

The discovery of the neural basis of these memories will in fact be one of the major issues for neurobiology in the next century. Here I will consider just one particular form of memory—spatial memory—which has not been extensively studied by neuropsychologists but which nevertheless constitutes an im-

portant model for understanding the relation between perception and movement.

Spatial memory is essential in representing space. Paillard defined its role
very clearly:[2]

> A basic assumption of our argument is that a *sensorimotor* mode of pro-
> cessing spatial information coexists with a *representational* mode and
> that both modes are generating and storing their own mapping of space
> . . . The *sensorimotor mode* concerns mainly that part of the physical
> world to which the organism is attuned by virtue of its basic sensori-
> motor apparatus. Local sensorimotor instruments entertain direct dia-
> logues with that world and thus contribute to the continuous updating
> of a body-centred mapping of extra-corpor[e]al space where objects are
> located and to which actions are directed . . . The *representational mode*
> derives from neural activities which explore and consult internal repre-
> sentations of the physical environment, that are embodied in memory
> stores. They include mental representations of local maps, spatial rela-
> tionships of routes relative to landmarks, relative positions between
> objects, and the position of the body itself in relation to its stationary
> environmental frame. The question arises as to whether the two pro-
> cessing modes operate in parallel, each using its own neutral [sic] cir-
> cuitry and generating its own mapping of space in two fundamentally
> different ways.[3]

Topographic Memory or Topokinetic Memory?

NAVIGATIONAL MEMORY

Suppose that I ask you to recall the trip you make to get to work. You will
have to remember both the way, that is, the topographic aspects of the route,
and the movements you make walking or driving.

Topographic memory allows us to locate a place and to find our way back
to it. In a sense, it is a form of procedural memory, because it involves a succession of places or local scenes and movements (turning left, right, and so
on). But it can also be a mental survey of the places that represent the trip as
on a geographic map. It is believed that this kind of memory is responsible for
anamorphoses, or distorted images. For example, if a subject is asked to draw

a map of Paris with his eyes closed, he has a tendency to make the center large and the periphery small.

Among these navigational abilities, there is one that shows up in all species of animals, from the fly to humans, and that is dead reckoning or, more generally, the ability to find one's way back. Darwin was interested in this function of navigation. He wrote:

> With regard to the question of the means by which animals find their way home from a long distance, a striking account, in relation to man, will be found in the English translation of the *Expedition to North Siberia,* by Von Wrangell . . . He there describes the wonderful manner in which the natives kept a true course towards a particular spot, whilst passing for a long distance through hummocky ice, with incessant changes of direction, and with no guide in the heavens or on the frozen sea. He states (but I quote only from memory of many years standing) that he, an experienced surveyor, and using a compass, failed to do that which these savages easily effected. Yet no one will suppose that they possessed any special sense which is quite absent in us. We must bear in mind that neither a compass, nor the north star, nor any other such sign, suffices to guide a man to a particular spot through an intricate country, or through hummocky ice, when many deviations from a straight course are inevitable, unless the deviations are allowed for, or a sort of "dead reckoning" is kept. All men are able to do this in a greater or lesser degree, and the natives of Siberia apparently to a wonderful extent, though probably in an unconscious manner. This is effected chiefly, no doubt, by eyesight, but partly, perhaps, by the sense of muscular movement, in the same manner as a man with his eyes blinded can proceed (and some men much better than others) for a short distance in a nearly straight line, or turn at right angles, or back again. The manner in which the sense of direction is sometimes suddenly disarranged in very old and feeble persons and the feeling of strong distress which, as I know, has been experienced by persons when they have suddenly found out that they have been proceeding in a wholly unexpected and wrong direction, leads to the suspicion that some part of the brain is specialised for the function of direction.[4]

Darwin's text was prophetic. As I noted in Chapter 3, neurons have an activity connected to the orientation of the head in space. Since Darwin's time, a large number of publications have been devoted to mechanisms of navigation.

However, the literature is less than clear on what we mean by the term "navigation," and it is worth trying to clarify it here.

By navigation we mean the ability to find one's way at sea and, by extension, on land and in the air. Navigation has been studied in connection with the flight of pigeons and bees, reproduction in salmon, the migration of Eskimo peoples in the Antarctic and, more recently, the voyages of Micronesian sailors, who traveled to far islands without instruments, using only cognitive maps and techniques for updating their position in relation to virtual landmarks.[5] It is important to make a distinction between far space that can be perceived directly (in a stadium or in coastal navigation, for example) and very great distances, which involve their own particular representation (as for airplanes and migrating birds). The ability to find one's way home or to memorize a route is not unique to humans; crabs return to the sea, bees to the hive, and desert ants to their nest using the sun to find their way. Each species has devised individual solutions to the same problem.

INERTIAL NAVIGATION

Darwin's account in 1873 aroused very interesting reactions at the time. To explain this navigational facility, some people proposed a mechanical analogy very close to the inertial navigation systems used on modern airplanes.[6] Perhaps the vestibular apparatus of animals served as an inertial detection system for navigating. Or perhaps the vestibular apparatus enabled the brain to directly detect the Coriolis forces and to use this information in navigating. It was even proposed that the Coriolis effect could be estimated by detecting the movement of blood in the vascular bed, a very controversial idea.

The assumption that the vestibular system plays an important role in navigating short distances was taken up by Beritoff, an Armenian physicist who also went by the name of Beritashvili. He elucidated these processes in dogs, cats, infants, and adult humans. He suggested that spatial orientation is the aptitude of organisms to locate the position of objects and to relate this position to themselves and to other objects.[7] From these perceptions, organisms form images of the distribution of objects in the world around them. These images initiate and guide movement oriented toward the objects, even if the objects are not seen or perceived by other sensory organs.

Beritoff conducted experiments with dogs. He blindfolded them before taking them into the laboratory from one side and leading them along a specific path to a food source located on the other side. Then he led them back to the point of departure before moving them to other parts of the room. He discovered that even blindfolded, the dogs had no trouble finding the part of the room where the food was located, whatever their starting point. He investigated the possible role of all of the sensory cues (auditory, olfactory, tactile) but found performance of the task to be impaired only after bilateral labyrinthectomy. Moreover, the animals found their way to the food just as well when they were carried to it during training as when they walked to it. These results seem to rule out kinesthesia (the set of cues supplied by the proprioceptors of the joint muscles and locomotor commands). Though he realized that the cerebral mechanisms called into play were extremely complex, Beritoff was the first to demonstrate a possible role of the vestibular system in dead reckoning.

A precise return to a point of departure in the dark and deterioration of performance following vestibular lesions have been observed in the rat.[8] Supplementary proof of this contribution from the vestibular system came from work on batrachians and desert mice[9] as well as the golden hamster.[10] In this animal, a lesion of the vestibular system causes deficits in navigational tasks that are reversible following recovery of vestibular function.[11] However, detecting linear movement using vestibular cues has been the subject of a major debate, for homing experiments with rodents suggest that these animals do not correct linear displacements or rotations as easily. There seems to be a dissociation between integration of linear and angular trajectories.[12]

ROTATIONAL MEMORY

The brain thus possesses mechanisms for memorizing displacements. Does the vestibular system really contribute to these mechanisms, as Beritoff suggested? One analytical approach separates out the components of rotational and translational displacements and asks the question: Can the brain memorize rotatory displacements and translations and reproduce them using the eyes, head, or whole-body movements? This is the question we sought to answer. I will describe these experiments in detail, because they led us to a fundamental concept: that vestibular memory is memory of movement, not of position.

One elegant experiment, called vestibular memory-contingent saccades, demonstrates the ability of the human brain to accurately detect the angle of a rotation from the acceleration signaled by the vestibular receptors.[13] The subject is seated in a revolving chair in front of two visual targets. One target is fixed in relation to the earth, the other in relation to his head. He is asked to look at the first target and to memorize its location in space. Then all the lights are switched off, except for the light on the target mounted on his head, which turns with him and keeps his eyes from moving during the rotation.

The subject thus has only vestibular cues about the displacement of his body to rely on, since he is in the dark and can make no movement of his eyes (which could otherwise inform his brain about his displacement by means of extraocular proprioception and motor discharge).

When the revolving chair is brought to a standstill, the illuminated target mounted on the subject's head is turned off, and he is now completely in the dark. Several seconds are allowed to pass for the vestibular effects to diminish, after which the subject is asked to recall the position of the earth-fixed target and to make an ocular saccade toward that spot. In other words, he is asked to displace his gaze from the place where the chair has stopped to the point of departure he memorized. This task amounts to having the subject make a visual oculomotor return to his point of departure. After this memory saccade, the light is turned back on. If the subject's gaze is on the target, his eye does not move; if his gaze is slightly deviated, he will execute a little correcting saccade that makes it very easy to estimate the error.

Normal subjects can make saccades toward memorized targets after rotation in the dark with extraordinary accuracy.[14] The brain is thus capable of estimating displacement of the head solely from information supplied by the horizontal semicircular canals. The subject is asked to wait 4 or 5 minutes in total darkness after the chair has stopped moving before executing the saccade. Despite this delay—which requires that the subject memorize the spatial information about the angle of rotation of the body, that the brain reconstruct this information, and that it activate the cortical centers that produce the saccade—the result is always excellent. In other words, vestibular information about displacements of the body can be stored in spatial memory. This is true for all orientations in space, which is only natural, since the vestibular system did not develop solely to execute horizontal movements but to enable complex movements, such as jumping from tree to tree.

Next a subject is placed on a turntable and rotated in total darkness. We ask him to reproduce the angular displacement in the opposite direction, which he can do using a lever that controls the movement of the table. A healthy subject can return to his point of departure with great precision.[15]

Which areas of the brain are involved in this ability to estimate rotations? Applying this test to patients with lesions of the cerebral cortex gives a partial answer.[16] Serious impairment has been observed in patients with damage to the supplementary oculomotor field, the prefrontal cortex, and the vestibular cortex. These findings suggest first and foremost that the prefrontal cortex is involved not only in visual memory but also in vestibular memory. This is no surprise, since in my opinion memory of space is memory of movement in space: thus it is essentially multisensory. Moreover, it seems to me perfectly natural to find deficits in patients with lesions of the vestibular cortex, given its importance in reconstructing movements of the head in space and its importance for transmitting to the rest of the cortex information about displacements of vestibular and tactile origin (see Chapter 3).

TRANSLATIONAL MEMORY

Man's aptitude for using otolithic vestibular cues to estimate translations in the course of linear movements was recently established. Recall that the otoliths are sensors of linear acceleration in their planes. We placed the subject on a trolley for translational movement.[17] This sort of experimental sled was constructed in our laboratory at the Ecole Nationale Supérieure des Arts et Métiers during the 1970s. It used a linear motor technology initially conceived to control the movement of high-speed trains, but eventually abandoned owing to the dissipation of heat that occurs with this mode of propulsion. The trolley made displacements of several meters with accelerations of up to about 1 meter per second per second. The subject was seated transversally, his displacement thus occurring in the frontal plane (that is, the plane of the two ears). Before beginning the move, the subject was shown a visual target fixed with respect to the earth—in this case a clothespin hanging on a line. Then the light was switched off, and the trolley was moved 60 centimeters to 1 meter. The subject's eyes did not move during the translation. Once the trolley had stopped, the light was switched back on, the effect of the accelerations on the otoliths was given a few seconds to dissipate, and the subject was asked—as for the rotations—to make a saccade to where he remembered the clothespin to be. The movements of his eyes adapted perfectly, whether his head was fixed or free. Additional support for these findings was provided by experiments made with the SLED (Figure 5.2) installed at the time in a glass shell at the Cité des Sciences in Paris. The subject was displaced in an anteroposterior direction, facing the movement. He was asked to signal his crossings under targets positioned all along the linear track.

Later, we adapted a mobile robot to study perception of displacements.[18] The subject was seated on the robot, in total darkness, and was subjected to noise to obliterate any acoustical cues resulting from his displacement. The robot followed a rectilinear trajectory whose profile was either triangular (acceleration, then deceleration), trapezoidal (acceleration, constant velocity, then deceleration), or square (sudden acceleration, constant velocity, then sudden deceleration). Once the passive displacement—1 to 10 meters—was com-

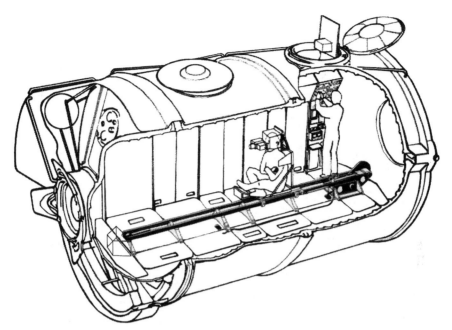

Figure 5.2. The SLED in the Spacelab laboratory. This space sled was used in the American space shuttle to study the influence of microgravity on the functioning of the otolithic receptors. It consists of a runway approximately 4 meters long on which a trolley moves by means of wires connected to a motor. The subject is placed in any of three positions (front, lying on his back, or crosswise) during translations. A digital camera, constructed by LETI (part of the French Atomic Energy Authority), records movement of one eye, and a miniature television screen projects visual scenes to the other eye, which stimulates the optokinetic reflex. Eye movements are also recorded with electrooculography. The subject uses a lever to indicate which way he thinks he is moving. A Pelletier effect system allows injection of hot air into the subject's ears to study the effect of microgravity on the vestibulo-ocular reflex. This apparatus was constructed by the European Space Agency. Our laboratory participated in its use with the French National Center for Space Research.

pleted, the subject was asked to reproduce the displacement, from memory, with the aid of a lever that controlled the velocity of the robot. All normal subjects executed this task accurately.

How does the brain go about this reproduction? Does it calculate the displacement in meters and memorize numeric values, or does it perhaps store other variables? We assume that it does not calculate displacements but instead reproduces a dynamic velocity profile. It simulates the movement carried out. Recall the words of Poincaré: "To localize an object simply means to represent to oneself the movements that would be necessary to reach it." Indeed, when different velocity profiles were imposed (triangular, trapezoidal, square), the subject tended to reproduce these profiles and not a standard profile that allowed him to optimally detect the displacement.

Which cues does the brain select among the available sensory data? A paraplegic subject managed a perfect performance despite the lesion to his spinal medulla that left only his head sensitive to muscular and tactile sensations. For him, the vestibular information seemed to be determinative. However, subjects presenting with bilateral severance of the vestibular nerves were not able to reproduce the velocity profile as accurately, though they succeeded in carrying out the task all the same, probably by using other combinations of information about or estimation of the displacement. The brain devises many ways of solving the same problem.

So the brain memorizes movement, not just places. Paillard speaks of velocity space and positional space. Reproducing a distance requires comparing an internal simulation of a memorized movement and information from the senses. The brain is a comparator that detects discrepancies between its own predictions based on the past and the information it selects about the world according to its purpose.

Locomotor Memory

Thus the role of the vestibular system in the memory of passively experienced displacements becomes apparent. What happens to this memory during self-displacement while walking? In this case, the information supplied by proprioception in the legs—length of step, signals of the motor commands—is very important for evaluating distances.

Anyone can do the following experiment, credited to Thomson. Look at a point on the ground about a dozen meters away; close your eyes and walk toward the point. You will see that you arrive at your destination pretty exactly. Another test is to look at a target point far away on the ground and walk there with your eyes closed but in a roundabout way. You will again observe that you reach your goal with good accuracy. You can also return to your point of departure with your eyes closed (what ethologists call dead reckoning). Finally, you can repeat your same route several times, always with your eyes closed, which means that the information has been stored in your spatial memory.

How do we explain this? There are only three possibilities: the distance covered may be calculated from proprioceptive data supplied by the muscles and joints of the legs; it may be derived from the memory of the motor commands for each step; and finally, it may be derived from inertial cues of vestibular origin. For the last of these possibilities, the brain would have to calculate the distance covered based on vestibular information about acceleration, a process that is called path integration. Poincaré had suspected this possibility: "Knowing the acceleration of rotational movements of the head at each instant, we can deduce through unconscious integration the final orientation of the head with respect to its initial orientation."[19] According to this theory, the brain calculates a distance covered. But we suggest that it dynamically updates spatial representations.[20] I have proposed the term "topokinetic memory" or "topokinesthetic memory" to indicate this dynamic type of spatial memory.

I return to Thomson's experiment described above. There are three ways of interpreting his results: Perhaps the brain has a memory like a map (in the geographical sense) of the environment and calculates its position on the map, or perhaps the distance is encoded in "action units," to use the expression favored among Gibson's psychologist disciples; that is, the number of steps, without there having to be a cartographic representation of the route in the brain. Perhaps a memory of movement (rotations, translations, and their combinations) is retained and detected by the sensory receptors.

When the target is situated 3 or 4 meters away, people who no longer have the function of their vestibular receptors still carry out the task, making only a few mistakes in selecting a path. Recently we reproduced this experiment by asking subjects to reach a target spot over a triangular path with their eyes closed. Each side of the triangle had a length of several meters. In every case, subjects who had no labyrinth, or who had unilateral labyrinthic lesions, demonstrated significantly impaired performance. So it is plausible that the aggregate of sensory messages contributes to this internal updating of spatial representations. Recent work by Reiser, Loomis, and Tresky in the United States, as well as by Thynus-Blanc in Marseille, shows that there are several possible mental strategies for updating.

The memory of displacements is probably a genuine dynamic memory that, when invoked, induces internal simulation of the path. The brain does not merely compare sensory information with memorized information, it also calls into play anticipatory mechanisms.

Recent observations confirm the anticipatory character of cerebral activity during navigational tasks. For example, 1 or 2 seconds before every bend, the gaze of a driver on a mountain road becomes fixed on the tangent inside the bend.[21] A calculation based on these data shows that the direction of this point in relation to that of the car makes it possible to predict the curvature of the road beyond the bend. In other words, the gaze of the driver is positioned on a point such that the information supplied by the optic flow allows him to predict the curvature of the trajectory. Thus the brain does not merely know the curvature of displacement of the car at a given moment based on visual and vestibular cues: it seeks to predict the curvature at a future time.

Similarly, when we turn a street corner, our gaze anticipates the rotation of our body. This anticipation appears in the infant in the course of its development, and I think it must be absent in patients with lesions of the parts of the brain that participate in anticipation.

It thus seems reasonable to suppose that in carrying out a navigational task, the brain in a way follows an internal representation, a model of the trajectory that anticipates the path, and not the other way around. This principle could be very useful in robotics. Instead of building machines that use so-called sensory receptors to guide their movements, we recently suggested to a team of roboticists that they control movements using gaze to anticipate the trajectory. This principle of guided navigation turns out to be much more interesting than that based on passive detection of displacement. Here, as in other tasks, the brain prefers a "go where you look" strategy, in which the object guiding a control strategy is placed close to the center of vision. I think that these anticipatory mechanisms reflect the fact that the brain carries out tasks of guided movement based on mental paths that it constructs and that enable it to make predictions. Navigation is no more than the execution of an internal plan based on past experience, and the senses are used, as in the case of the ski champion I mentioned earlier, to ensure that the plan unfolds and to make corrections.

The Neural Basis of Spatial Memory: The Role of the Hippocampus

What are the neural mechanisms of navigational memory for re-finding one's way in a city? The beginning of an answer to such a question may be found by turning first to neuropsychology to discover which lesions of the brain cause impairment of navigation or what is called place memory or topographic memory.

Although several neurologists since J. Hughlings Jackson have noted disturbances in locating one's position and navigating in familiar places, it was Milner who established the basic role of the hippocampus in spatial memory[22] (Figure 5.3) based on study over many years of an epileptic patient whose two hippocampi had been removed.[23] Generally, studies of lesions introduced in the rat and the monkey and sustained by humans following vascular injury or surgical excision suggest that the hippocampal region plays a determinative role in spatial memory. However, a recent reassessment of these lesions and

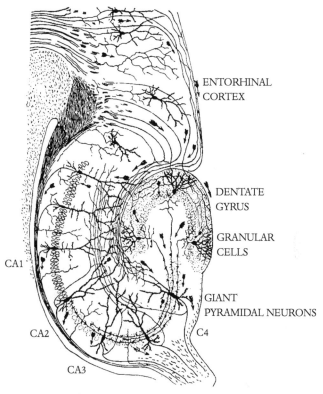

ENTORHINAL
CORTEX

DENTATE
GYRUS

GRANULAR
CELLS

CA1

GIANT
PYRAMIDAL NEURONS

CA2

C4

CA3

Figure 5.3. Anatomy of the hippocampus.

their effects indicated that the deficits induced might also be due to a small neighboring region that could have been damaged at the same time as the hippocampus—the entorhinal cortex. Nevertheless, several different conceptualizations compete with one another. They are worth summarizing briefly, because they explain why I maintain that the brain uses memory to predict the consequences of action.

Short-term Memory

In humans and in animals, damage to the hippocampus and neighboring regions of the temporal lobe impairs memory; so any plausible explanation of this phenomenon must be compatible with the physiological properties of the hippocampus. Marr was one of the first to propose that the hippocampus plays a role in short-term memory.[24] He was interested, among other things, in its possible role in the genesis of dreams. He posited that during paradoxical sleep (during the dream phase), the hippocampus fires stored memories at the parietal cortex. Later, it was hypothesized that the hippocampus contributes to medium-term memory. Mishkin suggested that the hippocampus is involved in "recognition memory."[25] He demonstrated this in a delayed recognition task used largely to test short-term and working memory: the monkey had to match the arrangement of objects in front of it before the delay with a new arrangement after the delay.

SPATIAL MAP OR MEMORY OF SPATIAL VIEW?

Another theory is that of O'Keefe and Nadel: the hippocampus is a "cognitive map" used by animals to update spatial maps of the environment based on "place cells" and to calculate the twists and turns needed to reach the destination.[26] This theory is based on the observation that if a rat moves freely in a test box, neurons of the hippocampus called pyramidal cells fire each time the animal passes by a particular spot in the box, whatever its orientation. Recording the action potentials that represent the neuronal activity reveals an increase in frequency each time the animal passes by or visits certain locations in the box. These neurons are called place cells. They cover a spatial domain that varies but that does not go beyond a radius of several dozen centimeters. The group of these place cells constitutes a spatial cognitive map.

The hippocampus thus stores spatial cues in allocentric coordinates, that is, geometric space independent of the animal, in contrast to the egocentric coordinates of the vestibular system. But perhaps this interpretation is insufficient.

In the monkey, it appears that space is encoded differently in the hippocampus. The neurons do not encode the spot where the animal is in the room but an area of space where the neuron is activated by a cue relevant for the animal. The activity of place neurons does no more than reveal an area of space where the neurons are involved in a wide variety of behaviors.[27] They are influenced by several sensory modalities (visual, auditory, olfactory, and so on). The neurons memorize conjunctions of these modalities, as well as their combinations in relation to gaze movements and perhaps elements of the repertoire of actions (turning, walking, and so on) in which the animal is involved at a given instant.[28] It is also possible that certain neurons are activated by objects important for the animal (such as a banana for a monkey).

When a monkey is given a task to perform; for example, to recognize whether a bowl placed on top of a wall is the one that contains food, certain cells behave like spatial view neurons; they respond to a particular view of the surrounding space. Similarly, other neurons are activated in connection with the direction of gaze of the animal toward a particular point in space. The development of the cerebral cortex in primates may have minimized egocentric encoding of place neurons ("where I am") in favor of allocentric encoding of topographical relations between landmarks in the external space and those of the place where the subject is in relation to these landmarks. For example, in an apartment, instead of encoding the bathroom, kitchen, and bedrooms in relation to you when you are seated at the dining-room table, what would be encoded would be the relationships between these rooms (the kitchen is next to the bathroom). This distinction is important, because it facilitates mental processes involving external space. Here is another example: In a village, each significant landmark (the train station, the post office, the park, the church, the synagogue) can be represented in relation to you if you are seated outside a café and you imagine the way you would have to take to go to the post office. But to find the best route, it is also very useful to be able to imagine the distances between each of these places and to calculate their relationships.

Understanding this perceptual decentering requires that the function of the hippocampal neurons be studied in new experimental situations. The hippocampus is sensitive not only to visual, auditory, and olfactory cues. Vestibular cues also can alter the activity of the hippocampal neurons. In the rat and monkey, these neurons are activated when the animal is passively transported to a place that it knows. But their activation is not connected just to the displacement of the body; these neurons are activated only when the animal is moved along certain walls, for example, or when it is moved toward a door or turns a certain corner. The same movement of the body in another location of the enclosure evokes no neuronal activity. So things become progressively more abstract as successive levels are crossed from the vestibular system to the hippocampus.[29]

In the vestibular nuclei, information about bodily self-motion is encoded in combination with visual and proprioceptive cues, and is influenced by the direction of gaze. But here encoding is exclusively egocentric (independent of the environment). In the parietoinsular cortex described above, movements of the head are situated in the more general context of movements of the body and the body schema. It is only at the level of the hippocampus that movement of the body is situated in allocentric space, and especially, that it is related to the particular use the animal might make of this or that part of its space. The brain labels its perceptions according to its intentions and goals, and the hippocampus appears to play a significant role in this process.

A major function of the hippocampus is thus to detect and memorize simultaneous combinations of multiple sensory messages. The hippocampus is critical in associating together simultaneous perceptual cues. This memorization of independent but simultaneous perceptual or behavioral events, which

Rolls calls "an episode," is thus a configurational memory of events and the temporal sequence in which these events occurred.[30] The hippocampus memorizes episodes in connection with other structures, the parietal cortex or the temporal cortex, where these episodes are stored over the long term. It also plays a role in association with the inferotemporal cortex in visual memory and identification of shapes.[31] For example, if one day we see two people at the same time, and then see one of them a month later, the memory of the other person comes to us. Or, if it is a couple, we might say: "Hey, isn't that the guy I saw with that girl a month ago?" Along with other structures in the temporal lobe, like the inferior temporal (IT) area, which is important in identifying visual objects, the hippocampus may participate in this associative visual memory.

But an episode is not only made up of images, sounds, and smells. The hippocampus does more than remember combinations of sensory information and detect new sensory signals. It is also involved in remembering and identifying combinations—what are called configurations—of perceptions and actions. And this is where the hippocampus probably plays a major role in organizing *sequences* of actions, such as a series of saccades or perhaps the steps in learning how to tie a shoe or knot a tie.

Proust's Madeleine

For me, one of the most fascinating aspects of Rolls's theory is that he identified, in the neuronal structure of the hippocampal network, properties that enable it to recall an episode or a combination of sensations given only a portion of the information initially memorized (Figure 5.4). This property is the result of the autoassociative structure of the networks of neurons of the hippocampus, owing to connections of hippocampal neurons that project to themselves or to neighboring neurons and endow the structure with a capacity for memory and for recovering memorized information (recall). This mechanism is used in certain machines to reconstruct an image from fragmentary clues, and it enables the use of past episodes as models for what an ongoing action might produce. In other words, to predict the consequences of an ongoing action, the memorized action is recalled, even if its context is different, to serve as a model for anticipating the consequences of action and possibly to alter the action. What is still not known is whether the hippocampus is mainly involved in storing new events or in recalling past ones. The right prefrontal cortex as well as the medial parietal cortex are also involved in recalling memories.

Rolls gives the following example: One day in Oxford he is crossing High

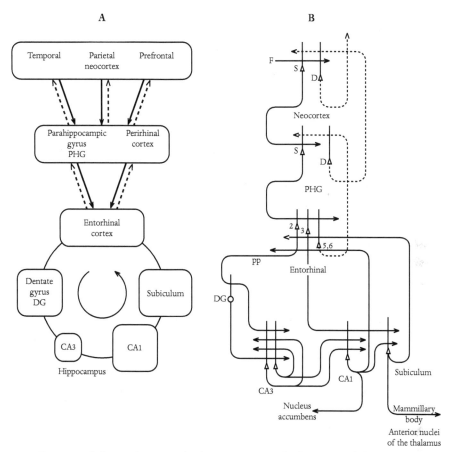

A

Temporal Parietal Prefrontal
neocortex

Parahippocampic   Perirhinal
gyrus   cortex
PHG

Entorhinal
cortex

Dentate
gyrus
DG

Subiculum

CA3   CA1

Hippocampus

B

F

S

D

Neocortex

S

D

PHG

2
3
5,6

pp

Entorhinal

DG

CA3

CA1

Subiculum

Nucleus
accumbens

Mammillary
body

Anterior nuclei
of the thalamus

Figure 5.4. Relations between the hippocampus and other parts of the cerebral cortex involved in spatial memory and planning action. *(A):* The main centers of the brain. *(B):* Axonal connections to neurons. *(Lower right):* A characteristic feature of the neurons of the CA1 and CA3 layers of the hippocampus is the projection of their axons back to the same cell or to other adjoining cells. This arrangement gives the network so-called autoassociative properties that allow it to memorize combinations of signals and to recall the entire combination even if some of the information the neurons memorized at the outset is missing.

Street when a bicyclist runs over his foot. The visual cues (the street, the bicycle arriving at the periphery of his visual field), proprioceptive cues (the bicycle on his foot), the motor action (the aggregate of information associated with the action of placing his foot in the street), and the auditory cues (the cry of the cyclist) constitute the configuration of sensory cues about this episode. One week later, he is just about to step out into another street. The situation contains enough similar cues for him to recall the bicycle episode. He suspends his gesture for an instant, fortunately, for a bus is passing by at great speed, and his hesitation saves him from being seriously injured. Memory of the past allows him to predict the consequences of his action.

This mechanism might also explain the brain's faculty for "filling in." By this we mean its ability to reconstruct episodes, shapes, words, and gestures from a few elements among a configuration of signs. For example, perhaps you do not always realize that a word is missing a letter, or an image some detail; the brain supplies the missing information. Phenomenologists, Husserl in particular, became interested in this capacity, which they studied in connection with caricature: how are we able to recognize a face in a simple caricature, or a man dancing, as Johansson showed, with only five or six points in motion? The episode of the madeleine, which Proust describes at length in *Remembrance of Things Past*, is a nice example of filling in. He has only to smell the madeleine dunked in a cup of tea to be able to recall the entire memory.

It is a general property of the brain, and among the most remarkable, not to need every bit of information to identify a memory, even a very complex one. Shepard, whose theories are discussed in detail in Chapter 8, proposes the idea of resonance between a repertoire of internal representations (preperceptions) and clues, even incomplete, from the environment, and here he parts ways with Gibson: "Instead of saying that an organism picks up the invariant affordances that are wholly present in the sensory arrays, I propose that as a result of biological evolution and individual learning, the organism is, at any given moment, tuned to resonate to the incoming patterns that correspond to the invariants that are significant for it."[32] A resonator can react to a signal only very slightly different from the one to which it is tuned, provided it has some relationship—for example, a harmonic—with the tuning frequency. Shepard uses this capacity of internal systems to resonate with stimuli that resemble normal stimuli to explain certain properties of perceptual filling in; that is, the capacity of the brain to continue to see the external world even if some portion of the required information is absent.

OSCILLATIONS THAT MAKE IT POSSIBLE TO PREDICT TRAJECTORIES

A recent development in hippocampal model theory attempts to explain how the brain can predict a trajectory during displacements in space. When a rat is active, its hippocampal neurons fire in bursts that repeat about eight or twelve times per second. When the global activity of the hippocampus is recorded, that is, the electrical sum of all the activities of the neurons, the result is an oscillating wave called θ rhythm. No one knows yet what this wave does, but we do know that when a rat moves in an enclosure, the discharge of the pyramidal neurons of its hippocampus is connected to θ rhythm. According to

O'Keefe and Recce, there is a relationship between the instant of firing of the neuron in each cycle of θ rhythm (the phase) and the position of the animal in space.[33]

But a second rhythm in the hippocampus of the rat that has an oscillating frequency of around 40 Hz also exists in parts of the cortex involved in processing visual, olfactory, and auditory messages. This is called γ oscillation. The two oscillations are superimposed in the hippocampus.

For each oscillation of θ rhythm, there are around seven to nine oscillations of the γ type; hence Lisman's hypothesis: Each θ cycle contains a series of seven to nine small packets of information—seven to nine episodic memories that will be replayed in this way six times per second.[34]

He takes the example of memory for telephone numbers. We cannot readily memorize more than about seven numbers. In Lisman's model, each figure corresponds to the simultaneous activation of a certain number of neurons from a population of hippocampal cells. With each γ oscillation, one of the figures is activated. Thus we can store a set of seven figures, but no more. What makes it possible to memorize them is that the brain perpetually replays the seven figures in the form of seven γ oscillations, eight or twelve times per second. In other words, external reality is broken down into packets of information contained in the simultaneous activation of subpopulations of neurons that are themselves activated every 25 milliseconds. Series of seven packets are repeated around every 125 milliseconds.

How does this mechanism enable prediction of a memorized trajectory? Lisman starts with the assumption that if the seven packets store seven locations (identified by the configuration of synapses activated by the different signals that constitute the episode) in a room—that is, the configurations of sensory information that correspond to the seven positions of the animal in the room—when the animal passes by the first of the positions, the automatic cycle of the sequence of seven positions is played at the γ frequency and thus predicts the course of anticipated positions.

It is not possible to describe this complex mechanism here, but what is essential is that the succession of the animal's positions in space will be replayed by the brain at least seven or eight times per second. In other words, seven or eight times per second, the animal will experience all the different sensory configurations that correspond to the positions he will occupy if he correctly negotiates the trajectory he has learned. I can think of no better example of what I mean when I say that perception uses memory to predict the future consequences of action. This interplay of past and future repeats on a time scale of several dozen milliseconds in the brain. It is also possible to imagine that slower mechanisms exist. These ideas are of course only models, but they are becoming increasingly plausible, even though the presence of θ rhythm and rhythm at 40 Hz in the hippocampus of the monkey have not yet been firmly established.

How Was Prehistoric Man Able to Draw So Well?

Recent findings on the projective mechanisms of filling in inspired me to a new theory to explain the extraordinary quality of the designs found in prehistoric caves. Although convention requires a scientist to support his conjectures with rigorous experimentation, I cannot provide any proof of my idea. But from time to time it is useful to listen to one's imagination.

Here, as I see it, is the cognitive mechanism that enabled Lascaux man, working in half-light, to draw the marvelous animals whose shape so accurately conveys their anatomy and movements. In the cave, Lascaux man's torch projected an evocative play of light and shadow onto the walls. You have only to spend a few minutes in a dimly lit section of these caves to see how the exceptional variety of shapes made by the limestone can stimulate the imagination. While a flashlight adds movement, candlelight considerably enhances the apparent reality of the forms.

Trying to recognize an animal in the shape of a stone or a rock or in the branches of a dead tree is a common experience. Country traditions are replete with rocks named after animals. It is also common to seem to recognize a familiar face in an anonymous crowd. Likewise, I think that prehistoric man, strolling about these caves, perceived on the walls the shapes of well-known animals.

The animals that Lascaux man saw or perceived were the recollections of animals in action that he had in his memory. He projected onto the walls the shapes of these familiar animals just as we project onto passersby the features of a face that we know and say, "She looks like so-and-so."

But the essential point of my theory is that the drawings of Lascaux are perfect because the painter was tracing, in some way, the image of the animal stored in his memory and projected onto the wall, just as one draws an image on a screen where a slide has been projected. The motion of the torch in the cave must have added considerable force to these impressions. Modern psychophysics tells us that perception of the curvature of a planar object can be caused by motion. This extrapolation of three-dimensional structure from simple motion has been studied in our laboratory and is probably made in the parietal cortex in association with the temporal cortex.

The perceptual theory of Empedocles is a naive form of the idea of a projective brain that I defend in this book. The brain projects the image of the animal on the wall of the cave; Lascaux man sees it there, external to him. Recall that Michotte wrote:

> The role of stimuli is not, as was believed for a long time, to give rise to "sensations" that are then combined, linked one to the other, and even altered by certain psychic processes under the predominant influence of acquired experience. Their role seems, on the contrary, to boil down to the simple initiation of endogenous constructive processes that obey the proper laws of organization, largely autonomous and independent of experience, and that lead directly to constructing the world of phenomena.[35]

It was enough for the shapes on the wall to suggest a part of the form of a buffalo; for example, to enable the brain of the painter to mentally reconstruct the entire body. I have shown in connection to my discussion of the hippocampus that neural networks of the autoassociative type permit construction of an internal representation of an episode from just a few clues.

At the same time, the person looking at the animals on the wall of the cave was probably filled with wonder at the supernatural existence, as it were, of the animal he normally saw in nature. Perhaps this wonder even gave rise to quasi-religious notions of a kingdom of the beyond where these creatures lived and where one could interact with them in sanctuaries. Be that as it may, prehistoric man was not drawing out of any mysterious symbolic impulse; he only fixed on the walls the shapes seen there. He must have been amazed by what he saw, even though his brain was recreating these shapes from various cues, just as I am present as a spectator at my own lectures, an impression produced by a brain whose expression I listen to with astonishment.

The example of cave paintings has a very general significance. Artistic creations seem to me to be projections of internal simulations of the world. I recently saw a fifteen-year-old sculptor crafting a horse, or maybe it was a dog, out of clay, without the aid of any model or scale, without even looking at what he was doing, essentially by feeling it. You could clearly see that the form emerging from the hands of this young prodigy was not even in his imagination; it was in the tips of his fingers, just as the pilot feels the landing strip under the wheels of the airplane as if it were his own feet, or as we feel the tip of a pencil on paper like the end of our finger. "Vision is the brain's way of touching," said Merleau-Ponty. He wrote:

The animals painted on the walls of Lascaux are not there in the same way as the fissures and limestone formations. But they are not *elsewhere*. Pushed forward here, held back there, held up by the wall's mass they use so adroitly, they spread around the wall without ever breaking from their elusive moorings in it. I would be at great pains to say *where* is the painting I am looking at. For I do not look at it as I do at a thing; I do not fix it in its place. My gaze wanders in it as in the halos of Being. It is more accurate to say that I see according to it, or with it, than that I see it.[36]

Perception is not representation: it is simulated action, projected onto the world. Painting is not a set of visual stimuli, but a perceptual action of the painter who has translated, through his gesture on a limited medium, a code that evokes the scene he perceives, not the scene represented. The painting moves us because it reproduces the miracle of the images of Lascaux in reverse. I look at the canvas in the place of the painter who has projected onto it his mental activity. A genius is someone who helps me to perceive things the way he does.

6

NATURAL MOVEMENT

> The most direct, and in a sense the most important, problem which our con-
> scious knowledge of nature should enable us to solve is the anticipation of fu-
> ture events, so that we may arrange our present affairs in accordance with such
> anticipations.
>
> —H. Hertz

The problems the brain has to solve are mainly problems of mechanics. Modern philosophy seems to have forgotten this, so captivated as it is by language
and so convinced that the higher functions of the brain have to do with formal
logic or can be explained by analogy to computers. But if the body is to be re-
habilitated in modern neurobiology, the rules that underlie its movements
have to be rediscovered. These rules[1] are intuitively understood by sculptors,
who are able to render the movements of the body and their relationship to
emotions, as are actors in Asian theater.[2] These actors demonstrate that posture is the first expression of movement. In other words, it is intended or suggested movement, the dynamic form of which Bernstein called "readiness to
move." They also demonstrate that the kinematics of movement conveys
meaning, and that the trajectory of a finger, the displacement of the head, the
swaying of the body must respond to laws that are at the crossroads of mechanics and neurology. Moreover, they confirm that natural movement is a
source of pleasure.

It might seem surprising that I open this chapter by speaking of pleasure.
Yet pleasure is a necessary element of perception and cognition. And the
source of this pleasure is in movement. The proof of it is the delight taken in
an elegant dance step, a nicely formed letter, a ball well thrown, or the joy got-

ten from certain movements that are possible only in particular situations. As I reported earlier, an astronaut once confessed to me his sadness at returning to "this sticky earth"—so happy was he to be freed from the constraints of gravity. He made me think of the poet Ronsard, a great master in the art of pleasure: "So let this muddy hide rot out, / Whose lot both destiny and fortune gambled over. / Be spirit only. Let the body be."[3]

One of the greatest pleasures of my own brain is to fancy that I am floating like a glider, free of gravity. I have extended dreams in which I fly over the mountains and along rivers. I particularly like to launch myself from the top of a hill and drift slowly, close to the meadows, teasing the goats, abandoning myself to the delights of effortless forward motion, impervious to mechanical forces. By day I also derive extreme pleasure in feeling vection, which is the illusion of advancing that you experience watching a river flow or watching clouds go by when you are lying on the sand.

Over the course of evolution, we learned how to make use of mechanics. In terms an author of the Enlightenment might have used, this book is a paean to the mechanics of the body in complex beings and the brain's accommodation to it. In this chapter, I will show how the brain managed to conquer mechanics.

It did it first by simplifying the problems, a common device in mathematics. You reduce a complex problem to a simpler one that you know how to solve. The work of the roboticist Slotine is a good example of this approach. Slotine developed a set of concepts from using robots that he taught to play tennis with remarkable dexterity.[4] In this way he demonstrated that making them work at what he called "composite variables" simplified calculations considerably and increased the robots' capacity for prediction and adaptation. The robots responded much more quickly and even solved problems for which they lacked sufficient data. The principle of composite variables itself is simple: instead of asking a robot to control position, or velocity, or acceleration separately, one asks it to work on a variable s that is a combination of all these variables and whose movement is defined by what is called a Lyapunov equation. This equation is chosen in an ad hoc manner. The remarkable advantage of this straightforward transformation is that nonlinear problems involving successive derivatives of the nth order (for example, velocity, acceleration, jerk, and so on) are replaced by linear problems (problems that can be resolved by much simpler, so-called first-order equations). Of course, actual implementation of these techniques is very sophisticated and does not concern us, but the idea is important because it developed over the course of evolution.

In the nervous system simplification was achieved by creating internal models of physical reality that enabled simulation of movement and that constrain perception. This dual effect is difficult to dissociate because its two aspects—perception and action—are so entangled. Neuropsychology, which grapples with dissociations in patients who have lesions, is in this respect indispensable, but it cannot explain the mechanisms. Thus new theoretical and experimental tools have to be constructed. I will provide a few examples of recent efforts.

A neuroethology of natural movement also will have to be constructed. It should be one that clarifies the relationships between movement and the emotions movement arouses or expresses. Darwin showed how posture expresses the emotions of the animal or human subject (see Chapter 11).[5]

Unfortunately, the neurophysiology of motor systems remained for a long time a neurophysiology of connections between nerve centers. Structure and function were not always associated. Theory was dominated by the stimulus-response paradigm of Pavlov and the cybernetic paradigm of Wiener. Few alternatives were entertained. The physiology of reflexes was predominant from the beginning of the century, reinforced by the discovery of numerous tools for analytical study of the nervous system—tools like electrical stimulation of nerves. Generations of neurophysiologists (a group to which I belong) duly measured reactions and responses resulting from the application of easily manipulable physical stimuli such as light, force, stretch, sound, and so on; they also constructed an analytical neurobiology. How is it possible to get from a physiology of reaction to a physiology of action, from analytical neurobiology to holistic neurobiology? To do so we have to study natural movement and abandon formulaic reductionism.

Pioneers

Aristotle, in *De animalium motu,* then Borelli, and finally the brothers Weber were the first to attempt to adapt the laws of bodily mechanics to the movement of animals. Then, in the second half of the nineteenth century, Marey was able to generate the first real descriptions of natural movement using chronography. At the inauguration of the chair in the natural history of complex organisms at the Collège de France, where he succeeded Flourens, Marey declared:

> When we stretch our fingers and we think about the sequence of actions that had to occur . . . , at the start we find the action of volition, a psychic action, then the transmission of this volition, a nerve action,

then the contraction of the muscle, a muscular action, and finally the movement of the organ, a mechanical action. In what order should we study these events? A philosopher of the past, a Spinoziste, would not have hesitated: follow the logical path; introduce the facts in the very order of their appearance. This is precisely the approach that our contemporary school rejects. The physiologists of today rethink the order of events by beginning with the crudest and the most visible, and working up progressively to the most refined and obscure.

Marey undertook to capture movement at the center of gravity with cinematography, calculating the work done and then the impulsive force, from which he deduced the amount of movement.[6] He used a dynamograph to infer the activity of the muscles. The limitation of his work, like that of his successor, Jules Amar, was in failing to go beyond this descriptive stage and come up with hypotheses about cerebral mechanisms. Bernstein did similar work first in Russia, then in semisecrecy in the Soviet Union between 1910 and 1950, owing to the official support of Pavlovian theory.

Today, rapid digital cameras, connected to computers capable of simultaneously capturing images and recording neural activity, enable us to envisage a physiology of movement that benefits from the power of kinematic analysis (see Figure 3.3). Why is this important? The measurements of the movement of the planets, painstakingly carried out by Tycho Brahé, made it possible for Kepler to infer their laws. Similarly, kinematic observation of natural movements reveals algorithms that the brain uses to control movement.

For example, these methods reveal that *all* natural movements (writing, drawing, complex movements, and so on) are organized in discrete segments.[7] Write the letter "a" on a piece of paper in different sizes. Then sketch this same letter "a" with your finger in space, or with your foot in the sand or on the living-room rug. These movements will all have approximately the same distribution of velocity and acceleration, in accordance with the principle of motor equivalence. Their trajectory, to all appearances continuous, is in reality composed of units along which tangential velocity is constant and proportional to the cube root of the curvature of the trajectory. This relationship illustrates a law called the two-thirds power law.

The anticipatory nature of the control of movement is also made evident by the simple analysis of bodily movements. Try, for example, to bend over to pick up something from the floor. Just before this flexion, an automatic synergy, discovered and studied by the neurologist Babinski, moves the body very slightly backward. Without this displacement, the projection of the body's center of gravity in front of your feet would cause you to fall down. Patients presenting with lesions of the cerebellum or disturbances of the basal ganglia, like those with Parkinson's, no longer display this anticipatory movement.

The analysis of movement thus brings to light solutions devised over the course of evolution to anticipate the consequences of action and to simplify the control of movement. I will take a look at some of these simplifications.

The Problem of Number of Degrees of Freedom

Bernstein was the first to formulate the question of how many degrees of freedom need to be controlled. "The first clear biomechanical distinction between the motor apparatus in man and the higher animals and any artificial self-controlling devices . . . lies in the enormous number (which often reaches three figures) of *degrees of freedom* which it can attain . . . Because of this there is no direct relationship between the degree of activity of muscles, their tensions, their lengths, or the speed of change in length." Bernstein gives the example of a ship at sea and an automobile, and continues:

> I draw my second example, for comparison, from the field of normal human motor co-ordination . . . Fasten the handle end of a ski-stick in front of the buckle of a subject's belt. Attach a weight of 1–2 kg to the far end and on the right and left sides of the wheel attach a length of rubber tubing long enough to allow the ends to be held in the subject's left and right hands. Instruct the subject, turning the stick point forwards, to stand before a vertical board on which a large circle, square or other simple figure has been drawn, and to try, manipulating the ski-stick only by pulling on the rubber tubing, to follow the contours of the figure with the point of the ski-stick. The stick here represents one segment of an extremity with two degrees of freedom; the tubing is analogous to two antagonistic muscles introducing a further two degrees of freedom into the system. This experiment (which is very useful for demonstrations in an auditorium) makes clear to all who attempt it just how difficult and complicated it is to control systems which require the co-ordination of four degrees of freedom.

He emphasizes that the basic problem is one of coordination: "The co-ordination of a movement is the process of mastering redundant degrees of freedom of the moving organ, in other words, making it a controllable system."[8]

The hundreds of degrees of freedom (Figure 6.1) that characterize the anatomical and dynamic organization of the skeleton would have made control

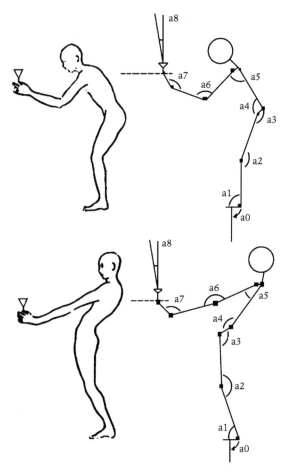

Figure 6.1. Two ways of holding a glass. These two natural postures result from controlling the number of degrees of freedom of the joints. Defining a posture simply requires defining the relationships between angles rather than the spatial position of each limb. This relative method simplifies control.

of movement impossible if, over the course of evolution, methods had not been devised for decreasing both the number of degrees of mechanical freedom by means of the geometric organization of the skeleton and the number of degrees of freedom the brain has to control. Roboticists, who to this day have still not found a way to build machines matching the complexity of the least little insect, know the extent to which any computer is quickly saturated, both in terms of its ability to make calculations and in speed, by the few degrees of freedom of the robots they construct. What are the tricks nature has found to reduce the number of degrees of freedom?

If you were to take a walk through the new gallery of the Museum of Natural History in Paris to see the display of animal skeletons patiently constructed by the systematic mind of French zoology in the last century, what might strike you is both the enormous diversity and a certain invariance of the collection. Despite the considerable efforts of zoology, the laws that govern the organization of the skeleton—its geometry—are not always related to the constraints of movement, and yet the purpose of this geometry is to reduce the complexity in controlling movement. Several recent findings have nevertheless attempted to link skeletal geometry to motor functions and to predatory behavior. Coppens and the paleoanthropologists thereby demonstrated that the skeletal anatomy of certain forerunners of modern man, like the famous Lucy, resembles both tree-dwelling quadrupeds and bipeds that walk upright.

In man, nonhuman primates, and other mammals, the anatomy of the cervical column was dictated by the need to reduce the number of degrees of freedom. Watch a stork flying, or a lion or gazelle or ostrich running. They keep their head held horizontally, such that the plane of the horizontal semicircular canals is perpendicular to gravity (see Figure 4.1). This constancy of position in relation to gravity makes the head a stabilized platform, which considerably simplifies processing of vestibular and visual information, as well as their coordination.

Indeed, suppression of the rotation of the head restricts the optic flow to alterations of the visual field connected to translations. These cues complement those supplied by the vestibular receptors of the otoliths which signal linear accelerations, but cannot distinguish translation or tilt of the head. Stabilization of the head thus simplifies the fusion of visual and vestibular information. Of course, it also reduces retinal slip and allows the vestibulo-ocular reflex to ensure the stabilization of the visual world on the retina.

Maintaining the head in this posture is achieved, in part, through the arrangement of particular cervical vertebrae.[9] In fact, curvature of the cervical column, of which the swan is a good example but which is a feature of all birds and all mammals, facilitates lifting of the neck, which places the head on a pivot that has the double advantage of positioning the semicircular canals horizontally perpendicular to gravity and creating a preferred plane of rotation.

A second property of the skeleton that simplifies biomechanics is limits on possible movements. Once more, movement of the head provides a good example. The anatomy of the cervical vertebrae allows only certain well-defined movements, familiar to specialists in kinesthesiology and rheumatologists. It is easy to make very quick orienting movements of the cervical column, which constitutes a pivot. Horizontal rotations are facilitated by a specialized muscular system, linked to horizontal movements of the eye. This architecture is what allows the sparrow you see on the lawn to turn its head in the horizontal plane with dizzying speed.

Look quickly straight up toward the ceiling. You will have difficulty doing it. Your center of rotation is located at the level of the first cervical vertebra. All the other vertebra are anatomically locked, which consequently forces a single center of rotation. You will note, moreover, that it is relatively easier to make an oblique movement upward and left, or upward and right, in the plane of the vertical semicircular canals.

Now bend your head quickly downward to look at the floor. This time, the cervical vertebrae are locked in such a way that the center of rotation is located at the junction of the cervical vertebrae and the thoracic vertebrae. On the other hand, the movement appears simpler because gravity aids the fall of your head.

In other words, nature used vertebral anatomy to reduce a very complex system to several basic movements, which simplifies its control. This description of anatomical constraints could be extended to the movements of the arms, legs, and so on. Biomechanics imposes geometric solutions that have been optimized over the course of evolution. For that matter, it is fascinating to note that the brain recognizes, so to speak, this organization of the possible movements of the skeleton.

The skeletal architecture of animals limits their movements in such a way as to considerably reduce the number of degrees of freedom. It is represented in internal circuits that allow mental simulation of movement. A new science is needed to understand this dynamic architecture.

Kinematic Phase Constraints

If you draw a figure eight in the air, the angles between the various parts of your arm will probably seem very different to you. But they are not. The relationships between them are very precise. A second trick devised by nature to simplify the number of degrees of freedom was to create kinematic constraints on the different parts.[10] When the angle of the arm increases in relation to the body, the angle of the arm in relation to the forearm decreases by an equal amount: they are in phasic opposition. This kinematic constraint uses a single parameter, varying only the relation of amplitude of the two angles, which considerably simplifies control. In this way, the brain controls global variables (elevation and azimuth of the movement from the end of the finger) and not the local variables (the multiplicity of the angles that make up the segments of the limbs themselves).[11]

Muscular Geometry

Yet another trick discovered by nature to reduce the number of degrees of freedom is to play upon the arrangement of the muscles. Muscular geometry complements the simplifications contributed by skeletal geometry. For example, most of the muscles are located between two joints (like the knee and the hip), but others connect distant joints (like the foot and the hip): these are called biarticular muscles, and people have been trying for a long time to decipher their role. The following example makes it easy to understand why: Take a mug of beer and set it on the table. Now, pick it up and bring it to your mouth. The interplay of angles made by your arm, your forearm, and your shoulder is such that had we only monoarticular muscles, some would do negative work, which is not very economical energetically. The fact that we have biarticular muscles enables us to recover this negative work and to reconstruct it in a positive form at one of the joints.[12] Of course, it takes a bit more mathematics to demonstrate than can be shown here. The point is that nature found an elegant way to resolve a problem of mechanics using geometry, just as carpenters do in solving problems of static distribution of weight.

Relationships between Kinematics and Geometry

Relationships or covariations between geometry and kinematics are another way of reducing the number of degrees of freedom. Take, for instance, the mechanical system formed by the hand, the forearm, and the arm. This system has a geometry, a shape. In general, the geometric characteristics of a mechanical system cannot impose a law of movement a priori along a given trajectory. The trajectory depends mainly on the temporal organization of the motor commands.

Mathematicians tell us that if a principled relationship exists between geometric and kinematic variables—for example, the radius of curvature of a trajectory (a geometric feature) and the tangential velocity along the trajectory (a kinematic feature)—"which is invariant for a class of movements generated by one controlling system, this relationship must be the reflection of a general rule that the system follows in planning the forces. Ultimately, any consistent pattern of covariation between quantities related to geometry and kinematics is likely to provide a clue for understanding the logic of the controller."[13] Two covariations that meet these criteria have been discovered.

Two-thirds power law. When you draw an ellipse on a sheet of paper with a natural movement, you may have the impression that the speed with which you draw is totally independent of the shape. Well, that is not the case. There is an extraordinarily precise relationship between the curvature of the form you are drawing and the tangential velocity along the curve. If you try to draw an ellipse, you will see that the speed of your pencil increases at the parts where the curves are greatest. The movement of the pencil along the ellipse can be described in at least three ways: first, describing the motion in time of the Cartesian coordinates $x(t)$ and $y(t)$ of the tip of the pencil; second, using equations to describe the shape of this trajectory and the law of movement along the trajectory; third, describing the movement by specifying two parameters: radius of curvature $R(s)$ of the trajectory and tangential velocity $V(s)$ of a point along the trajectory (s represents the curvilinear coordinate of movement). The trajectory has been shown to be completely defined by these two parameters.

By measuring these parameters, it is possible to demonstrate the relationship between curvature and the velocity of movements of the hand. In 1983 Viviani and his co-workers discovered a simple relationship between curvature ($C = 1/R$) and angular velocity ($A = V/R$) : $A = KC^{2/3}$, which they called the two-thirds power law.

This law held only for a certain class of movements. Subsequent investigations of the law in relation to three-dimensional movements resulted in a reformulation that involves a greater variety of natural movements than drawing; it even holds for points of inflection.[14] The two-thirds power law gradually becomes operative in infants as motor function matures. It connects the radius of curvature at any point of the trajectory to the tangential velocity V, as well as angular velocity. It is expressed by the following equation:

$$V(s) = K(s)\left(\frac{R(s)}{1+\alpha R(s)}\right)^{\beta}$$

Factor K depends on the length of the trajectory and not on its shape. When $\beta = 1/3$, $\alpha = 0$, and K is constant, the new formulation is equivalent to the original equation. Is this law connected to the mechanical properties of the limbs, or does it reflect the general principles of motor control by the brain? It appears to be due to neural mechanisms and not to geometric or mechanical constraints connected to the muscles and limbs. Natural movements are often complicated. They can be broken down into parts, and this law then applies to each separate part.

The principle of isochrony. This second principle establishes that the velocity of movement from one point to the other tends to increase with the distance between the points. If you have two objects before you on a table, the principle of isochrony predicts that you will cover the distance between one object and the other more quickly if the distance is greater. Go ahead and do the experiment. It is also true for movements of the eyes: the velocity of a saccade (see Chapter 10) increases with its amplitude. A saccade of 10 to 15 degrees is made at around 30 degrees per second; a saccade of 50 degrees is made at more than 500 degrees per second. According to this principle, which holds for all natural movements, the duration of a movement is relatively independent of its linear extent, hence the term "isochrony" (fixed duration).

Recently, Viviani and Flash noted that modulation of the average velocity along a trajectory depends on two factors: the length of the trajectory and the distribution of the curvature along this trajectory.[15] These relationships disclose a more fundamental property that involves the prediction of movement. "In both cases, the fact that velocity is modulated by a global geometrical quantity (the linear extent of the path), even before the trajectory is fully executed, suggests that an estimate of this quantity is available to the motor control system as part of the internal representation of the intended movement."[16] This theory assumes that the motor control system is equipped with a spatial blueprint even before the movement begins, which fits with the assumptions of motor program theory; it also assumes that the geometry of the planned trajectory limits the kinematic and temporal aspects of action.

The Invention of the Eye

Kinematic constraints do not only involve movements of the limbs. An extraordinary property of the movements of the eye testifies to the work done by nature to simplify the number of degrees of freedom. The eye is one of the most remarkable inventions of evolution. First it was a simple pivoting retina, such as crustaceans have; then, in fish, this little orbit also became an organ that could quickly direct gaze in three dimensions toward prey or predator in just several dozen milliseconds.

I maintain that in the fish the eye is not only a retina that pivots but a genuine organ of anticipated simulation of movement. The fish must maneuver a complex and heavy body in an environment that allows it only a single move. Catching prey or escaping from a predator affords the fish no room for error. Displacement of gaze by a simplified mechanical organ (six muscles operating in pairs produce rotations of the eye in three perpendicular planes) allows the fish to confirm that a movement it is planning to execute is the correct one.

What a fantastic invention this little biological machine is! With a single movement of the eye, the brain can visually pinpoint prey and check whether its gesture is adapted to the reality of the external world. The movement of the eye simulates the one the predator must make to capture its prey. Thus the eye is more than just a pivoting retina: It is a very simple physical model of the action needed to secure a meal. If I am a predator, all I have to do to figure out how much I need to move my body to capture my prey is to duplicate with my body the movement specified by my eye. Nature first had to simplify the number of degrees of freedom of the problem and then solve it with a very simple instrument.

Next, nature invented the principle of relative control of movement. It is actually not necessary for the fish to calculate its position in space (and it would have a hard time doing it in the water); its brain has only to minimize the distance from its present position to its goal. Borrowing the language of cybernetics, all the brain has to do is to minimize motor errors.

In primates, this internal simulation of a change in direction of gaze was refined to an even higher degree by the invention of the displacement of attention. Look straight ahead and turn your gaze quickly to the right. You will notice that just before your eye turns, you feel your gaze move forward. In Chapter 10 I discuss briefly the anticipatory displacement of the receptor field made by the neurons of the parietal cortex before the saccade of the eye and the mechanisms that enable internal simulation of gaze movements.

Transforming 3D Movements into 2D

Biology devised remarkable ways of enhancing the efficiency of this anticipatory action of the eye and of simplifying neural processes.

Consider the eyeball. It consists of an orbit moved by six muscles. In the fish, these muscles work in pairs and guarantee the horizontal, vertical, and torsional movements of rotation that you observe in an aquarium. Over the course of evolution human eyes migrated in their sockets and moved from a lateral position to a frontal position. At the same time, the fovea developed (the area of the retina that enables fine spatial resolution and highly sophisticated color vision). The fovea was a late find in evolution. The rabbit has only a horizontal furrow of greater visual resolution, and in the cat a retinal area of approximately 10 degrees (the area centralis) has a higher concentration of retinal cells. Only primates and man have a real fovea.

In man, horizontal eye movements are produced by the pair of horizontal muscles, but vertical movements require the cooperation of the vertical muscles and of torsion, and ocular torsion has very limited amplitude. The work of the physiologist Hering, in 1879, gave rise to this question: How does the brain encode rotations that enable the eye to go from one position to another? Suppose that you are looking at a point on the wall in front of you. Displace your gaze by an oblique saccade to the right and upward. From a geometrical point of view, this rotation of the eye can be broken down into three rotations, horizontal, vertical, and torsional (Fick coordinates), or vertical, horizontal, and torsional (Helmholtz coordinates).

The problem is that this change in the sequence of rotations does not return the eye to the same position. The axis of gaze is appropriately oriented toward the target, but the torsion observed at the end of the two movements is not the same. Indeed, the rotations do not constitute what mathematicians call a commutative set.

Figure 6.2 illustrates this noncommutativity with dice. Take a die and turn it over a few times, making the same rotations but in different order. You will

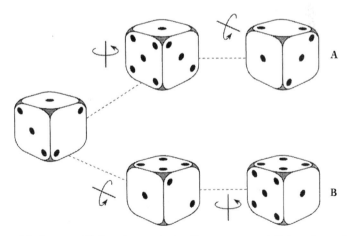

Figure 6.2. A game of dice demonstrates the noncommutativity of rotations. Reading left to right: Position a die with the number 1 uppermost. Turning it first around a vertical axis, then around a horizontal axis will result in a position of the die with a 2 uppermost. Now, start from the same original position and invert the order of the rotations. You will end up with a 4 uppermost. Thus, the order of rotations about an axis changes the result: they are not commutative.

see that the final position of the die is not the same. The rotations are not commutative. The same is true for the eye. If no simplification had occurred, the images of the same part of the environment would project to the retina in different ways. The result would be perceptual confusion and major difficulty in constructing a coherent image of the visual environment based on local images.

How did nature resolve this problem? Listing[17] derived a law, initially proposed by Donders,[18] who formalized the biological solution: Every movement of the eye can be represented by a vector whose axis is perpendicular to the plane of rotation and the chosen amplitude in a particular relationship with the angle of rotation. Is it possible to provide a straightforward illustration of this vector? Simply imagine a top: the vector of rotation of the top is the handle you use to make it turn, and that is perpendicular to its plane of rotation. Now, suppose that we make the length of this axis proportional to the velocity of rotation. It will be quite short when the top turns slowly and very long when it turns very fast. This use of vector of rotation is a very elegant way to describe the movement of the top in three-dimensional space. All that is necessary is to define the direction of the vector and its amplitude. The three-dimensional problem is then reduced to two dimensions.

Listing's law states a remarkable property: all the vectors of rotation of the eye are in the same plane, the frontal plane of the head, called Listing's plane. All the saccades you make in reading this page have vectors of rotation in this plane. When Listing's law is satisfied, there is no torsion of the eye, which constitutes experimental proof of the law. Many laboratories have now confirmed the validity of Listing's law for eye movements.

It was believed that command of eye movements could be expressed using mathematical rules that govern these rotational vectors. The representation of movement by rotational vectors presents another advantage: it is possible to characterize a vector by its constituents in space with the aid of a mathematical construct called a quaternion. A sequence of rotations can be represented by calculating the product of their scalar constituents. It was thus suggested that nature had discovered this property of quaternions and that the brain uses it to encode movements in the neurons. However, this theory has yet to be confirmed.

The Law of Two-thirds Power and Ocular Pursuit

Listing's law may not be the only mechanism for simplifying the control of eye movements. Following a moving target (a passing car, for instance) calls into play the neuronal subsystem of ocular pursuit (see Figure 3.2). This slow movement of the eye appeared late in the course of evolution and followed the appearance of the fovea, which, as I mentioned, exists only in primates and in man. You can see how it works very simply by performing the following experiment. Hold a finger out in front of you and fully extend your arm. Move your finger slowly from left to right and from right to left, ever more quickly. You will notice that your gaze follows your finger for slow movements, whose cyclic frequency is about once per second (1 Hz). But as you increase the speed, your gaze will start to lag behind your finger, and you will get to the point where you cannot follow it anymore. In fact, the tracking system is a slow system (up to 100 degrees per second)—unlike the saccade, which is extraordinarily rapid (up to 800 degrees per second)—probably because it involves complex pathways. Ocular pursuit follows the law of two-thirds power (this is also true of manual pursuit),[19] but it remains a slow movement. Nature's way around this slowness is prediction.

Oculomotor pursuit is essentially predictive. Indeed, it is not only the target itself that is tracked, but an internal simulation, an internal model, of the trajectory predicted by the target.[20] If you move your finger in front of you like a metronome, one oscillation per second, you will observe that after about two or three oscillations your eye is ahead of the finger it is tracking. Your brain is anticipating the trajectory.

The Form of a Drawing Is Produced by the Law of Maximal Smoothness

Most theories about the control of movement assume that a trajectory is shaped by a motor program that defines a kind of image, a sort of drawing of movement. But one can also imagine that the shape results from the simple laws of mechanics, just as the trajectory of a shell can be explained principally by its velocity leaving the cannon, the force of gravity, its mass, and the laws of Newtonian mechanics.

It has been suggested that the motor system follows a principle of minimum energy that leads to movement produced with *minimum jerk*. Thus we can define a cost function (CF) proportional to the mean square of the jerk, jerk being defined as the derivative of acceleration.[21]

According to the principle of minimum jerk, shape results from properties of morphogenesis (creation of shapes) that are determined by the value of this single variable. The principle is very satisfying; it predicts the trajectory

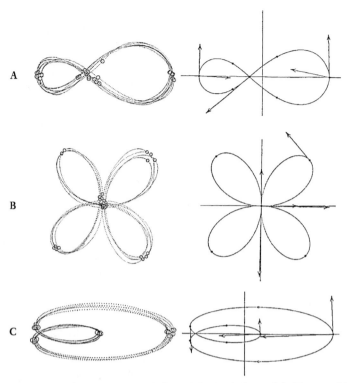

Figure 6.3. Natural movement and the minimum-jerk model. Three subjects were asked to draw three different shapes *(left: A, B, C)*. To the right are traces of the trajectories predicted by the minimum-jerk model.

and the velocity based on the total duration of movement, and the position, velocity, and acceleration at points where it starts, ends, and through which it must pass. Experiments have shown that natural movements (drawing a figure eight, a spiral, and so on) made by subjects could be modeled very precisely using the law of minimum jerk (Figure 6.3). However, this model does not take into account the global isochrony of movement. One compromise is thus to search among the theories that take into account global isochrony and relationships between geometry and velocity. The synthesis proposed by Viviani and Flash assumes that the brain first constructs an abstract representation of the intended movement, encoded as a sequence of boundary conditions, or velocities and positions of departure and arrival. These conditions are then transformed into trajectories by the laws that constrain movement (minimum jerk, point of equilibrium, and so on).

Current theories of motor function thus adopt two completely opposite points of view. One starts with a geometric representation of the trajectory and assumes that the brain organizes a sequence of basic actions to follow it; the other suggests that the brain ignores the shape of the trajectory, which results quite simply from high-order constraints such as the stability of paired oscillators operating in tandem or the minimum jerk of a movement. Taking it one step further, some would say the body follows the form, whereas for others, the form follows from the functioning of the body.

7

SYNERGIES AND STRATEGIES

In olden times, children might be given a splendid Harlequin doll made from cardboard as a present. Its back concealed a network of strings, connected in such a way that a single string made the arms and legs flex at the same time. This invention is a legacy of Chinese shadow puppets and of Asian puppet theater, and it illustrates a basic form of synergy. The principle involves coordinating the limbs so that a single command sets in motion a network of combined actions that constitute this synergy. What remarkable simplicity! Indeed, in puppet theater, different gestures are created by manipulating various combinations of strings. Musical automata relied on the same principle.

The word synergy comes from *syn* (together) and *ergos* (work). This concept was proposed by Bernstein to support the idea that, since the nervous system cannot control all degrees of freedom, evolution selected a repertoire of simple and complex movements, which are called natural movements. These involve groups of muscles working together. I have already mentioned the constraints the skeleton imposes on the variety of movements possible at each joint. Moreover, this repertoire is not very extensive. Watch a dancer, and see just how meager the motor repertoire of the human body is. What is recognized as the genius of choreography and the expressive richness of dance is the combination in time and space of the elements of movement and the interplay of partners.

Motor synergies are the basis of movement. The remarkable flexibility of neural networks makes it possible to manipulate these synergies to produce what I call strategies: the selection either of a particularly well adapted synergy, or of a sequence of synergies that constitute a complex movement oriented toward a goal. In fact, movements are organized in sequences of synergies on which behaviors are based, as Lorenz demonstrated superbly in his experiments with greylag geese. How synergies are chosen depends on the goal of action.[1]

The challenge is to construct with ethologists what I call a neuroethology of motor function, and to work with molecular biologists to ascertain whether some of these synergies are genetically determined by knocking out specific parts of the genetic code in the rat or the mouse.

I will first describe the contribution of anatomy, especially the branching of axons, to the construction of synergies; then we will analyze data from research on neurons of the motor cortex with reference to problems discussed at the end of the preceding chapter on the laws governing movement of the arms.[2]

The reader would be well advised to resist the temptation to skip this passage on anatomy. The digression is worthwhile. You do not need to understand the architecture of a flower to enjoy it, nor do you need to analyze how the branches of an oak are arranged to benefit from the shade it provides. But anyone who seeks the principles of physiological functioning must find them first in anatomy. My objective is not to review the wealth of anatomical facts accumulated in the aftermath of the pioneering work of Ramón y Cajal, owing to the diversity of techniques anatomists have at their disposal. Rather, I will discuss how the branching of axons determines motor synergies and how it links motor function and perception, in short, how anatomy comes to the aid of physiology.

Vestibular Axon Branching and Gaze Stabilization

Look straight ahead and turn your head upward while keeping your gaze fixed on an object. Your gaze will remain stable. The vestibulo-ocular reflex, described in Chapter 2, contributes to this stabilization; vision is not necessary. Indeed, try the following experiment: Look straight ahead, letting your gaze rest on an object. Close your eyes and think of the object. Lift your head while fixating behind your closed lids on the object you memorized. Open your eyes. You will see that your gaze has remained where it was. In the dark your eyes moved in a direction contrary to that of your head as a result of the vestibulo-ocular reflex. Several muscles in each eye had to be jointly activated to make this compensatory movement. This is a very simple example of motor synergy.

Actually, this reflex is complex. It is not the simple movement of a Harlequin doll. Try another experiment: Again, look straight ahead and close your eyes. But this time, concentrate mentally on the end of your nose. With your eyes closed, lift your head, fixating, in the dark, on the mental image of your nose. When you open your eyes, you will see that this time you raised your gaze. The vestibulo-ocular reflex was blocked by a signal related to the goal. So this reflex is both stereotypical and flexible. Its stereotypical character is due to your anatomy, which composes a synergy. Its flexibility is due to a wide range of mechanisms that involve the cerebellum, the cerebral cortex, interactions with other elements of the oculomotor repertoire, and so on. As with the Harlequin doll, organization into synergies simplifies control by obviating the need to keep track of all the muscles separately. Each synergy constitutes a unit of movement. But first I will explain how these synergies are organized.

In the fish the problem of the relationship between vestibular receptors and the muscles of the eyes is simple. The eyes are lateral, and three pairs of extraocular muscles produce, respectively, horizontal movements (directly lateral and medial), vertical movements (straight up and down), and torsional movements (inferior and superior oblique). An arc of three neurons connects the vestibular receptors and the muscles of the eyes. However, in the course of evolution, in higher organisms the eyes migrated and became frontal at the same time that the fovea appeared. A vertical eye movement in the monkey, as in man, requires an activation synergy of the vertical muscles and the oblique muscles. This synergy implies that synaptic activation of these different muscles is very precisely distributed. This distribution is achieved through the connectivity of each second-order vestibular neuron. So-called collateral axon branches innervate the muscles that participate in this synergy. Figure 3.5 shows an example of these connections in the cat.[3] Axonal connectivity and efficient synapsing of the axonal endings in each motor nucleus of a second-order vestibular neuron thus encode the geometry of the movement by synergistic activation of several muscles.

When someone jostles you and your vestibular reflex simultaneously rights your head and your eyes to stabilize your gaze, the synergy is much more complex, since your neck has about thirty muscles whose insertions are extremely varied. This coordination of the movements of the eyes and the head is also facilitated by axonal connectivity. Some second-order vestibular neurons project both to the motor nuclei of the eyes and to those of the muscles of the neck, which are located in the caudal-most end of the medulla. Thus, it seems plausible that this axonal distribution underlies the eye-head synergy that facilitates movement in various directions.[4]

Remarkably, the extent of the axonal projections of these neurons is not limited to the motor nuclei. Collateral axons seek out very precise targets that are required for the more complex aspects of this synergy, as well as for its perceptual and behavioral context.

First, axonal collaterals project to the vestibular nuclei of the other side of the brainstem to keep the eyes and head from moving in the opposite direction. This is called reciprocal innervation. Next, other collaterals project to neurons whose target is the cerebellum, which plays an essential role in controlling the plasticity of reflexes. Collaterals branch off to areas of the reticular formation where the cell bodies of the reticulospinal neurons play a role in controlling posture; this connection ensures coordination between gaze and posture. Finally, other collaterals project to interneurons, which feed back to the colliculus, the cerebral cortex, and different parts of the nervous system associated with perceptual stabilization, like the thalamus and the vestibular cortex. These organs influence cortical processing of the perception of movement.

A copy of the motor command is sent to multiple centers,[5] contributing to the various mechanisms involved in gaze stabilization. The question really is one of central anticipation, since the nervous system does not wait for sensory cues to reorganize its functioning in this reflex activity. Note also that this so-called reflex is a complex entity integrated in a behavior, arising from the most basic level of the three-neuron arc. Axonal branching allows a single neuron to activate different levels in the synergy and to participate simultaneously in the motor command and in perception. The notion of simple sensorimotor assemblies with an entry and an exit disappears, to be replaced by a concept of a neuronal character that is both extremely specific to the nervous system and completely integrated with a set of behaviors and perceptions.

Interpreting the anatomy of the neurons can sometimes lead to varying conclusions depending on the implicit and explicit theoretical conceptualizations of the observer. Thus, at the beginning of the century, the neurons of the reticular formation were considered to be hairy neurons that projected everywhere. Following on work by Morruzzi and Magoun, these neurons were attributed diffuse functions of activation or inhibition. This was certainly true for the ascending neurons of the dopaminergic and serotinergic systems, for example. But now we know that their axons and collateral axons have exquisitely precise targets that correspond to differentiated functions (motor, reciprocal innervation, corrective, perceptual, and so on). Accordingly, synergies have to be understood as genuine perceptuomotor actions.

But there is more. Axonal organization and the structure of connections are alterable, at least at certain stages of development. I have emphasized that connectivity controls the geometry of movement. For example, in the flatfish the brain can use anatomy to change this geometry. Here is the story of the baby flatfish that wanted to swim flat on its belly.

When it is born, the baby flatfish, like all other fish, swims vertically upright with its eyes on each side of its body (Figure 7.1a). A distinctive feature of this fish is that it undergoes an extraordinary metamorphosis when it is young. It suddenly flips onto its side and starts moving around flat at the bottom of the sea (Figure 7.1b). One of its eyes is thus on the under-

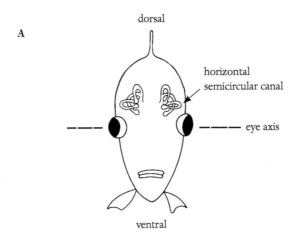

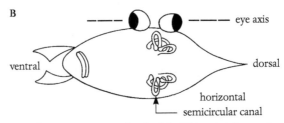

Figure 7.1. Eyes change position in the flatfish. When it is born, the flatfish has eyes on each side of its head and swims upright like all other fish (A). Soon after, the fish flips onto one side and swims around flat at the bottom of the sea. As it develops, one of its eyes migrates to the other side of its head. (B): Because the vestibular receptors do not change position, the vestibulo-ocular reflex no longer functions correctly; in response, the anatomy of the vestibular neurons is reorganized.

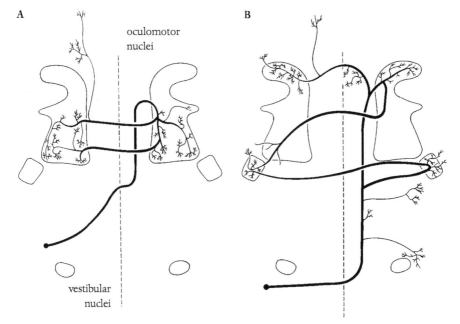

A oculomotor nuclei B

vestibular nuclei

Figure 7.2. Reconstruction of the second-order vestibular neurons (involved in the horizontal vestibular reflex) in the flatfish after metamorphosis. *(A):* An excitatory neuron; *(B):* an inhibitory neuron. At first these neurons, which receive signals from the horizontal semicircular canals, activate the vertical and oblique muscles. The projections of their axons are subsequently reorganized, resolving the problem of the new geometric relationship between the horizontal canal and the eyes after metamorphosis.

side of its body. To compensate for this disadvantage, the lower eye migrates to the other side of the fish's head. The two eyes end up right next to each other.

This daring transformation of the position of the visual sensor risks completely changing the functioning of the vestibulo-ocular reflex that stabilizes the image of the world on the retina. But in fact, the semicircular canals and the otoliths do not move in the skull. When the animal turns, the precise connections that I have described displace the eye, which has migrated in a direction that no longer corresponds at all to the correct geometry. A remarkable reorganization of the axons of the vestibular neurons then follows, which can be seen by injecting these neurons with a tracer (horseradish peroxidase) that makes it possible to reconstruct the anatomy of the neuron (Figure 7.2).[6] Following the transformation, the anatomy of the neurons is completely changed: the projections are reorganized and now ensure gaze stabilization with the new arrangement of the eyes.

Thus axonal anatomy does not determine only the geometry of the eye. It facilitates adaptation to changes in the body. This biological solution to the problem is extraordinarily elegant because it spares the animal complicated neural calculations to reorganize its perceptual and motor space.

The Neural Bases for Encoding Movement of the Arms

If the neurons that underlie local reflexes are organized in a way that produces synergies, how is the cortical control of movements organized? Do the neurons of the cerebral cortex that direct our movements control the muscles one by one, or do they use the same principle as the vestibulo-ocular reflex, that is, does each neuron drive a specific synergy? Consider the case of a part of the cortex up to now considered simple: the motor cortex located at the apex of the skull in the central gyrus, also called Brodmann's area 4. The early work concerning neural activity in the motor cortex was done by the American physiologist Evarts in 1968. It was first assumed that each projection neuron of the motor cortex (large pyramidal cells that project to the medulla via the corticospinal pathway) activates a muscle or possibly a group of antagonist muscles at the joint.[7] Further studies contradicted this approach. The axons of the corticospinal neurons, like the vestibular and collicular neurons, actually branch out systematically into various levels of the medulla. Activation of any one of these neurons simultaneously activates several groups of muscles throughout the body; this contributes to a precise movement (for example, leaning forward); that is, to a synergy. In other words, once again, anatomy underlies synergy.

The connection between the pyramidal neurons of the cortex and the various muscles of a motor synergy is, moreover, specific to a function and not to the muscles that constitute their target. This functional specificity is illustrated by the fact that the neurons of the motor cortex can be active during precision gripping, as when you are picking a strawberry, but the neuron is silent if the same muscles are activated with great force, in what we call power gripping.

How these neurons control force is still a mystery. Each pyramidal cell maintains a variable relation with the force exerted by the finger that excludes a simple causal relationship. Certain neurons of the motor cortex are activated only when the frequency of the movement is increased, which suggests that these neurons control stiffness (the connection between a force exerted and the resulting displacement) of a given joint.

The Theory of Vectorial Encoding by a
Population of Neurons

A major surprise awaited researchers once they turned to arm movements more complex than simple grasping. They expected that many types of discharges would be associated with local variables of particular joints. Instead, they discovered that the cortex was involved in the overall movement of each limb, such as the complete trajectory of the finger. The activity of the *population* of these neurons is associated with the direction of the finger in space (this is also true in varying degrees for other cortical areas).[8] If a monkey is asked to point its finger toward targets located in various directions in front of it, specific neurons are activated for a particular direction of the pointing movement. In other words, certain neurons fire more when the movement is made, for example, toward two o'clock on the face of a clock. But these preferential directions are somewhat indistinct, and some neurons fire for movement in all directions. If the activities of a population of neurons (about a hundred) are simultaneously recorded during the same movement toward two o'clock on the clock face, some neurons will fire frequently because the gesture is made close to their preferred direction; others will fire less because their preferred direction is farther off. Taking the average of all these discharges will reveal that the resultant vector of the population is very precisely oriented in the direction of the movement of the finger toward the target.

Neurons of the Motor Cortex and Mental Rotation

The motor cortex also participates in anticipating movement via a signal that changes over time. Suppose that I ask you to point to a target in front of you. Then, suddenly, just before you start toward the target, I indicate another point to the right of it. Your brain has to reorient the gesture; it must make a mental rotation in the new direction.

When a monkey is asked to make a movement toward a target, but the target is moved just before the monkey begins his movement, the animal must redefine the direction of its movement. The vector representing the activity of the population of neurons in the motor cortex also changes direction during the preparatory period, as if it were participating in a genuine *mental rotation* of the planned movement.[9] This change in direction of the vector signals participation of the motor cortex in anticipating and reorganizing movement, not just in its execution. The motor cortex does not only control force, or even only synergies; it is also involved in a complex way in the spatial planning of trajectories. But the debate on the true mechanisms of control of trajectories is still open. As I explained in Chapter 6, defining an optimal trajectory may require no more than specifying points of arrival and departure, and adhering to the law of maximal smoothness (minimum jerk). The role of the cortex may thus be to define trajectories by modulating the constraints imposed by this law. Yet again, movement probably results from a reciprocal adjustment of neural commands and biomechanical properties.

I would like to reflect a few moments longer on the profound significance of this concept of synergy. If a neuron of the motor cortex controls a motor synergy (for example, a pincer motion of the fingers), this neuron can also be used by the brain during an internal simulation of the movement. Suppose I ask you to imagine that you are plucking a strawberry. Look at your fingers— they have not moved. Yet the brain can still fire the neuron that causes this synergy, blocking its execution via an inhibitory mechanism. (I will discuss later how these mechanisms are organized, taking gaze as an example.) An axonal collateral of this neuron can transmit a copy of its activity to other structures in the brain. The idea of plucking the strawberry is thus realized when the neuron fires. This discharge does not *represent* the synergy; it *is* the synergy.

Coordination of Synergies

The organization of movement is thus based on a repertoire of synergies that make up a multitude of possible actions. But it is not enough to have a library of movements that are easily activated and mutually compatible because they use the same frames of reference or are bound by the same geometrical principles. They still must be selected.

Locomotion is one of the best examples of this type of mechanism. Horses must move from trotting to galloping, from walking to jumping. How are such synergies selected? How are strategies organized? There are so many mechanisms of selection that to describe them is beyond the scope of this book. The most interesting modern concept is that of the *functional flexibility of networks*. For example, in the case of locomotion, a network in the medulla (described in detail for swimming in the lamprey by Grillner and his team) organizes a locomotor rhythm. Alternative modes of swimming are selected at centers located at critical spots in the network, which reorganize the movement through neurons in these centers with the aid of neurotransmitters. Some connections are inhibitory and others are excitatory. A basic mechanism of selection thus facilitates reorganization of the same set of neurons. This

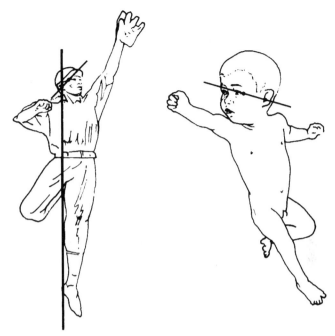

Figure 7.3. This movement is a synergy. It consists of a rotation of the head that orients gaze to one side, while the arm on that side is extended and the arm on the other side is flexed. The leg on the same side as the gaze is extended, and the other leg is flexed. This synergy is present in the baby; it is an integral feature of many gestures, like that of an athlete catching a ball or a swordsman fencing.

type of functional flexibility has also been described by Moulins and co-workers in invertebrate animals such as the snail. A network of neurons produces rhythmic contraction of the stomach muscles of this animal and is also implicated in an additional motor activity. The change in neural configuration between these two modes of functioning is assured by specialized neurons, independent of the network. Moulins and Clarac showed that several synergies can share the same network of neurons.

Thus, genetically determined local synergies—different kinds of locomotion, ocular motor systems (saccades, vestibulo-ocular reflexes, and so on), sexual behaviors, postures (Figure 7.3), and so forth—make up the sensorimotor repertoire of each species. These synergies are organized as behavioral strategies guided by global mechanisms. In higher animals and in man, these strategies can be anticipated, selected, and internally simulated before being executed, using the same neural structures as those of the action itself. I will show examples of these processes in gaze orientation and postural control.

The cerebellum probably plays a fundamental role in organizing synergies. The scope of this book prevents me from examining the role of this structure in the coordination of gestures, for I would also have to compare its function with that of the basal ganglia, and so on. I will merely take note of the following: It is well known that stimulation of muscular, tactile, and joint receptors activates the Purkinje cells of the cerebellar cortex. These sensory projections are distributed over microzones, each receiving projections from sensors located on the tongue, neck, arms, hands, feet, and so on. For several years it was surprising to find that the same part of the body projects to several microzones. Why, for example, are several projections of the head required?

It was discovered that the projections corresponding to the parts of the body involved in a precise motor action are grouped together. In other words, they compose a synergy.[10] For example, projections for whiskers, mouth, and front paws are found side by side; they correspond to the action of cleaning. Here again anatomy reflects activity, an element of the motor repertoire organizing actions and not isolated movements. "In the beginning was the Deed," says Faust. The choice of a strategy for action is thus simplified: all that is needed is to sequentially activate these sets of contiguous neurons that control the repertoire of actions. Similarly, the very same sets of neurons can be used to internally simulate movement because these sets *are* the neural mirror of action. It is known that cortico-ponto-cerebello-thalamo-cortical loops exist, within which internal simulation of movement can occur completely independent of its actual execution. These loops contain . . . here I do well to hesitate, like Faust. "Representations" is too vague; "models" is modern but probably vague as well; "images" is too visual; "schemas" is the term perhaps most common in the literature; "kinesthetic series" would make Husserl happy.

There is still a long way to go before it is understood how synergies and strategies are assembled in the various parts of the brain that control movement. How the rigidity of their repertoire is compensated by the plasticity of the rearrangements, as revealed by recent findings from neurobiology, is still unclear. A tantalizing scientific adventure awaits.

8

CAPTURE

Perception is active exploration; it is a question put to the world, a wager, pre-selection—it is also capture. Consider gaze. Visual perception is possible only by actively exploring the environment through gazing, through the changes in perspective enabled by movements of the eyes that I call stationary locomotion. Each gaze cast constitutes capture, especially if the object of regard is in motion. This capture is anticipatory, predictive. For example, try to read this page aloud. You will be struck by the coexistence of two coordinated actions: your voice, which articulates the text, and a silent reading preceding it. Bernstein said that we have two texts in our heads: the one we read aloud and the one that we look at in advance to prepare our reading. This phenomenon is well known to musicians, who are always reading the score one or two measures ahead.

According to Bernstein, planning a motor action, however it is encoded by the nervous system, necessarily involves recognizing patterns of what will be but does not yet exist.[1] Planning requires exploration of the future. Just as the brain constructs an image of the real external world, it must be capable of planning in advance.

What happens when an object is in motion? Frogs manage to catch flies and cats mice only because they make use of anticipatory mechanisms. Observe a tennis player. To initiate coordinated activation of his muscles and to assume a posture that will enable him to hit the ball, his brain needs information about the ball's velocity. A certain amount of time (several hundred milliseconds) passes between the start of movement and the instant in which the racket is in the desired position. During this time, the ball will have traveled several meters. Under these conditions, the player is perpetually condemned to miss it. This is the problem beginners have. The movement of the racket in the direction of the future position of the ball has to be programmed. To do that, the direction of the ball and its trajectory must be predicted.

This ability to predict the trajectory of objects in motion plays a fundamental role in the survival of species, even the most primitive. At a very early stage of evolution, the nervous system developed the means for detecting potential predators or prey and predicting the direction of their movements. The lizard thrusts its tongue toward the future position of the fly it hopes to catch.

At the beginning of this book, I proposed that in the course of evolution the brain was invented essentially to facilitate capture of prey or escape from predators; in other words, to allow decisions to be made almost instantaneously based on perception. Before I look at several striking examples of this kind of anticipation in detail, I would like to describe the neural mechanism that enables a toad to decide if a moving shape is prey or predator.

The Toad's Decision

When a hungry toad sees something moving, it has to decide whether the thing is an earthworm (prey) or an eagle (predator). In the first case, it must spring up and capture it; in the second, it must hide. If you have the time to sit by a stagnant pond, you will notice that toads, despite their nonchalant appearance, are capable of making these decisions in a flash. The naïve psychology of representation would claim that the toad recognizes the worm and the eagle as such, and keeps central representations of them that it classifies as prey and predator, such that identification of the object induces a reaction of fight or flight.

Actually, the mechanism is much simpler and remarkably efficient. Specific neurons are responsible for the behavior.[2] They are located in the optic tectum, a brain structure in the frog that is sensitive to visual cues from the retina indicating the position and movement of objects. This structure is the ancestor of the superior colliculus in mammals and primates. The neurons of the optic tectum are sensitive to a particular configuration of visual stimuli: longitudinal movement of any elongated object. This is obviously a characteristic feature of earthworms, so a lure such as a little rectangle of paper will also serve, provided that it moves in the direction of its length. This assembly of neurons in the optic tectum is responsible for orientation and capture.

Other neurons located in the posterior portion of the thalamus—an important neural center for multisensory convergence as well as a center for relaying sensory information to the cerebral cortex—are sensitive to another visual shape, which Ewert called "antiworm" because, unlike the worm, this configuration corresponds to a large target, like an eagle or a large animal. These neurons are connected to motor assemblies that induce avoidance behavior. An essential property of these systems is this: each inhibits the other. The two behaviors of capture and flight are thus mutually exclusive.

Of course, one could say that the earthworm is represented at the central level by neurons sensitive to the movement of an elongated object. However, a genuine physiological interpretation of the brain would emphasize the fact that this neural system is very simple. Receptors and first-order neural processing detect the relevant variables directly. A few of the properties of the neural networks—classification (the ability to recognize two sensory configurations), neuronal connectivity, and a clever combination of elements of synaptic inhibition and excitation—are sufficient to work out the mechanisms that sensitize the animal to what I call expected sensory configurations, or neural hypotheses. Schemas as Schmidt understands them guide the selection of sensory configurations recognized by the system.[3] Compared with more highly evolved animals, the toad lacks the flexibility or capability to recognize a fake worm. You can catch frogs using a bit of red cloth, but you cannot catch a cat with a stuffed mouse.

The classification of perceptions (for example, of prey or predator) is determined by the repertoire of possible actions (in this case, capture or flight). Hence a further illumination of my proposition, *perception is simulation of action*. But there is more. We capture first of all with our gaze. To study how mechanisms of gaze control anticipate targets in motion, we constructed a visual game in the laboratory like the ones at a country fair where you have to shoot at a line of ducks or pop multicolored balloons that an air current blows around at random. Despite their simplicity, these games are difficult, which is how the fairgrounds people manage to stay in business.

In our experiment, the subject first has to fixate on a spot in the center of a screen on which, from top to bottom, a second spot appears that falls suddenly, like a ball. An auditory signal set off at various moments during the fall instructs the subject to catch the falling target with an ocular saccade. During the 150-odd milliseconds during which the eye remains fixated, the brain estimates the velocity of the target, which moves toward the periphery of the eye. You can do this experiment yourself by focusing on a point in front of you and holding an object in your right hand above the horizontal. Let the object fall, and then try to catch it with your gaze during its fall.

We observed that at the instant the auditory signal sounds, the eye moves not to where the target is but further on in its trajectory. The brain guesses—

predicts—the future position of the target. How are perception and action connected to be able to program a saccade toward the future position of the target? We still do not know.

The Art of Braking

The toad thus exhibits two kinds of behavior that depend on prediction: capture and evasion. A second avoidance behavior is collision prevention. When you are in an automobile, when do you brake to avoid running into a stalled truck in the middle of the road? Lee's hypothesis is that the brain uses an optic variable, which he calls "time τ-to-collision."[4] In fact, Lee observed that if the speed of the automobile is constant, τ is equal to the relation between the observed diameter of the car in front and which you are approaching, and the speed with which this diameter expands.

Time-to-collision is thus defined as the function of a derived distance. Speed equals distance divided by time. If distance (the observed diameter of the car on the retina) is divided by speed (the speed at which this diameter expands), the distance units cancel out, and pure time in seconds for small angles is the result.

If movement proceeds at constant velocity, time-to-collision can thus be evaluated by the speed with which the retinal image of the object expands. It is a genuine optic variable. This expansion is detected by specialized neural elements. In the monkey, neurons sensitive to this expansion of the image were found in the parietal cortex, which as it happens was predicted by the Dutch physicist Koenderink, an optic flow theorist.

What is new about Lee's theory is that he estimates time without ever calculating distance. Time τ to contact is given solely by retinal cues. I use the word "given" to reiterate an expression of Husserl. It is a good example of what can be *given* to perception in contrast with what has to be *calculated*, of what is better explained by a *biological*, not *computational*, theory of the brain. This does not mean that these processes cannot be represented using mathematical methods or that the processes that take place in the nervous system are not equivalent to mathematical operations. But a model remains a formalization of a biological reality, accounting for some aspects of it.

This model was originally developed to describe an approaching obstacle, but it obviously holds for estimating the time-to-collision (here, time-to-contact) with a ball thrown toward someone (or an eagle diving toward a toad!). If the motion is accelerated—a falling ball, for example—the time-to-contact

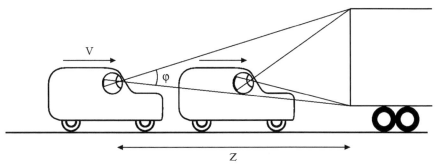

Figure 8.1. Lee's (1976) theory about the perception of time-to-collision during a movement. Without having to estimate the distance, the driver of a car can directly work out time-to-collision with a stalled truck based on optic flow. As the car approaches the truck, the image of the truck expands on the retina at a velocity proportional to that of the proximity of the two vehicles. If velocity *V* is constant, time-to-collision is equal to the apparent diameter of the image of the truck on the retina divided by the velocity of expansion of this image. These two pieces of information are supplied by the retina and require no knowledge of the distance.

depends on the current velocity of the target and its acceleration. Lee proposed an extension of his theory to take this case into account. He assumes that the brain uses what he calls the "τ margin."

Returning to the definition Lee gives, suppose that I am driving a car at velocity *V* (for example, 110 kilometers per hour) and that a truck suddenly appears in front of me (Figure 8.1). Given my velocity *V*, the time-to-collision, if I do not brake, has a certain value τ. Now I must ask what margin of time (τ_m) I need to brake and not run into the truck. If I need a certain time τ to begin braking, and if the time required to avoid a collision is τ_m, the distance to the collision is $(\tau_m - t)V$, and the constant deceleration to avoid the collision is $V/2(\tau_m - t)$. Lee's calculations show that if the driver needs 2 seconds to begin braking, it would be reasonable to use a margin τ_m of 4 seconds for speeds up to 110 kilometers per hour, subject to the car's ability to decelerate at about 4 meters per second per second. However, this deceleration is so high that it is uncomfortable, and if one wishes to decelerate at less than 2.5 meters per second per second, τ_m must be approximately 8 seconds.

Thus the question is whether the brain uses these optic variables, which are directly measurable on the retina. Two experiments suggest that the brain uses either τ or the margin value τ_m—that is, optic variables—and not the physical variables that constitute the angles, velocities of targets, and their accelerations.

PLUMMETING GANNETS

Picture a gannet diving from different heights to the sea to catch fish. It begins by adopting a posture that enables it to effectively battle air resistance while still allowing it to direct its flight. But if it were to reach the water in this position, wings spread wide, it could get hurt. So just before entering the water, the gannet folds its wings and assumes another aerodynamic—or, rather, aquadynamic—posture. How does it know the precise moment to assume this posture?

Studies of films suggest that whatever the altitude above the sea from which it begins its descent, the bird begins to fold its wings using the τ margin given by the expanding retinal image of the surface of the water.[5] The higher the bird starts, the higher the altitude at which it reaches this value. The films show that the higher the altitude at which it begins its descent, the sooner the gannet begins to streamline. This strategy allows the bird a greater distance, as it is falling, to ready itself for contact with the water, and it has two advantages: it is simple, because it uses only one optic indicator on the retina, and it is useful, because it reduces the risk of injury.

CATCHING A BALL

Does this theory apply to catching a ball? Early studies on catching objects in motion borrowed their ideas from information theory. But investigation of natural movement quickly revealed to experimenters the existence of anticipatory mechanisms whose very nature defied description in terms of information processing. For example, well before contact, the shape of the hand adapts to the apparent shape of an object; the muscular activity of the arm is always initiated at the same time (before contact), no matter what the velocity of the target. The brain seems to extract information about the ball's movement from sampling the retinal slip during a segment of the trajectory and processes this information. Geometrical features, such as asymmetries, can be important for predicting impact. Although this apparent distortion of the object may indicate a change of trajectory, that is certainly not the case when the object moves at a precise distance from the body. In an attempt to resolve these difficulties, several research groups applied Lee's theory to games such as baseball, tennis, and volleyball.[6]

Originally, the concept of τ margin was applied only to the case in which the ball was aimed at a person, as in baseball. But what about when the ball goes quite far into the so-called extrapersonal space, as is the case in tennis?

Studies reveal that here, too, the movement of the racket is remarkably constant and independent of the approach velocity. Thus the beginning of the movement may in fact be adjusted by the value of the τ margin. But sometimes the movement begins after a relatively flexible delay, and a connection between perception and action enables a person to adjust the stroke or catch according to the trajectory of the ball.

Now imagine that a volleyball player is leaping up to reach a ball in free fall, thrown very high by the opposing team. Lee studied this case: He asked the player to bend his knees, then to jump right up to the ball and strike it. He measured the angles of the elbow and the knees to evaluate the temporal course of the movement and connected that to the height from which the ball fell. The results demonstrated that the simplest relationships between the action of reaching the ball and the magnitude of the fall were again obtained using the τ margin and not the actual time-to-contact.

If the brain does extract dynamic information from the appearance of the image in motion, it is helpful to know how much time is needed to gauge these variables and to estimate the distance, or the time, to contact. For catching a ball, numbers of 240 to 300 milliseconds have been posited. To catch a ball travelling at a speed of 10 meters per second, the athlete has a temporal window of 50 milliseconds during which his hand must remain closed. If he closes it too early, he will miss the ball; if he closes it too late, the ball will bounce off his hand. A solution roboticists like Slotine have found is to assign the hand a speed equal to that of the ball and to have a robot hand catch the ball in this moving frame of reference.

An experiment was conducted to determine whether subjects do use the τ margin to catch a ball. This experiment works best with a rubber ball whose size changes as it approaches the subject (it deflates). If the τ variable triggers the closing of the hand, the subject must close his hand later, when the ball deflates, for its relative size diminishes and its speed of expansion is less. This hypothesis has since been confirmed. The distance of the ball from the body can also be perceived in units of object size. The brain operates not by seeking the most perfect predictive information but rather the most useful information for regulating the action. In this case, the information does not specify "when to be where" but "how to be in the right place at the right time" without worrying about where that place will be.[7]

Even if these ideas are disproved in the future, they are interesting because they suggest that the brain does not construct Cartesian space or topographic space, but units of space connected to action.

When we are trying to get close to an object to grasp it, we have to curb our movement to reach the object at zero velocity relative to it. This is also true for birds in flight who wish to alight on a tree branch or on the ground. Lee recently became interested in the gentle deceleration that makes it possible to reach a target without perturbing it thanks to these graceful postures. This achievement amounts to maintaining a constant deceleration of the moving body until it reaches zero velocity at a precise location. Pigeons landing on their feeder, who have to slow down their flight to land gently; acrobats executing perilous jumps, who have to control their landing so they do not fracture their ankles; a person carrying a strawberry to his mouth, who wants to grasp the fruit delicately—anyone or anything executing a gentle braking must evaluate time-to-contact.[8] Time will judge the correctness of this theory. There are other alternatives, most recently ideas inspired by the theories of advanced dynamics systems for the control of movement of Kelso[9] and Schöner[10] (see the end of this chapter).

What If Newton Had Wanted to Catch the Apple?

The theories of Lee and his successors give short shrift to gravity. Does the brain use this basic acceleration, which acts on all moving bodies, to predict their trajectory? Have we internalized certain properties of the gravitational field? For example, what if instead of simply having watched the apple fall Newton had tried to catch it? How would his brain have resolved the problem?

Imagine the following experiment: You have waited impatiently for autumn, and now here you are, seated beneath an apple tree in Normandy, anticipating the gust of wind that will knock an apple from the tree. You hold out your hand, palm toward the sky. The wind rises and the fruit falls straight into your hand. This problem is simple compared with our tennis problem—you have only to place your hand within the vertical trajectory of the apple. You do not even have to brake gently, just adjust the dynamic parameters of your muscles to adapt the stiffness of your hand and arm so the apple does not bounce. It will, however, be necessary to predict the time-to-contact and the force of the impact. How does the brain do it?

An experiment provided the answer.[11] A seated subject, forearm positioned horizontally and hand turned toward the ceiling, was asked to track and catch a falling ball. The trick of this experiment was to place a motor at the level of the elbow that subjected the forearm to small, very short rotations that stretched the biceps. The subject responded to these impulsive forces by a reflex contraction. This made it possible to test the amplitude of the reflex, stiffness of the muscles, and so on before, during, and just after the ball dropped. The results showed that catching a ball is preceded by an anticipatory muscular activity that prepares for impact and is followed by a reflex activity induced by the impact. If Lee was right and if the subject's brain uses a τ margin, the longer the duration of the fall, the earlier the anticipation of contact and initiation of the muscular reaction must be. The start of anticipatory activity of the muscles is fixed, but subjects seem to base their response, whether anticipation or reflex, on a prediction of the dynamic properties of the ball under the influence of gravity.

Thus the brain anticipates the dynamic properties of the ball based on several kinds of cues: its velocity (detected by vision), its mass (estimated based on past experience and remembered relationships among its angle, mass, and inertia), and the influence of gravity on both limb and ball. Here we encounter the role of memory. Lacquanti concludes:

> This then suggests that . . . an internal model of the dynamic interaction that is expected to occur at impact is built based on *a priori* knowledge and available on-line information (such as visual information on the velocity of the ball and the limb). The response of this model to the actual perturbation is compared with the response of the limb to produce an error signal. This error is subsequently used to calibrate the parameters of the neural controller of the plant [the bones and muscles] and to update the internal model. Thus, if the model does not accurately predict the desired performance, possibly because of a faulty estimate of the properties of impulsive impact, kinesthetic and cutaneous information obtained with the first trial can be used to correct the estimate.[12]

So this model includes an internalization (whose mechanisms are still mysterious) of the effect of gravity on moving limbs and objects. Newton's laws are identified in the course of development and written in the neural circuits that make up the internal model. This idea confirms the possibility of an internal simulation of movement without execution, since our nervous system would contain models both of our limbs and of the effect of gravity on our movements. It also explains why we dream of actions that seem to unfold in a perfectly Newtonian universe even though no force opposes them, except, of course, for dream flight, which shows that our dreaming brains do not simulate all the effects of gravity.

We proposed that NASA carry out this experiment in space. This was accomplished in April, 1998, during a flight of the space shuttle Neurolab. We questioned what strategy the nervous system chooses when movement is not accelerated by gravity but has a constant velocity.

The idea of internalization of gravity in the nervous system is not new. In the 1950s, the Soviet school had already assumed that the brain internalizes properties of the physical world in the structure and activity of its neurons. I will come back to this point. Perception and action are thus connected through these mysterious internal models of the properties of the limbs and objects in the physical world. The consequences of movement can be simulated and predicted by the brain using these internal models.

Because it is worthwhile to delve more deeply into this idea of internalization, I will suggest yet another example.

The Problem of the Waiter

The problem of the waiter can be formulated in the following way: when the waiter arrives with a tray full of bottles and lifts one to give it to a customer, he reduces the weight supported by his arm by about 1 kilogram. Yet his tray remains perfectly stable. But if a customer takes the bottle without warning him, his tray becomes unsteady.

The fact that the tray does not move when the waiter removes the bottle means that he has an internal model that enables him to predict the effect of the unballasting on his arm. Before removing the bottle, he decreases the tonicity of his muscles. This anticipation probably requires a motor memory. In this case the frame of reference used by the brain is the horizontality of the tray in relation to gravity. This ability to predict appears in infants approximately 1 year old.

The concept of the internal model even turns up in cybernetics. In fact, engineers were the first to discover that performance of servo systems improves significantly if instead of waiting for the sensors to detect errors and make corrections (which takes time and leads to oscillations), a calculator is included in the circuit that estimates the initial state of the sensors and their variation (or mathematical derivative) based on their current value.

It was an engineer, Kalman, who developed the predictive filter that bears his name.[13] The Kalman filter is the optimal filter, the best-known predictor of the state of a linear system, and so it is used in many automatic guidance systems. Young, from MIT, was among the first to use it in a formal model of postural control based on the idea that cues from the sensory receptors (vision, proprioception, inertial vestibular sensors) are compared with estimates made by a Kalman filter. The motor command for correcting posture during a perturbation is thus not the simple response of a servomechanism or a measure of the discrepancy between the state of the system and the desired state. It is dictated by comparison of the state of the sensors with the prediction made by the Kalman filter.

Is the brain equipped with Kalman filters? We have no idea.

THE SHORTEST PATH FROM ONE POINT TO ANOTHER

Has the nervous system internalized the physical properties of the world? Many great thinkers have embraced the idea of a cerebral image (in modern terms, an internal model) of the world. Among them is von Uexküll, who had a profound influence on modern physiologists of the corollary discharge school like von Holst and Mittelstaedt. Von Uexküll defined the concept of *Umwelt,* or environment.[14] According to Merleau-Ponty, "for von Uexküll, *Umwelt* signals the difference between the world as such *(Welt)* and the world as the realm in which this or that living thing exists. It is a transitional reality between the world as it exists for an absolute observer and a purely subjective domain."[15] The sea urchin is an example of an animal that is no more than reflexes and a repertoire of independent synergies; the sea urchin, says von Uexküll, is a "republic of reflexes" that exists in an *Umwelt* that represents things that are often dangerous but to which it is so well adapted that it truly lives as if there were only one world. Its nervous system contains no image of the world. On the other hand, for higher animals, *Umwelt* is not closure but openness. The world is possessed by the animal. The external world is distilled by the animal which, differentiating sensory data, is able to respond to them with fine motor actions; and these differentiated actions are possible only because the nervous system is equipped like a replica of the external world *(Gegenwelt),* like a copy. It is a mirror of the world *(Weltspiegel).* The *Gegenwelt* is itself divided into *Merkwelt,* which depends on how the sensory organs are arranged, the world of perception, and the *Wirkwelt,* which is the world of action. Today it is clear the sea urchin is more than a republic of reflexes, but von Uexküll's ideas have the benefit of setting the debate.

The idea that internalization of the properties of the physical world constrains perception, that we perceive the external world through the laws of the physical world integrated within the functioning of neural networks, is also supported by psychologists. Shepard, for example, writes:

> I believe the external constraints that have been most invariant throughout evolution have become most deeply internalized, as in the case of

the circadian rhythm. Such constraints may be extremely general and abstract: The world is spatially three-dimensional, locally Euclidean, and isotropic except for a gravitationally conferred unique upright direction, and it is temporally one-dimensional and isotropic except for a thermodynamically conferred unique forward direction.[16]

Several properties are subject to internalization by the nervous system, in particular the laws of geometry and kinematics, which control relationships between movements of rigid objects or various parts of nonrigid objects. For example, despite the countless ways of superimposing two identical objects located separately in space, there is always a simple rotation that can do it. This optimal path property was established between 1763 and 1830 by Italian (Mozzi and Giorgini) and French (Chasles) mathematicians. Suppose that we wish to move an object in space from point A to point B by combining a rotation and a translation. This is the movement a helicopter makes when it takes off. It is also the twisting movement of a screw. According to Chasles's theorem, there is only one axis in space such that the object can be moved from A to B by both a rotation around the axis and a simultaneous translation along it. Shepard maintains that the brain has internalized these kinds of properties.

If the brain does know these laws, the proof should be in the way it splits up perception of movement into rotations and translations. The vestibular system actually seems to do this for movements of the head, as evidenced by the semicircular canals, which detect rotations, and the otoliths, which detect translations (see Chapter 2). In the monkey, this geometric segregation is found in the analysis of optic flow by the neural networks of the cortex: in the visual cortical areas of the superior medial temporal sulcus some neurons are sensitive to a combination of rotation and expansion of visual objects, but others analyze the components of rotation and translation separately. In the accessory optic system, geometric segregation of visual movement is carried out very early, as I discussed in Chapter 3.

Shepard also considered the case of Chasles's planar theory: for two positions, A and B, of an object in a plane, there is always a single axis such that the object can be rotated from A to B (imagine a top with points A and B indicated on the circumference of the circle the top makes, for which the handle would be the axis). To confirm that the brain really has internalized this property, Shepard showed a subject two identical polygons in a row, each inscribed within a circle, the second having undergone a rotation relative to the first. The succession of the two images gives the impression either that the polygon is moving or that it is spinning. The polygon gives an illusion of movement that follows Chasles's theory to the letter.

Several other experimenters have demonstrated that perceptual illusions of movement appear to respond to the constraints of kinematics. For Shepard, when the brain predicts a trajectory, it does not take into account specific anatomical characteristics; it predicts the trajectory according to a hierarchy of simple presuppositions. The suppositions at the top of the hierarchy assume the rigidity of the moving object. However, when the brain has to work quickly, it takes shortcuts and can no longer stick to this rigidity hypothesis. If different trajectories are possible in a three-dimensional space, the brain will choose geodesic curves—the shortest on the surface in question, just like airplanes going around the earth from Paris to New York or from Paris to Tokyo.

What is the neural basis of this internalization? Shepard has no answer. He does, however, propose that neural mechanisms enable genuine simulations in the internal circuits, and in this he differs with Gibson, who denies any autonomous simulation activity that is not connected to reality.

The fundamental laws of geometry and kinematics are not the only ones to be internalized. If when we are looking at a series of photographs of a person jumping, like the photographs by Marey or Muybridge, we are shown another photograph and are asked whether it belongs to the series of jumping movements, we can identify it all the more quickly if it pictures a stage that corresponds to the continuation of the natural (predictable) trajectory of the jump. Similarly, shown two photographs, we can very quickly recognize the movement of a person's arm if the motion required to move from one posture to the other is physiologically possible. Here the principles of the motor theory of perception, introduced in the first chapter, again come into play.

Proof is accumulating that the laws of motion constrain the perception of movement. Recently, Viviani showed that if a subject is shown a point of light moving on a screen in an elliptical trajectory, for the subject to have the impression that the point of light is moving at a constant velocity, in reality it must move with a velocity that varies with the curvature of the ellipse, as the law of two-thirds power predicts. Thus the point has to move as the finger would naturally move. The laws of natural movement constrain perception![17]

INTERNAL MODELS AND THE INVERSE PROBLEM

I have discussed the concept of internal models extensively. Before going any further, I must explain the so-called inverse problem. Say that you want to make a drawing, and that the neurons of your motor cortex fire, activating the motor neurons of your arm so they bring about the movement required to draw a circle. This neural command activates muscles that have a particular geometry, stiffness, and viscosity that will activate the arm, which has an inertia, leverage mechanisms, and so on. The motor command is thus mediated by the biomechanical properties of the limb, and the movement runs the risk of being completely distorted. For the motor command to be faithfully transmitted, it would have to be distorted in advance so that it could somehow mirror the distortions the limb will impose on it. This is called the inverse problem, because the question is one of subjecting the motor command to an inverse transformation of what it is about to undergo. Roboticists have devised many ways of making this inverse calculation, because it is a problem that comes up with all robots: it is easy to program a circle into the computer and to send commands for drawing a circle to the motors in the robot's arm. But if this arm is elastic, subject to frictional forces, heavy, and so on, such properties have to be represented in a model, and the inverse of the command signal has to be applied before it is sent to the motors.

To plan a defined trajectory, it is thus necessary to account for this filtering of the limb in advance. One solution is to have an internal model in the nervous system, before the final motor command, that modifies motor signals according to a transformation exactly the inverse of the modification imposed by the arm, that is, the controlled object. When the movement is executed, the internal (inverse) model and the arm are linked. The result would be the following sequence: "desired trajectory–inverse properties of the arm–properties of the arm." And what you would have at the end would be the desired trajectory undisturbed by the filtering of the arm.

Recently, Kawato proposed an additional concept, also used in robotics: a model of object control.[18] This feedforward model is used as a substitute for a limb or object controlled by the brain. Setting a desired trajectory, an inverse model, and an internal model in a series results in a simulated virtual trajectory, which it is possible to correct without having to execute the action in the external world.

The different operations of these models can be represented schematically, ranging from the least to the most predictive:

Initial situation
Planned trajectory → motor command → controlled object → distorted trajectory

Addition of an inverse model
Planned trajectory → motor command → inverse model → modified motor command → controlled object → ideal trajectory

Addition of a feedforward model

Planned trajectory → motor command → inverse model → modified motor command → model of the object → virtual trajectory → comparison with the planned trajectory → error → internal correction of error → modified motor command → controlled object → ideal trajectory

For the brain, the advantage of working with these kinds of internal models is that if one takes into account the properties of neurons, processing time is on the order of 30 milliseconds—much less than a return using proprioception or vision (on the order of 100 to 200 milliseconds). This mechanism is also capable of learning and adapting. In fact, it is possible to correct these internal models by sending them error signals detected by the senses or by the internal neural circuits that monitor relationships between command and execution.

More generally, how is it possible to imagine that the networks of neurons in the brain can be adapted to the conditions of the physical world? This process must be implemented in the baby. Walton and Llinás made an important observation.[19] They noticed that baby rats tremble when they are with their mother in the first days of life. This trembling is the result of electrical oscillatory properties intrinsic to the motor neurons of the medulla that involve all the limbs. Movement, or the trajectory of the limb, results from the filtering of these neural oscillations by the mechanical properties of the body. The rat's sensory receptors detect these movements and relay the information back to the medulla. This return oscillation can be used to specify the difference between central activity and actual movement and to calibrate the neural networks by constructing an internal model. The implications of this idea have not been studied, and how these internal models might be constructed is not well understood.

Children's first movements and play thus function both as training for motor programs and also as a way to construct these internal models. Their significance and especially the variability of the skills that children acquire depend on the internal models that they have managed to construct.

DYNAMIC THEORIES

This theory of internal models of mechanical properties of the body is one way of accounting for the capacity of the brain to simulate relations with the environment that it can use in anticipating. But it is not the only theory, and its working assumptions (for example, that the brain contains a representation of the trajectory of the arm) are controversial. For several years, another school of thought has taken a different tack, that of the theory of dynamic systems, which holds that the behavior of the nervous system is integrated with nonlinear interactions of oscillators.

The main objective of this theory is to define a paradigm for studying the neural basis of movement. It seeks to substitute a paradigm of paired oscillators for the stimulus-response paradigm. When these oscillators function interactively, they have a tendency to revert to a stable state. This tendency to revert to a preferred mode of oscillation when there is a deviation from its functioning is considered a reflection of internal properties of the system.

The dynamic systems approach refers to this property.[20] It considers diversion of the system from a stable state to see how it reverts to that state or how it reorganizes itself to reach another stable state. Stable does not mean motionless. Thus, the rhythm of walking is a (dynamically) stable state. Tap on the table with one finger of each hand at the same time. You will see that you have a natural tendency to adopt a particular rhythm of tapping; you establish a dynamically stable state. Now try to tap a bit more quickly with the right hand: you are upsetting this dynamically stable pattern. Moreover, it is known that a particular region of the brain—the supplementary motor area—is immediately activated when you make such a change.

Actually, dynamic theory assumes that the variables controlled by the nervous system are not only intrinsic variables of the elements of the body (length of the muscles, muscular force, angles of joints), but also variables connected to the task and to the kinematics of movement. These last two are the most important; the nervous system organizes motor activity by controlling these so-called coordination variables.

The biological relevance of this approach remains to be demonstrated, but it is supported by many arguments. In fact, internal neural circuits do exist, like those that connect the cerebral cortex, the basal ganglia, and the thalamus (which projects in turn to the cerebral cortex), or that connect the cerebral cortex to the nuclei of the brainstem (called pontine nuclei), which themselves project to the cerebellum and, from there, to the cortex. Internal oscillation generators cause rhythmic activity that sweeps through these circuits (see Chapter 1). For example, the cortico-striato-thalamo-cortical circuit is swept by an activity oscillating at 40 Hz, other motor circuits involving the inferior olive are swept by bursts of activity at 10 Hz, those involving the hippocampus and the limbic system by activities of 6 to 9 Hz, and so on (see Chapter 5).

9

THE LOOK THAT INVESTIGATES THE WORLD

I have demonstrated that the critical issue is not so much a matter of cataloging sensors, but what questions the curious brain asks the world based on the assumptions it develops and tasks it proposes to carry out. The brain not only formulates hypotheses and tests them using the sensory receptors. Evolution made the brain a machine for predicting, not just a machine to take account of situations. It also made it an organ for detecting, predicting, and interpreting movement, for there is no action without movement. Gaze *orientation* was one of the first functions that required the development of a brain that could predict, a brain that was curious, and a brain that could simulate action.

Gaze Orientation

Orientation toward a sensory source is a behavior that turns up in the simplest organisms. Ethologists call it taxis, and distinguish phototaxis, heliotaxis, thermotaxis, and so on. It often involves locomotion, flight, or swimming toward the source of attraction. In insects such as the fly, movements of orientation result from very subtle but relatively rigid automatic mechanisms. In the toad, as I mentioned in Chapter 8, movements of orientation are associated with capturing prey (an earthworm, for example) or with evading capture.

Lorenz studied the repertoire of orienting movements in mammals and birds. He emphasized that some of the synergies involved in these movements are completely independent of sensory inputs and "physiologically they [motor patterns] are distinguished by the fact that their fixed pattern of movements, contrary to what one would expect, is not activated by a succession of reflexes but by processes which take place in the central nervous system itself, without receptors being involved."[1]

But let us consider gaze orientation. Gaze orientation is projection, questioning, prediction. It is stationary locomotion. It is the model of choice for understanding the physiology of perception. Chapter 2 described the mechanisms that stabilize the visual world on the retina owing to the vestibulo-ocular reflex. And Chapter 3 explored how the analysis of movement in the visual world through the pathways of the accessory optic system (responsible for optokinetic nystagmus) is carried out via geometric segregation of retinal slip in three planes that are the same as those of the semicircular canals. Also in Chapter 3, I wrote of the late appearance in the course of evolution of the mechanism by which an image is maintained on the fovea.

The idea that gaze is a projection of the brain on the world is not new.[2] Even in Babylon, gaze was either male, and projected, or female, and receptive to light. In Grecian times, Empedocles (Figure 9.1) tried to explain sight as follows: the interior is fire, around which are water, earth, and air, through which fire passes by virtue of its subtlety, like the light in lanterns. The pores of fire and water are arranged in staggered rows; through the first, white objects are perceived, and through the second, black objects. According to the so-called theory of extramission, this internal fire produces light that in turn is reflected by the object and returns to the eyes (perhaps the origin of the expression "a burning gaze"?).

Plato proposed his own theory of interaction. Visual rays emanate from the organism and lead to interaction with ambient light to form a cone of vision whose vertex is in the eye and whose base is at the object. This cone makes the object vibrate and transmits this vibration to the eye. This signal activates the cognitive components of the soul that are located in the brain. Later, Alahazen (965–1038) proposed another interactive theory of gaze: effer-

Empedocles (~490–430 B.C.)

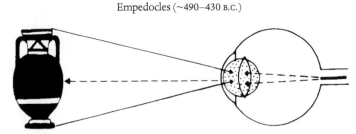

Figure 9.1. The projective gaze. According to Empedocles' theory of extramission, a light projected by the eye grasps an object, which reflects the light back to the eye.

ent visual signals produced by the brain *(spiritus visibilis)* at the level of the optic chiasm penetrate the eye and interact with the visual waves produced by the objects. This interaction is then sent back to the brain, where it combines with that of the other eye to produce a perception.

The art of the Renaissance proceeds largely from the discovery of perspective. In contrast to the stilted view of earlier paintings in the East as well as in the West, gaze came to life and became a way of looking at the world from various vantages. It estimated proportions and guided architectural relationships.

The theory of corollary discharge is a modern version of a theory of interaction (see Figure 3.4). This theory is often attributed to von Holst and Mittelstaedt,[3] but it can be traced to Bell and to Purkinje (1823–1825). Bell and Purkinje observed that external objects seem to move when the eye is made to move, whereas they remain constant during active gazing movements (for more on this subject, see Chapter 10).

A copy of the motor command is sent to the center of perception when the movement is actively triggered by the brain. This copy makes it possible to update representations of the object and to compensate the displacement of its image by a virtual image that remains constant. Within the framework of cybernetic theories of servo control, which have dominated thinking about gaze mechanisms for 20 years, the copy of the motor command corresponds to the concept of *feedforward* as opposed to *feedback*.

The capacity for actively exploring space via orienting movements, produced not in response to environmental stimuli but according to a person's intention, is important. Lorenz even suggested that the organization of a progression, that is, a locomotor trajectory—based on constructing a representation of space and requiring an associated sequence of orienting movements— is the most highly evolved form of orientation reaction.

This idea turns up again in the development of gaze in infants. When seated in a highchair, a baby first directs his gaze using his own body as a reference point. He turns his head and eyes about the axis of his body, an egocentric frame of reference. But once he begins to walk, he can no longer explore the world with his eyes in a fixed reference system and must develop a new strategy for keeping objects constant in space. Before 18 months, he fixates on a landmark (for example, a corner of a table) to which he anchors his progression around the room. This strategy is probably the first sign that he is constructing allocentric relationships (locating an object in the environment). If his anchor point is hidden from him, the infant reverts to an egocentric strategy (related solely to his own body). At 18 months the infant adopts a third

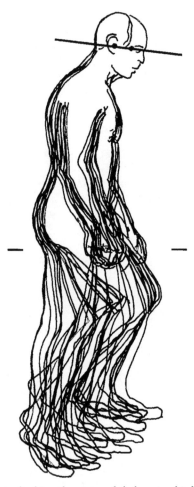

Figure 9.2. The man is climbing the stairs while keeping his head stabilized—his head does not turn in the vertical plane. The angle of his head is dictated by the direction of gaze. The straight line drawn on the head joins the meatus of the ear to the external corner of the eye socket. It indicates roughly the plane of the horizontal semicircular canal. The direction of gaze is lower than this line. (After photographs by Muybridge)

type of functioning. He no longer fixes his gaze, but is able to mentally update his position in the room while he is walking, employing a mechanism of mental rotation and translation.

In adults, gaze is not directed at random during locomotion, as shown in the images I drew based on photographs by Muybridge (Figure 9.2). The gaze of a person mounting a staircase or running, or of people carrying out a task, is oriented toward a precise goal. It can be readily stabilized in space to enable processing of optic flow and combining of information about movement (see

Chapter 4). It can be anchored to an object, even an object someone is carrying. Enormous functional flexibility succeeds egocentric organization in the infant.

"Go Where I'm Looking," Not "Look Where I'm Going"

Although we are freed from the constraint of having to fix our gaze, we continue to guide locomotion, or, rather, our direction indicates anticipation of the trajectory. You can do the experiment yourself. Clear a space of around 3 meters by 3 meters in your living room and draw a circle on the floor with chalk. Place yourself anywhere in the circle, and look at the circle. Then close your eyes, and make two complete tours trying to stay within the circle.

You can do the same experiment by drawing a triangle or square 3 meters on a side, which obliges you to walk in a straight line and then to turn corners. During these different tasks both in the light and in the dark, a person's gaze turns toward the interior of the concavity about 300 milliseconds in advance of the feet and the trunk. In other words, we head toward the place we are looking, and not the contrary. The brain does not develop a simple motor program that enables it to proceed along a circular trajectory. It follows an internal model of the trajectory just as you pursue a moving target using your eyes or your hand. In other words, we mentally simulate the trajectory, and we compare the actual movement executed by our feet with the predicted movement.[4]

Eye-to-Eye Contact

Gaze orientation not only enables us to fix on a target selected from the environment, it is also required to construct a coherent representation of space in primates and in man, whose vision is fragmented owing to development of the fovea. Analyzing the evolution of vision in vertebrates, Gibson reminds us that fish have a fisheye, simulated today by camera lenses, which can take in wide surroundings with a single glance. In the course of evolution, the appearance of foveal vision was accompanied by migration of the eyes from a lateral to a frontal position, which enabled perception of depth, tracking, and so on. But the disadvantage of frontal vision was that animals lost their panoramic vision and had to reconstruct the visual environment based on a series of restricted views. Orientation thus was not used solely to guide the mouth or the body, but to explore the visual world. Coordination of exploratory movements ensures the acquisition of successive images that, according to

Gibson's example, is essential to a coherent representation of the entirety of the room in which one happens to be.

But gaze movements do not just serve for capture or for coordinating points of view; they also govern relationships with others. One of the functions of gaze is data gathering, but it also plays a fundamental role in interactional equilibrium, for example, the exchange between a mother and her child. I will come back to this example in relating the development of ocular movements in the infant. This exchange has fascinated the research groups of De Ajuriaguerra, Minkowski, and Lésine, and we are indebted to Saint Anne Dargassies for one of the first reliable descriptions of it. De Ajuriaguerra calls it "eye-to-eye contact." It has long been known to so dominate the act of eating that infants are sometimes totally distracted from suckling. This eye contact may be among the innate triggers of maternal response. But it is not enough to be able to direct gaze; one must also be able to turn away from the target that attracted it.

> From studies of precommunication or communication modalities between mother and infant, with reference to gaze reciprocity, we find a whole set of ambiguous problems in their mutual understanding: the eye that touches and that encompasses, that takes and rejects, that hears or that listens, that licks and devours . . . the infant can follow the other through ocular scanning without being able to leave her; fascination gives way to intrusion. For this the two gazes must encounter each other. And thus vision becomes sorcery . . . The infant feels more comfortable once this fixation becomes reversible, movable, and relative; and when, in the love-hate battle that is created over the course of life, exchanges that safeguard control and intentionality are established. This happens as soon as the infant is able to break away from the magnetic hold of another's gaze by turning his head and closing his eyelids.[5]

It is now known that development of the frontal cortex enables this evasion, because patients presenting with lesions of these regions persevere in their fixation.

The eye of the other is not always benevolent. All Mediterranean civilizations are familiar with the "evil eye." In North Africa and in the Near East, it is experienced as an ambivalent organ, both receptor for the physical world and transmitter of a living force. It flows out like water from a spring for which the eye is named in Arabic. First fascination, then desire, it brings on misfortune. As early as the third century, Heliodorus of Emesa had summed up the essentials of these Mediterranean beliefs: "After that, it is enough for anyone to look at a beautiful object with an envious eye to communicate pernicious qualities to the surrounding air; his breath spreads, full of bitterness, over his neighbour, and, thanks to its subtlety, penetrates to the bones, even to the very marrow. Thus it is that, frequently, this species of envy which has been called by the name of 'fascination,' affects the health of those whom it attacks."[6] So the glance exchanged may also be one of horror; the orientation reaction may turn into an avoidance reaction.

This exchange also mediates the social status of the individual in a group. Bourdieu describes the role of gaze restraint on the habitus of the Kabyles in this way:

> The virile man who goes right to the point, bluntly, is also one who, ignoring looks, words, gestures, contortions, faces up to and looks straight in the face of whomever he is about to welcome or toward whomever he is heading; always on the alert, because always under threat, he lets pass nothing of what goes on around him, a glance lost in the air or glued to the ground being the mark of an irresponsible man who has nothing to fear because he is of no consequence within his group. In contrast, a properly brought up woman, one who does nothing out of turn "neither with her head, nor her hands, nor her feet," is expected to go about slightly bowed, eyes lowered, to avoid any gesture, any unwarranted movement of her body, head, or arms, looking nowhere but the place where she will put her foot, especially if it happens that she has to pass in front of a group of men . . . In short, truly feminine virtue, that is, "lah'ia," modesty, restraint, reserve—*orients* [italics mine] the entire female body downward, toward the earth, the interior, the house, whereas male excellence—the "nif"—is confirmed in movement upward, outside, toward other men.[7]

In this exaggerated interpretation of the orientation reaction, the stimulus is no longer simply a sensory configuration. As Bourdieu makes clear in his definitions,

> in practice stimuli do not exist solely as objective truth; they operate only as conditional and conventional triggers, dependent on encountering agents that are conditioned to recognize them. The *habitus* can only produce the response written objectively in its "formula" insofar as it brings to the situation its effectiveness as a stimulus by constructing the situation according to its principles, that is, in causing it to exist as a relevant issue in relation to a particular way of questioning reality.

To my mind, this text is fundamental. The sociologist, just like the physiologist, concludes that the brain does not just passively experience a set of sensory events in the surrounding world. Quite the opposite, it questions events based on its presuppositions. A genuine physiology of action would be based on this principle.

Gaze and Emotion

To conclude this overview, which is intended to show the complexity of the orientation reaction, I must mention the relationships between gaze-orienting movements and emotion. More may be needed than a biochemistry of passions. Facial expression is a powerful indicator of emotions. The essential role of gaze, reinforced by general body posture (see Figure 11.3), was illustrated wonderfully by Darwin in *The Expression of Emotions in Man and Animals*.[8] The physiology of orienting movements is thus also a physiology of expression. In other words, it is *projection* and not only reception. When I direct my gaze toward a target, I am trying at the same time to act on that target. Movement is an integral part of perception.

Theater provides excellent examples of this intimate link between the body and its movements, perception, and the expression of emotions. In the West, especially in France, attempts were made to dissociate textual theater from the theater of expression through a Cartesian intellectualism that separated the intelligent mind from the emotional body and language from action. However, great directors like Kantor, Grotovski, and Mnouchkine discovered the essential role of movement and of the body in the actor's acting, in his ability to make language and ideas come alive in a direct way. One of the remarkable examples of the theatrical use of gaze is that of *kathakali*, which has influenced many contemporary directors. One of the rules that masters of this choreographic art teach their students could not be more relevant for us: where the hands go to represent an action, the eyes must follow; where the eyes go, the intellect must follow; and the action represented by the hands must give rise to a specific feeling that is reflected on the face of the actor.

The repertoire of an actor of *kathakali* consists of a complex alphabet of gestures that are composed of facial expressions, gesticulations, and pre-established bodily kinetics. Thus, there are nine movements of the head, eleven ways of looking, six movements of the eyebrows, four positions of the neck, and sixty-four movements of the limbs. The gestures of the hands and the fingers have a narrative function based on fixed shapes, the *mudras* (in Sanskrit, "symbolic gestures"). Thus *kathakali* plays on two registers: the face, which expresses the emotion of the character in the situations he finds himself, and the hands and the body, which communicate the narrative aspect of the episodes.

Gaze plays an important role in many ritual activities. Here is an example practiced by the Chakyars, a community in Kerala famous for its excellent actors. On the very first day, as soon as the brand-new moon appears, the student sits down to train his eyes upon it. His eyes are anointed with clarified butter. He moves his eyes over the moon continuously until the heavenly body disappears. The first day, this exercise lasts about an hour, the time it takes the moon to cross the sky. The second day, the student sits down at the same time and applies himself to the same sort of exercise, this time for twice as long, because that is the period between the appearance and disappearance of the moon. The third day proceeds the same way. Thus he continues to exercise his eyes every night, always increasing the length of the exercise. On the fifteenth day, at the full moon, the student is seated from 6 o'clock in the evening to 6 o'clock in the morning, moving his eyes up and down, left and right, in a circle and diagonally, from one quadrant to the other, without interruption. Only at dawn does he stop. The butter has a refreshing effect on the continually rotating iris. This system is known as *nilavirikkuka,* literally, "seated by the light of the moon."

The actors train at least one hour per day for eight years. Orientational training toward an object is as follows: the eyes are wide open, and the head turns while observing, as though the eyes were moving the head. Suddenly, the head stops abruptly, and the eye fixes on an object that is not yet its final goal. The head remains in this stationary position, while the eyes move (slowly or rapidly according to intention) toward and then reach the predetermined goal. Only then does movement of the head follow; and with this movement completed, the face assumes a particular expression (hatred, contempt, joy, and so on).

But gaze is not only capture, an instrument of coherence, or an expression of emotion. It can also be absent, as it is, for example, in autistic children. Uta Frith describes Peter, an autistic child, as follows: "At last, Peter started to speak. But language did not open the doors to communication as everyone had hoped would surely happen. Strangely, he often echoed what other people said. Peter was quite indifferent to make-believe play or simple group activities . . . Often the family felt as if there were an invisible wall preventing them from making proper contact with Peter . . . Most of the time it seemed as if he was looking through people, not at them."[9]

The historical definition of the orientation reaction must thus include processes much more general than simple movement of the eyes toward a target. Pavlov provided one of the first physiological descriptions of the orientation reaction.[10] According to him, the appearance of a new stimulus immediately evokes a searching reflex; the animal trains all its relevant sensory receptors toward the source of the perturbation: pricking up its ears, directing its gaze toward the source, and sniffing the air. For Sokolov, who devoted many studies to the neural basis of the orientation reflex, the question is one of a generalized state of alertness that is not specific to any sensory modality.[11] In the Soviet literature, the orientation reaction is thus more preparation than execution; moreover, it involves structures, like the hippocampus, that process information at the highest level of the brain and that are particularly sensitive to novelty.

First off, the orientation reaction involves suspension of any ongoing movement through an increase in muscular tonicity,[12] which follows from Bernstein's sense of the word, regarding posture as a readiness to move. It also involves a lowering of sensory thresholds, probably connected to the nature of the source and to the selective redistribution of tonicity, to prepare to move. This lowering of thresholds for certain inputs preselects the senses that will be used for orienting movements. Two kinds of simultaneous processes occur next: one is inhibition of cortical resting activity, indicated by characteristic slow waves in recordings of electrical potentials on the surface of the skull. (Indeed, researchers recently demonstrated the importance of cortical and thalamic rhythms associated with states of alert attentiveness and emphasized alteration of these rhythms during transitions between waiting states and orienting movements in the cat and the monkey [see Chapter 1].)[13]

At the same time as these changes occur in the oscillatory activity of the brain, and perhaps sequentially, activity increases in numerous centers (for example, the cerebral cortex and the colliculus) activated by the hippocampus— that detector of novelty that compares configurations of actual stimuli with stored configurations. The excitability of other specialized receptors (for example, vision and hearing) is also increased. Moreover, alterations occur in visceral functions and in functions of the vegetative nervous system.

In this way the entire organism prepares itself for orientation. This response to the orienting stimulus disappears with repetition. To explain this extinction, Sokolov introduced a basic concept in his "neuronal model of stimulus." That concept continues to play a role in many theories under the names "internal model," "central estimation," "prediction," "intrinsic hypothesis," and so on. "This neuronal model of stimulus can be conceptualized as a matrix of potentiated synapses that encode the system with reference to properties of repetitively applied stimuli." This model registers not only the basic properties of stimuli, but also their recurrence and co-occurrence.

The orientation reflex thus involves not only short sensorimotor loops, but also a connection between the neocortex (a basic mechanism for analyzing signals) and the hippocampus, which detects novelty. The reflex results in a mismatch—today it would be called an error—between the neuronal model of stimulus and the external signal. The core idea of this theory is that the orientation reflex must not be considered simply as a motor reaction that directs gaze or attention toward a target, but as a mechanism that establishes a transition from one state of the organism to the other.

10

Read this line and answer the following question: Does this sentence contain the letter *w?* To answer, you have to scan the line with rapid movements of your eyes, which are called saccades. These gaze movements are guided here by a high-level cognitive task; they are not simple orientation reflexes toward a spot of light. Now, look at the word below and focus your gaze on the letter N in the middle. Without moving your eyes, see if you can find the letter Z in the word:

ANEMELECTROBACKUPPEDALOCUTWINDSHOWERSHEDNAILPROTECTEDCYCLE

You will note that it is possible to examine the letters of the word, at least those that are close to N, thanks to displacements of attention.[1] These displacements of attention are in reality ocular saccades whose mechanisms are in part the same as those for exploratory saccades, but whose execution is inhibited. Is it possible to understand the neural mechanisms of these gaze movements? I will proceed by detailing the neural basis of ocular saccades.

But before going into the details of how saccades are organized, I will first examine the role of inhibition.

The Brain Is a Fiery Steed

The brain is like a spirited horse that inhibition handles much the way a horseman handles his reins. Refinement of sensorimotor function is in part connected to the appearance, over the course of evolution, of subtle inhibitory mechanisms, the source of new motor competencies. It is easy to think that the brain contains only excitatory neurons. Rarely is any other role accorded to inhibition than that of suppression. This is borne out by the current expression "to be inhibited" in the negative sense of not being able to do something. Yet neural inhibition is actually one of the basic mechanisms of production of movement and its flexibility, and probably the main mechanism of sensorimotor training. It is also the source of perceptual mechanisms of filtering and selection and plays a decisive role in certain cognitive functions such as decision making.

MacKay assigns inhibition the leading role in his theory of mental nodes.[2] Autoinhibition is necessary to ensure that centers of integration, called mental nodes, are not superactivated by the echo of signals coming from nodes located downstream in the network that controls action, or by sensory messages. So stuttering, for example, corresponds to an inability to inhibit such an echo, leading to repetition or disruptions in language.

Inhibition takes extraordinarily varied forms in the nervous system. Local inhibitory neurons were the first to be described. For example, the motor neurons of the medulla (which activate the muscles) are inhibited by small local neurons, called Renshaw neurons, that are activated by the motor neurons themselves, whose recurring inhibition is their mode of regulation. Also in the medulla, inhibitory interneurons called Ia inhibitory neurons combine information from several sensory receptors with commands from the brain. They play a fundamental role in organizing movements. In the cerebellum, inhibitory neurons called Golgi neurons or basket neurons control the internal activity of the organ. In the thalamus, inhibitory neurons contribute to the production of oscillations, and so on. The full list would be long, but it would be worth drawing up.

One of the first functional roles attributed to inhibition was reciprocal innervation, that is, braking or blockage of antagonist muscle during a gesture. This mechanism is very simple. When you want to lift your arm, you have to activate the motor neurons of the biceps, but you also have to inhibit the activity of the antagonist muscle, the triceps. Reciprocal innervation is assured throughout the nervous system by inhibitory neurons. These neurons are also found in mechanisms of gaze control. For example, they are found in the vestibulo-ocular reflex: projection neurons called inhibitory second-order vestibular neurons induce reciprocal innervation of the muscles of the eyes (see Chapter 2 and Figure 2.4).

Ito's discovery of the inhibitory nature of the Purkinje cell in the cerebellum was a major finding.[3] Eccles and his students had made a detailed study of this remarkable cell, which is the *sole* neuronal exit from the cerebellum. This neuron (whose morphology was revealed by the work of Sotelo in Paris, among others) projects to several structures important in regulating and coordinating movement. Moreover, it is now thought that the cerebellum also plays a role in certain cognitive activities.[4] The fact that the main projection neuron of an organ as important as the cerebellum is inhibitory demonstrates the importance of a form of neural activity that researchers of the time thought could do no more than brake, or simply reverse, the direction of a command or a signal.

Another form of inhibition probably plays a fundamental role in the selection of sensory messages that accompany the planning of action: inhibitory neurons, located in the spinal medulla, inhibit sensory fibers just before they synapse on the target neurons of the medulla. This presynaptic inhibition is a gating mechanism for sensory input or modulation of its amplitude. Thus there are at least two major mechanisms for controlling messages from the proprioceptive receptors: γ and β commands on the spindles (see Chapter 2) and presynaptic inhibition. The brain has many ways of selecting sensory messages.

Cellular mechanisms of inhibition are also extraordinarily complex. Twenty years ago, this inhibition was thought to result from regulating transport of chloride ion across the neuronal membrane. The inhibitory neuron produced a hyperpolarization of the neuronal membrane, which distanced it from its firing potential and, consequently, prevented it from firing. It is now known that numerous neurotransmitters are responsible for synaptic inhibition (GABA, glycine, and so on), that several can be present in the endings of a single neuron, that inhibitory action can take diverse forms via blockage of numerous ion channels, that two successive inhibitions can be expressed by an excitation, that a combination of inhibition and disinhibition can have a multiplier effect or be used simply to introduce a delay, and so on. Molecular biology is currently completely overhauling our ideas about the mechanisms of inhibition. The work of Korn and his group in Paris also highlighted the powerful role of inhibition in modulating the excitability and plasticity of neural activity.[5]

It is impossible to broach this immense domain in just a few pages without falling into one of two traps, either a surfeit of details or neural tourism. But I will look at one example: the control of gaze, more particularly, the ocular saccade, to show how inhibition operates as a mechanism of selection and in what Piaget called "internalization" of sensorimotor behavior, and in motor imagination.[6]

A saccade is a change in the direction of gaze when the eye turns. It is rapid (20 to 150 milliseconds) and attains angular velocities of 800 degrees per second. It is the fastest movement humans are capable of producing. Each saccade consists of a rotation, followed sometimes by a slight shift to adjust the axis of gaze on the target, and sometimes other small correctional saccades if the gaze has not reached its objective.

Look at a point straight ahead of you and make a big saccade toward an object in the room. You will perceive neither movement nor change of position in the room, even though its image has shifted on your retina. This perceptual invariance is due to a mechanism called saccadic inhibition, which blocks the neural messages from the retina during the saccade and prevents you from being aware of the shift, seen but not perceived by the retina, because it remains blocked at the first visual relays. Moreover, a mechanism of perceptual shift maintains the positional stability of the environment.

I have made the following observation, which I invite you to confirm. If I focus on an object in the room for about 20 seconds, and then make a big saccade to focus my gaze on another object in the room, at the start of the first saccade, I actually have the impression that the world is moving slightly. But if I repeat it several times, my perception of the world is remarkably stable. I think that in the latter case, I anticipate the displacement of my gaze, which spares me the slight impression of shifting. I will examine the mechanisms at the level of the parietal cortex that may account for this stabilizing anticipation.

The results of the following experiment reproduce the observations of pioneers such as Purkinje, Bell, Helmholtz, and numerous other physicists and biologists who were all fascinated by these phenomena. Press laterally, rather hard, on your eyes to displace them. Don't be afraid—eyes aren't fragile! If you look straight ahead of you, you will see that during this passive movement of your eyes, the room seems to move in front of you. In other words, when your retinal image of the world is displaced, you have a perception of displacement of the world. In contrast, when you are actively gazing, the world looks stable despite retinal slip. Feedforward neural processes associated with the production of movement thus ensure perceptual stabilization. Many mechanisms have been suggested to explain this phenomenon. The most plausible is the use of an efference copy or corollary discharge (see Chapter 3 and Figure 3.4).[7]

These examples show what an interesting model the saccade is for studying the most highly developed mechanisms of active perception. But it gets even more interesting. Yarbus, a Russian psychologist, was the first (followed by Stark at Berkeley, Jeannerod in France, and the neurologists Lhermitte and Chain of the Salpêtrière Hospital) to show that the oculomotor path (the sequence of saccades) followed to explore a face is completely different depending on what the observer is thinking: whether she thinks that the individual is rich, sad, or well-coifed, that his ears are protruding, and so on. Similarly, in front of a landscape, the oculomotor path taken depends on what a person is seeking. To explore a face or an environmental scene requires complex cognitive decisions, so it is easy to see why the saccade is an interesting model for studying motor selection and decision processes. These paths are altered in patients presenting with lesions of the brain.

Finally, the production of saccades, like all our actions, requires access to memory. Visual memory is the first type: I can ask you to look at an image that follows a precise oculomotor path with a defined rhythm, then, in darkness, to repeat the same path.[8] Vestibular memory is another: I can ask you to focus on an object in front of you on the table, then to turn in your chair in the dark and not move your eyes; then I can ask you to make a saccade toward the memorized position of the object. In both cases, the saccades will be perfectly executed, which demonstrates the saccadic system's access to spatial memory.[9] I touched on these mechanisms in Chapter 5.

Next, I will consider the principal stages in producing a saccade. I will not go into particulars, but will roughly describe the general organization, as a physiologist does, even though it means losing a little of the fine grain of the analysis. I will start at the end—the motor centers of the brainstem—and move little by little toward the cortex. This method, from the bottom up, is contrary to the apparently more natural approach of examining the successive stages of image analysis. But I want to show that saccades are indeed executed hierarchically, which enables gating at different levels by inhibitory mechanisms, and that the same neural structures are brought into play in imagined movement of the eye as well as in actual movement.[10] The brain does not only carry out sensorimotor *transformations;* at several levels, motor commands influence how sensory data are processed. So, starting from the motor command comes back to thinking of action as an essential element of neural functioning and enables us to study how it organizes perception instead of simply trying to find out how perception defines action. A saccade is a decision to act, not a response to a stimulus.[11]

Figures 10.1 and 10.4 summarize some of the neural operations involved in the control of ocular saccades. My account is, of course, not exhaustive. I will review the saccadic system briefly and then discuss the individual elements. I use the word "neuron" to characterize each type of neuron, but what is involved is populations; that is, groups of diffusely or centrally localized neurons, or nuclei.

Let us begin with the *last* stage of execution: the ocular saccade is actually produced by the firing of motor neurons, called motoneurons (Mn), which activate motor fibers in the muscles of the eyes. These discharge with a very particular modulation of frequency made up of two successive parts.

First, a very intense burst of phasic activity (that is, a series of action potentials of high frequency and short duration) penetrates the viscous eye. This allows the saccade to be executed in a few dozen milliseconds. The premotor neuron, which produces the phasic activity that moves the eye toward its new position, is called the excitatory burst neuron (EBN) and is located in the brainstem for horizontal movements (EBNH) and in the reticular formation of the mesencephalon for vertical movements (EBNV). There are thus two separate centers for the two types of movement. This is the same geometric segregation I have mentioned in other chapters. The discharge of EBNs precedes the start of the saccade by around 10 to 15 milliseconds and projects directly to the ocular motor neurons. Their immediate frequency is connected (depending on one's interpretation) either to motor error or to the velocity of the eye. Generating an accurate copy of the displacement of the eye simply requires integrating the activity of these neurons (that is, adding up the number of action potentials). These phasic neurons also project to the local inhibitory neurons (called interneurons) situated in the vestibular nuclei. The functional significance of this projection is important: it blocks the vestibulo-ocular reflex during orienting movements. If not for this inhibition, the eye and head could not be directed toward a target, because the vestibular-ocular reflex tends to displace the eye in a sense contrary to that of the head (see Figure 2.4).[12] This is an example of the way in which inhibition achieves automatic coordination between two motor subsystems that have opposing functions.

Second, a sustained tonic discharge is necessary to maintain the position of the eye in its socket. The more the eye moves, the higher the tonic frequency: this is the signal of movement that is sent to the motor neurons. The two saccadic generators that receive a dynamic signal from the brain trans-

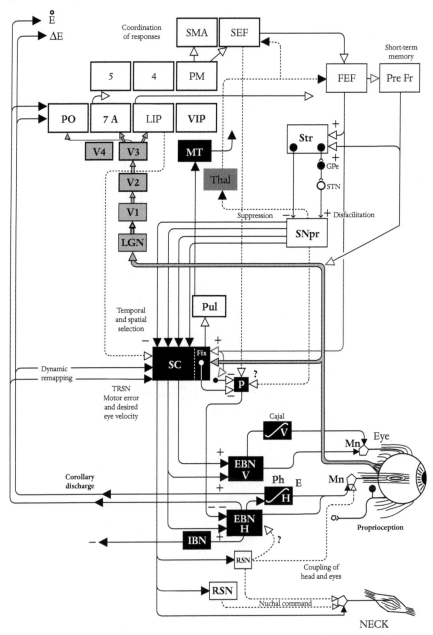

Figure 10.1. Saccadic movement of the eye *(lower right)* is produced by phasic excitatory neurons located in the pontine and mesencephalic reticular formation. The position of the eye is controlled by integrators. Copies of signals about movement of the eye are sent to the colliculus and to the cortex *(arrows at right)* and cause changes in the maps where position is encoded and where the saccade is initiated. The saccadic generator can be inhibited at a first level in the brainstem by the pauser neurons *(P)* and at other levels by a cascade of inhibitions.

form it into continuous activity owing to a local neural mechanism called the integrator. There is assumed to be one integrator for each of the two types of movement, horizontal and vertical. These are indicated by the symbol $\int$ in Figure 10.1.

As with the integrator of the vestibulo-ocular reflex, the mechanisms at work are still not well understood. They may involve properties of the neural membrane as well as those of neural networks. It is known that if a lesion is introduced to a small area close to the vestibular nuclei, the eye no longer maintains its horizontal position after a saccade. By analogy with electronic circuits, it is said that the integrator leaks. This appears to be a local mechanism. The final position of the eye is thus developed very late in the process, probably using internal models of the eye. You will see why this is interesting for imagining movement.

INHIBITION BLOCKS SACCADES

This premotor neural circuit, each neuron of which is well defined, is under the permanent control of a remarkable first-order inhibitory system. Neurons that fire continuously, whose cellular body is located in the median region of the reticular formation, inhibit EBNs and normally prevent saccades from occurring. For the saccade to take place, these neurons themselves must cease firing, or pause, to induce disinhibition of EBNs. Indeed, every time a saccade is made, these neurons pause briefly. Thus they are called pause (P) or omnipause neurons.[13] These pausers are a mechanism of *temporal selection;* they control when the saccade will occur and for how long. Up to this point, the signals from the brain are only those of *intended movement.* Removal of inhibition enables them to crystallize into orienting and gaze movements.

To remove the inhibition induced by the pause neurons, these neurons themselves must be inhibited. This inhibition is the job of neurons located in other structures of the brain that follow two parallel pathways, one descending directly from the cerebral cortex (prefrontal cortex, frontal oculomotor fields), the other most probably descending from the frontal (so-called rostral) pole of the superior colliculus.[14]

But saccades are not only controlled at the level of the brainstem, close to the ultimate motor command neurons. Execution of saccades is also blocked at a higher level, in the superior colliculus (SC). I explained in Chapter 3 that the colliculus is an important neural structure connecting visual and auditory space. The neurons of its superficial layers form a sensory map of visual space in retinal coordinates. The intermediate and deep layers contain tecto-reticulo-spinal neurons (TRSNs) that project to the brainstem and the medulla (see

Figure 3.6). These cells participate in producing orienting movements of the eyes and the head toward visual or auditory targets.

The precise role of TRSNs in generating saccades is still unknown, but it is clear that at least some of them fire in a burst of activity that precedes and accompanies the saccade.[15] They project directly to the EBNs, or alternatively to other intermediate neurons located in the brainstem, as well as to reticulo-spinal neurons involved in producing the saccade.

The firing rate of each TRSN determines the dynamics of the saccade,[16] and the connectivity of its axon determines the saccade's direction. Horizontal and vertical saccades are produced by TRSNs that project respectively to the neurons of the horizontal and vertical motor nuclei; an oblique saccade is produced by neurons whose axons project to both. Thus the motor synergy required to produce a saccade is determined in part by anatomy, as the example of vestibular neurons in Chapter 7 demonstrates. *Geometry is encoded by anatomy.*

The colliculus constitutes the second level of potential blocking of execution of the saccade because the neurons of its intermediate layers are under the inhibitory control of another formation—the substantia nigra pars reticulata (SNpr)—which is part of the basal ganglia at the center of the brain.[17] It has recently been established that projection of the substantia nigra to the superior colliculus is organized in a very precise way: some of its neurons inhibit other very specific populations of neurons. Each collicular neuron that projects to the saccadic generators induces a movement of the eye in a precise direction and thus orients gaze toward a defined location in space. This mechanism implies genuine *spatial selection,* since it is possible to inhibit saccades made in a precise direction in space. Another group of neurons, located in the rostral part of the superior colliculus, fire when the eye fixates. Thus, an independent mechanism exists for fixing the eye in its socket.

What might be called a spatiotemporal decision is thus worked out at the level of the superior colliculus. It enables displacement of visual attention from one area of space to another, since, depending on which neural population of the colliculus is disinhibited, a person's gaze will fall on this or that area of space. This attentional mechanism constitutes motor preparation, as the expression "motor theory of attention," originated by Rizzolatti, emphasizes.[18] This attentional orientation is tremendously complicated; it involves many multisensory processes and is probably influenced by mechanisms at the level of the cerebral cortex and even the hippocampal formation.

The third level of inhibitory action is also located in the basal ganglia. The neurons of the substantia nigra that inhibit the collicular neurons[19] cannot suspend their activity and trigger the saccade unless they are themselves inhibited by the neurons of the caudate nucleus in the striatum.[20] These neurons fire during voluntary saccades and are also active during saccades toward memorized targets. They are most likely involved in cognitive control of saccadic initiation.

But the process of saccadic selection is even more complex and subtle. Recent findings reveal that, concurrently with the cascade of inhibitions involving the caudate nucleus and the substantia nigra, a second cascade of inhibition and excitation accompanies and reinforces selection. This pathway, whose target is the colliculus and whose terminus is the subthalamic nucleus (STN), is made up of an inhibitory neuron, located in the striatum and projecting to the globus pallidus externus (GPe), which projects to another inhibitory neuron that in turn projects to the subthalamic nucleus. The target neuron in the subthalamic nucleus is excitatory and projects to an inhibitory neuron in the substantia nigra that suppresses activity in the colliculus, just as the neurons of the first pathway do. The STN also projects to the colliculus.

What is important is that the basal ganglia are not just components of simple command pathways: they form part of the internal circuits that connect the cortex, the basal ganglia, and the thalamus (Th), and that feed back to the cortex. In Chapter 1 I mentioned the possible role of these circuits and the sustained oscillations that traverse them in the development of mental representations. Hikosaka recently asserted that signals connected to a reward can influence them (Figure 10.2).[21] According to Hikosaka, the basal ganglia play a coordinating role among the different cortical areas involved in developing motor strategies. These areas are functionally connected in the course of training based on reward. The limbic system signals the basal ganglia as to whether the chosen movement will lead to a reward. This method of selection is thus even more complex and subtle than a simple gating mechanism at the level of the brainstem or spatiotemporal selection at the level of the colliculus (Figure 10.3).

The pleasure taken in looking at something, or the fear of what might be seen, or how interesting an object is—all contribute to the selection of saccades by these parallel cascades of inhibition and excitation.

It is not possible to describe here the totality of structures that contribute to the development of saccades. For example, it is clear that the cerebellum plays an essential role in regulating motor commands that modulate the amplitude, velocity, and duration of saccades. Lesions of the cerebellum cause

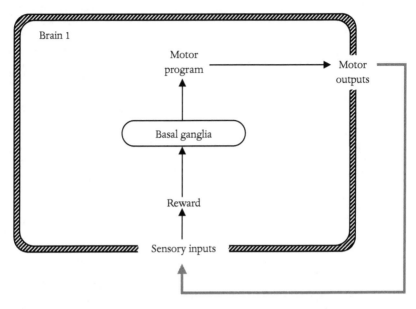

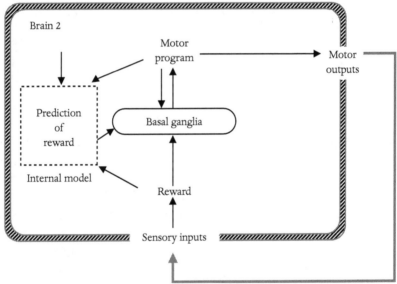

Figure 10.2. Evolution of the basal ganglia. In primitive animals *(Brain 1)*, the basal ganglia activate the elements of a genetically programmed repertoire of movements. The efficiency and success of the movements are enhanced by incentives that adjust these behaviors under similar conditions. In higher animals and in humans *(Brain 2)*, the behavior is more complex. A new mechanism (which probably involves the prefrontal cortex) makes predictions based on an internal neural model that simulates the different possible movements and evaluates their chances of success. The consequences of the resulting behavior are compared with this prediction and may modify the internal model.

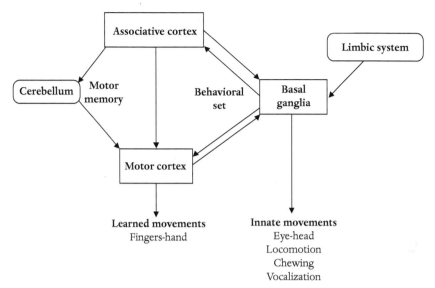

Figure 10.3. The basal ganglia coordinate interactions among cortical areas. When first learning new movements, the different cortical centers act independently. Their activity converges in the basal ganglia, which are sensitive to their interaction. When a movement has been rewarded, the limbic system promotes the successful combination in the neuronal synapses of the basal ganglia. As the new combination of actions produced by the cortical centers is reinforced, a new behavior is learned.

many oculomotor signs. The reticular neurons that control saccades are modulated by the cerebellum, itself involved in regulatory circuits that originate in and return to the cerebral cortex. Remember that the motor exit of the cerebellum is inhibitory (the Purkinje cells are inhibitory). Once again, inhibition is involved in the control of movement.

Thus, concurrently with the hierarchical cascade of inhibitions of the pausers and basal ganglia, there exists a supplementary regulatory mechanism based on inhibition. Moreover, subdivisions of the cerebellum play different roles in controlling the movement of eyes. The vermis (the median portion of the cerebellum), which dates far back in evolution, probably regulates automatic movements, and the lateral portions (the neocerebellum), which are connected to the frontal cortex, probably regulate more complex gaze movements, though this is pure speculation. If, as recent theories suggest, the cerebellum contains internal models of motor apparatuses and also facilitates the temporal coordination of movements, this structure, together with the cerebral cortex, is crucial in preparing gaze movements.

To examine the contribution of the cerebral cortex in controlling saccades, I will start with the retina. Two major pathways transmit information from the retina to the colliculus: one goes directly from the retina to the superficial layers of the colliculus; the other projects the image of visual targets toward the visual cortex through the intermediary of a relay called the lateral geniculate nucleus (LGN) (Figures 10.1 and 10.4). Interestingly, neural activity connected to the saccade itself has been found in this first visual relay. As with the vestibular nuclei, the motor command modulates sensory inputs. In other words, *action modifies perception at its source.* The brain is not constructed of simple systems that transform sensory signals into motor commands, but of closed circuits.

This reentrant information—these copies of the motor command that represent eye movements—is perhaps used as a substitute for sensory inputs. The movement of the eye is equivalent to that of a target, since moving my eye 20 degrees to the right amounts to the same thing as moving the target 20 degrees to the left.[22] Information on the position of the target is next transmitted to primary visual area V1, from whence it activates the neurons of areas V2, V3, then the parietal cortex, a very important relay in developing saccades.

The *parietal cortex* is an area of the brain that is essential in representing space and relationships between the body and the external world.[23] Although from the early part of the century the parietal cortex was conceded a basic role in constructing the body schema, opinion is still divided between theories that further define that role as directing visual attention and others that see it as planning movements. The parietal cortex is part of the dorsal system of visual information processing that is concerned with localizing objects, their movements, and the relationships between objects and intended action. The parietal cortex is not only involved in localizing objects (knowing where they are) and their relationship to the body; it also contributes to preparing actions (knowing how to grasp objects and handle them). Lesions of the right parietal cortex cause hemispatial neglect (see Chapter 3).

Neuronal recordings reveal the involvement of the parietal cortex in visuospatial behavior. Some authors posit that a change of coordinates occurs in the parietal cortex that reconstructs the position of the target in space by summing its position in relation to the eye with that of the eye in relation to the head and of the head in relation to the body, and so on.

The parietal cortex is divided into specialized zones connected to particular areas in the frontal cortex. Some are more specifically involved in preparing

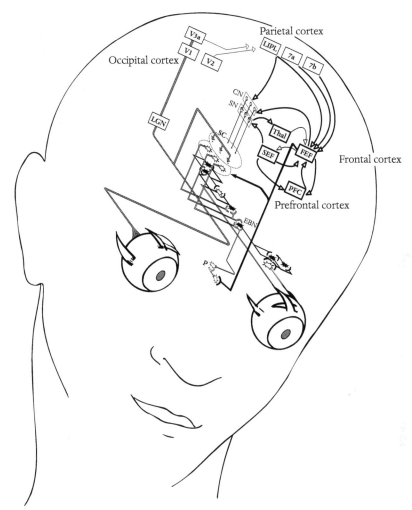

Figure 10.4. The principal production pathways of the ocular saccade. The first goes directly to the superficial layers of the superior colliculus (SC), where it is transferred to the deep layers whose neurons project to those that produce the saccade (excitatory burst neurons, EBN). A second pathway traverses the lateral geniculate nucleus (LGN) in the thalamus and projects to the visual areas of the occipital cortex (V1, V2, V3). From there, the information is transmitted to the parietal cortex, where it is combined with information concerning the movements of the head and body in space and with internal signals (memory, general context of action, and so on). The centers of the parietal cortex themselves project to the supplementary eye field (SEF), which coordinates sequences of saccades, and to the frontal eye field (FEF), located in humans in the precentral gyrus. The precentral gyrus projects to the SC and also directly to the brainstem. The prefrontal cortex (PFC) plays a role in the most cognitive aspects of the production of saccades, and a mechanism containing a cascade of inhibitions facilitates their selection. It involves the caudate nucleus (CN) in the striatum and the substantia nigra pars reticulata (SNpr), which projects to the colliculus.

gaze movements, in particular saccades or ocular pursuit. The intensity of their neuronal responses depends on the direction of gaze.[24] So a signal of detection of the position of the eye in its socket from the depths of the brain— from the neural integrator of the brainstem—is added to the activity of the neurons caused by vision. The position of the target in space is also encoded by other visual areas.[25] Recent findings have shown that this area is the center of multisensory convergence where visual, vestibular, and proprioceptive messages are combined. The neurons of these areas contribute to preparing and directing the saccade; they fire just before its initiation. Some say that the parietal cortex forms and maintains spatial coordinates that enable localization of objects and determine their position in space,[26] and that the prefrontal cortex (PFC) uses this knowledge to direct a response; others say that movements are encoded in coordinates connected to the head, in particular to the eye itself, and do not require any internal reconstruction of the position of the target in spatial coordinates.[27] Nevertheless, the question of the frame of reference in which this system functions has yet to be resolved.

The lack of activity of certain areas of the parietal cortex during voluntary saccades in the dark shows that they are mainly involved in visuospatial processing. However, using positron emission tomography (PET), we have demonstrated for the first time that they are also activated (together with frontal areas) in humans during tasks of memorized saccades carried out in the dark.[28] Successfully performing this task requires accessing the spatial memory that recorded the position of the targets, and recalling their order; thus it means producing a mental image of their positions to make a sequence of saccades. Patients with lesions of the parietal cortex make less accurate, delayed saccades.

Without summarizing the properties of the parietal cortex, which may also be involved in the memory of movement and in spatial memory,[29] in decisions connected to the perception of movement,[30] and so on, I will once again consider the essential role of anticipation and prediction in the organization of the brain. Indeed, I have shown from the outset that the senses detect the derivatives of sensory signals, that in this way they can anticipate future states. They can do this because motor signals modify sensory inputs, which alters perception based on action, and inhibition selects messages based on intended movement. But there is something even more remarkable: the nervous system can ensure the stability of the perceived world during movement by updating visual frames of reference in advance.

When the eye is stationary, visual objects are picked up by various neurons of the visual system according to where the objects are in space. It is said that these neurons have receptor fields (see also Chapter 3). These fields are small for the first relays of the visual system (area V1), but become increasingly larger over the course of successive stages of visual processing (areas V2, MT, MST, and so on). Up to now the question was thought to be one of the rigid properties of the neurons of the visual cortex, with fixed coordinates in the retinal space. The localization in space of the receptor field of a neuron of the parietal cortex is abruptly displaced before the saccade, in the direction of the gaze movement about to be made.[31] In other words, before the saccade that leads the gaze from one point in space to another, the brain displaces the receptor field as if it wished to anticipate the consequences of movement.

The advantage of this transitory anticipation is that the retinal representation of the world remains stable during movement. In other words, the brain places itself in the final reference system, adopting the vantage it will have on the visual world even before making the movement.

More than 20 years ago, an experiment done in our laboratory predicted these findings, which were obtained recently in the monkey.[32] You can try this experiment yourself. Look at the upper left corner of this page and make a single saccade toward the right lower corner. If you hold the book in front of you, the saccade will have an amplitude of about 40 degrees. Your eye will move about 400 degrees per second and thus cannot make use of vision because of the considerable shifting of the image on the retina. We observed that subjects could grasp words, although their eyes always moved faster than 200 degrees per second (which we recorded precisely using a magnetic field technique that I introduced in France) even when they slowed down in the middle of the page. We concluded from this that an active process of displacement of the receptor fields enabled the brain to compensate for the image's shifting on the retina via an anticipatory mechanism connected to the intended movement. In the 1950s, Verguiless and Zinchenko in Moscow spoke of the "functional fovea."

This is why I keep saying that the boundaries between sensation and motor function are disappearing. For the same reason, an internal copy of a movement command signal can be used as a signal given by the senses, and is itself influenced by movement from the first central neurons and, even, at the periphery.

It does appear that the information needed to produce simple saccades toward a visual target are transmitted from the parietal cortex to the frontal eye field (FEF), which projects directly to the colliculus.[33] Ferrier,[34] then Penfield,[35] showed that electrically stimulating the frontal eye field produces eye movements. In the monkey, the frontal eye field contains several types of neurons that fire in connection with saccades, ocular fixation, pursuit, and so on. In humans, its activation during saccades was demonstrated through several pioneering efforts.[36] Its localization in the precentral gyrus is now well established.[37] Generally, this area is activated during all voluntary saccadic movements.

The importance of this area is not due only to the fact that it projects directly to the colliculus. It is part of an internal circuit that I have mentioned several times as being one of the neural bases important for the internal simulation of movement. In fact, a circuit connects this cortical area with the basal ganglia and the thalamus.

Another area important for anticipation and prediction in developing movements is the supplementary motor area (SMA), located in the median portion of the frontal lobe, in front of the actual motor areas. Its location was studied by recording electrical (EEG) and magnetic (MEG)[38] activity and by PET.[39] It is also important to remember that it is part of the circuits that connect certain structures of the basal ganglia and the thalamus. Its function is thus associated with theirs, and even with that of the cerebellum. It also functions in connection with the parietal, prefrontal, and cingulate cortex.

The neurons of this area are arranged in a sort of map of the body as in the motor cortex (somatotopy). For example, in humans and in monkeys, facial movements are obtained by stimulating the anterior portion. This area may command groups of muscles and complex motor synergies. It is involved in global actions, since it initiates, for example, activity of the muscles of the left hand only when the movement mobilizes both hands. It also comes into play when the subject must make a choice, or decide whether to make a move (go–no go). Moreover, it is activated during sequences of movements.

In humans, lesions of the SMA cause alternating movement deficits of the limbs, mutism, disordered language, and impairments in gestural sequencing, bimanual coordination, memorized movements, and especially endogenously produced movements (movements that are freely constructed by the brain even in the absence of external stimulation). These findings have been confirmed by EEG recordings taken on the surface of the human skull. In the area of the cortex that corresponds to the SMA, a slow electric potential (readiness potential, discovered by Kornhuber and Deecke), precedes a simple finger movement by about a second and a half. This area is strongly activated when the subject is asked to do a bimanual tapping task with a different rhythm for each hand, as in playing a piano piece.[40] It scarcely registers in simpler movements, which mainly activate the motor cortex.

It has been suggested that the motor cortex contributes primarily to endogenous movements, whereas the premotor cortex contributes to movements caused by sensory signals. This distinction could be profound and may go far back in evolution. In the mammalian family tree, the SMA has its origin in the hippocampus and a limbic root in the anterior cingulate cortex. The premotor cortex probably had its origin in the piriform cortex and its root in the insula. Subcortical inputs from the SMA mostly come from the basal ganglia via the thalamus, whereas the premotor cortex receives its inputs from the cerebellum. The SMA connects to several areas of the cortex involved in perception of space and of the body. This distinction indicates a possible specialization of these areas in endogenous movements for the SMA and exogenous movements for the premotor cortex.

Whereas the posterior portion of the SMA is clearly involved in coordinating movements, the anterior portion contributes to more cognitive aspects. The anterior portion (pre-SMA) is connected with the prefrontal cortex and receives inputs from area 46 (anterior premotor cortex) and the anterior cingulum involved in intended movement and attention, and perhaps in training sequences of new movements (playing a Bach fugue, or dancing choreographed steps). Electrical or magnetic stimulation of the pre-SMA arouses an intention to move; subjects say, "I feel like moving." Recent findings obtained by PET suggest that the anterior portion of the pre-SMA is activated during imagined movement of the fingers, and the posterior portion during the actual movement.

There appears to be a further subdivision of the SMA itself unique to gaze movements. Stimulating this area produces ocular saccades that instead of having a fixed direction, all converge toward a single point.[41] In the monkey, this area is called the supplemental eye field (SEF). Patients with lesions of the SEF present with problems in executing sequences of two or three saccades toward memorized visual targets,[42] as well as saccades toward a memorized target after a rotation; that is, after vestibular stimulation.

The SMA is also connected to the basal ganglia and to the thalamus via a neural circuit that is probably involved in establishing new movements and in assessing the rewards or outcome of an ongoing action (see Figures 10.2 and 10.3). A special circuit for eye movements exists among the basal ganglia-thalamo-cortical loops specifically concerned with different motor functions.

Under the frontal cortex, the motor and parietal cortex are connected to a region of the cortex called the cingulate cortex, whose anterior portion is involved in attention and in perceptual decision making. It is associated with the prefrontal cortex and perhaps also is involved in emotion. The median portion has more to do with motor function, and its neurons probably contribute to organizing movement of the hands, eyes, and head. In humans it is activated during saccades[43] and during endogenous (that is, produced by the subject) movements of the fingers even in the dark, and thus contributes to creating movements based on memory. The posterior portion is involved in perceptual reorganization after a movement, or even following corrective control. In fact, the neurons of the posterior cingulate cortex fire after ocular saccades, and their activity depends on the characteristics of the preceding saccade.[44] Thus this region may contribute to integrating the consequences of action in a sensory or motor context.

I have examined the role of the brainstem as an ultimate controller, the colliculus as a visual-oculomotor interface, the parietal cortex as a predictor and spatio-temporal transformer, and the frontal areas as preparatory centers. I will now consider the role of the prefrontal cortex in the higher cognitive operations that control gaze orientation. Findings from neurophysiology show that in the monkey, neurons of the *prefrontal cortex* (PFC) are involved in organizing ocular saccades that are simple but are carried out with a delay.[45] They are also involved in training for sequences of movements or actions that obey a rule,[46] or require either memorizing a context or deciding among several movements, which necessitates selective attention combined with memory.[47] This function of representational memory and of work is thought to be one of the main functions of the prefrontal cortex.[48]

Patients with lesions of this portion of the cortex have trouble refusing requests—controlling themselves, one might say. They also have trouble making decisions and especially changing the rules of conduct once they have decided to act according to a certain rule or have established a plan of action. These patients are often deficient in performing tasks that require making saccades toward visual targets they must memorize[49] or when they are expected to make a saccade that corresponds to a rotation in the dark; that is, one based on vestibular information (see the protocol for vestibular saccades described in Chapter 5).[50]

They also fail at so-called antisaccadic tasks, which consist in making a saccade in a sense contrary to that of the illuminated target.[51] (Try it yourself: Ask a friend to strike a match on your right and force yourself to direct your gaze with a single saccade in a direction opposite to the match; for example, to the left if the match is on your right.) Moreover, in the monkey and in humans, prefrontal regions are involved in the command *not* to execute a move (no go).[52] Thus, there are probably mechanisms for selection and control of gaze orientation at this level that could be identified by imaging. Connections of the prefrontal cortex with the hippocampus, the parietal cortex, the SMA—all the areas involved in the voluntary production of saccades—suggest that a cortical system is involved in the construction of exploratory paths taken when we look at a visual scene, but probably also in spatial memory in general.[53] In any case, the prefrontal cortex surely contributes to the mental simulation of movements and displacements.[54]

So the prefrontal cortex is involved in processes of decision in the course of "voluntary" actions,[55] and more specifically when we alter our gaze based on affective or social constraints. Impairment of gaze organization observed in persons with schizophrenia (for example, distraction and deterioration of the ability to make predictions) are perhaps attributable in part to dysfunctions of the prefrontal cortex.[56] Understanding these mechanisms, which are fundamental to the development of thought and reasoning, will require new experiments using brain imaging in humans and neural recordings in animals.

Imagined Movement and Actual Movement

Now I will present the arguments supporting the idea that inhibition plays an essential role in internal simulation of action. But first I will summarize. First to appear were mechanisms that block saccades by means of pauser neurons; that is, a blocking mechanism of the collicular orienting reflex to prevent the animal from making a saccade the instant a visual target appeared. Then inhibition enabled blocking of the vestibulo-ocular reflex. Both mechanisms were required for a saccade to occur.

Next came inhibitory control by the substantia nigra of discharges of the neurons of the colliculus. This control was more subtle, because it allowed genuine selection of the part of the visual field toward which a saccade would be made. Moreover, inhibitory cascades in the striatum probably enabled working out the decision to make a saccade based on many other criteria important for the animal. At this stage of evolution, it was already possible to simulate saccades without having to make them. Thus, displacements of attention are saccades that are blocked at the lower premotor levels. I have proposed a theory of the hierarchical organization of inhibitory mechanisms to explain the common mechanisms of executed and imagined or simulated movements, and to account for the possibility of selectively blocking the execution of movement at several levels of the sensory motor system. This theory suggests that there is not a simple dichotomy between imagined and executed movements but a hierarchical superposition of different levels of possible blockage. This theory extends the motor theory of perception advocated for eye movements by Rizzolatti and colleagues.[57]

Development of the cerebellum probably enabled modulation as well as a choice of coordination between the saccade and other motor systems. Finally, the entire neural machinery of the cortex was put into place. When the colliculus is inhibited and the pausers are active, the set of cortical structures can prepare and simulate oculomotor paths: no saccade is observed. Inhibitory gating of execution thus permits the brain to imagine displacements of gaze without carrying them out. You can see for yourself that you do indeed have this option. Focus on a point in front of you and displace your internal gaze— your attention, as some would say, your "functional fovea," as Verguiless would say. You will have the undeniable impression that your gaze is displaced from one point of the room to the other.

The same neural structures thus ought to be used for imagined movements and for actual movements, since it is possible to gate execution at different levels without suppressing the functioning of internal circuits in which gaze movements are developed and simulated. To check this assumption, we compared the cortical areas activated during actual saccades and imagined saccades.[58] We measured ocular movements to make sure that the amplitude of the saccade remained insignificant during the imagined movement. The subject was asked to focus on a point of light in front of him and, first, to make voluntary horizontal saccades, then, with the target obscured by the dark, to make imagined saccades similar to the actual ones. Under these conditions, the oculomotor field, the SMA, and the cingulate cortex were activated during the imagined movements just as they had been during the actual saccades.

So perhaps the motor theory of attention is true. When you look at the long word that appears at the beginning of this chapter, focusing on the N in the middle and looking to see if there is a Z, you bring into play the same circuits as those that are active during an actual saccade. But there may be alternative interpretations. Only experimentation will tell. I mentioned earlier that there were possibly differences in the roles of the pre-SMA and the SMA.

Dynamic Memory and Predictive Control of Movements

Although the structures involved in orienting gaze are beginning to be understood, their description in terms of general algorithms that describe the processes involved remains controversial. In particular, it is obvious that the firing rate of many premotor neurons is connected, to different degrees, to acceleration, velocity, or the angular position of the eye. Moreover, the problem has been restricted to one dimension and to small movements to ensure the validity of a linear approximation of the processes. Finally, and more generally, the question of neural mechanisms that underlie the control of movement as simple as the displacement of the arm from one point to another has not been resolved: Which is controlled, the final position of the arm, or the balance of the contractions of agonist and antagonist muscles?

The solution to these various problems extracts from them a general concept that makes it possible to account for observed neural activity and at the same time to explain the control of orienting movements. It derives from a model conceptualized by Droulez called dynamic memory, mentioned in Chapter 5. This model is too complex to describe briefly. But I can give some idea of its essential properties.

The novelty of this model is twofold. First, it shows that the brain does not need to calculate spatial coordinates of the targets toward which saccades are made. Automatic mechanisms that use the motor command itself update the neural maps on which the image of the target is projected. Moreover, the saccade can be predicted even if the target disappears, owing to a mechanism of neural memory that maintains neural activity after the disappearance of the target.

It also predicts the results of a critical experiment, during which the authors diverted the course of a saccade unbeknownst to the animal by means of central electrical stimulation.[59] Despite this perturbation, the eye reached the target, which suggests that the brain encodes movement in spatiotopic coordinates (connected to space). We have shown that dynamic memory makes it possible to perform such manipulations without reconstructing a representation of the target's location in space in the brain.

Dynamic memory has another property. Its output can be used to make a move using different effectors. Indeed, orienting movements or movements of capture can be achieved with the eye only, or with the head, the trunk, or even the whole body. The model predicts that the orienting movement can be specified in terms of velocity in a rather general way, before it is executed by the eye, head, or a limb, owing to local integrators that determine changes in position of the effector based on its biomechanical characteristics. This property is very general, since it is easy to write a letter using a finger, hand, or foot. Very few current models include this property of motor equivalence. Our attempt was, of course, only a first step, but its formalism complements the empirical study of sensorimotor processes because it makes experimental prediction possible. In particular, it assumes that the brain is not a Cartesian machine that reconstructs the world in spatial coordinates. There is no little geographic map inside the brain that details places; instead, the brain notes discrepancies and updates the activity of its own internal maps in the course of movement. The model of dynamic memory is also a model of dynamic updating.

Was Piaget Right?

It is interesting to compare the facts and theories explained in this chapter with the theory proposed by Piaget in 1949.[60] "The entire evolution of intelligence in the infant is characterized by the internalization of effective actions into represented actions and reversible internal processes." He notes that between years 2 and 7 mental images remain strictly dependent on movements and preserve a motor component. The first mechanism that he considers is straightforward inhibition of action. He cites Bain: "To think is to keep from moving." I have shown, in the case of the ocular saccade, how inhibition can block execution of action. This first level, which enables internalization of action, is in fact present in the neural basis of gaze.

"A second stage," says Piaget, "is reorganization of motor function at around 7 years. Until then, after each representation, the child acts on objects and perceives the results of his action. This is a semi-internalization." Between 7 and 12 years, the question is no longer one of connections between afferents and efferents, but between associative pathways. Piaget predicts the appearance of combined mechanisms of orientation and of inhibition (which he also calls "channeling" and "blocking"). In this description, I see a possible analogy with the high-level mechanisms that enable selection of collicular neurons by the inhibitory cascade of the substantia nigra and the caudate nucleus; the difference between imagined movement and actual movement is thus not a dichotomy. A gaze movement, when executed, also entails a component of internal simulation, as our scheme from 1985 (see Figure 1.5) suggests. Imagined movement can bring into play several levels of the cascade of hierarchical mechanisms. There is not just one kind of imagined movement; there are several nested degrees of imagination and of execution.

These findings confirm the fact, discussed in the first chapter, that to perceive an object is to imagine the actions its use will involve. Similarly, listening to music has the same effect as singing it or playing it. In 1983 I had a series of talks with Melvill-Jones, who pointed out that as a member of a choir he had learned hymns simply by listening to them. Following these talks, and after having observed my daughter Rebecca learn the violin by the Suzuki method, which enables a child to play pieces just by listening to them on cassettes and singing them—without having to figure out the notes—I joined a team of psychologists in a study of how violinists are trained. I proposed the idea that singing provides the brain with a general dynamic template of the musical melody, which can then be transferred to the coordination system of the hand. This work is in progress, but preliminary results do indicate a greater facility in going from singing to violin playing than the reverse.

11

BALANCE

What a proud look the young child displays to his admiring relatives the first time he stands on his own: in an instant he has summed up several million years of evolution, definitively freeing his hands for the use of tools and feats of manual dexterity and art. Suddenly, escaping the caves of early infancy, head in the stars, Caliban turned Ariel due solely to the play of muscles that control posture, he triumphs over gravity, which up to now held him to the floor and obliged him to sit down to touch objects and beings. What mechanisms underlie this achievement?

The discovery by Sherrington and his school of the mechanism that causes a muscle to contract when it is stretched led to a view of balance as the linking up of a multitude of local reflexes: when the body bends under the effect of gravity, muscle receptors as well as those of the joints, vestibular system, and even vision detect this bending and cause a muscle contraction that rights it. In the West, this conceptualization flourished until the 1950s, concurrently with the spectacular growth in techniques of analytical neurophysiology that described links between sensory receptors and muscles. Each time neurophysiologists stimulated a sensory ending, they obtained a muscular contraction with a more or less prolonged latency, which reinforced the idea of postural organization based on servomechanisms that detect errors through the sensory receptors.

In reality, control of posture and its coordination with movement also bring into play very different mechanisms of these reflex loops. Remarkable biological solutions simplify this eminently complex problem. Mechanisms of anticipation and internal simulation were devised and make maintaining balance a superb model for studying reflexive and cognitive cerebral functions.

Yet my teacher and friend, the neurologist Pierre Rondot, used to say that as late as 1970 the postural manifestations of the major neurological illnesses were still interpreted by imperial fiat: in the consulting rooms of the Salpêtrière Hospital, the Mecca of French neurology, the "prince," that is, the senior consultant, explained the signs of this or that disturbance of balance based on the wealth of findings established by the French neurologists Duchenne de Boulogne, Déjerine, Thomas,[1] and so on. This *qualitative* semiology is still viable, for it is remarkably effective. It would be interesting if today, in cooperation with clinicians, a *quantitative* semiology of disturbances of posture and coordination of posture and movement were to be established.

A Physiology of Reaction

Engineers were the first to be interested in the quantitative study of posture and balance. They applied methods of analysis derived from servo systems developed during the Second World War by specialists in radar and electrical motors. A new science of automatics was born that provided a basic theoretical framework for reflex theories of postural control.

This approach is good for so-called linear systems—those whose output fluctuations (for example, tilting of the body) are connected, in a straightforward way, to disturbances of input (for example, displacement of a movable platform on which a subject is positioned). Relationships between inputs and outputs are expressed in the form of equations called transfer functions. This is how gain (the relation between the magnitude of variation of input and that of output) and phase (the delay in variation of output compared with that of input) are defined. In Chapter 2 I discussed how these techniques are used to describe the anticipatory property of the myotatic reflex, which compensates the slowness of muscle reaction by phasic anticipation. Similarly, some 30 years ago the English physiologist Merton developed the concept of servo-assistance to model the control exerted by γ motor fibers on the neuromuscular spindles.

The use of movable platforms was also a powerful tool in understanding the mechanisms of balance under natural conditions. First studied was the contribution of proprioception to the control of balance. For example, selective anesthesia of the nerves that transmit tactile information from the feet helped to demonstrate that a dog placed on a movable platform had difficulty resisting perturbations of balance. Thus it was possible to study posture from a functional point of view using established methods. The path lay open to a new physiology of balance.

Researchers then examined the contribution of vision to balance. Although its role was certain, its mechanisms were unknown. A decisive step forward was taken the day Lee hung a large box from the ceiling of the amphitheater of the University of Edinburgh.[2] The box measured 3 meters on a side, and was open at the bottom and on one side. He placed a subject upright within the box and gently oscillated it like a large pendulum, front to back. Instead of perceiving the box moving around him, the subject perceived it as unmoving and the amphitheater to be oscillating. In the present of a *relative* movement between the box and the ground, the subject's brain concluded that the ground was moving, not the room represented by the box. He made a genuine perceptual decision among the possible interpretations.

Lee observed a second phenomenon: simultaneous with the subject's perception of self-motion in relation to the visual world, judged to be nonmoving, the subject's body began to oscillate from front to back with the box. These oscillations could have been the result either of a direct effect of vision on muscle tone, or of a reinterpretation of the subject's perception of his body's vertical in space. Lee opted for the first possibility and suggested that vision plays a proprioceptive role.

These results were subsequently confirmed to be rotatory movements of the visual world.[3] Subjects were placed before a disk spinning in the frontal plane, at the center of which was a small line that could be tilted in relation to the vertical. You can easily try this experiment yourself with a bicycle wheel placed at the height of your eyes. Turn it with your hand or with a small motor. If you stand upright in front of the wheel, you will perceive a rotation of your body in a sense contrary to the movement. This circular vection in the frontal plane is mitigated somewhat because the otoliths and the tactile messages from the soles of your feet contradict the visual messages. You can remove this contradiction by lying down on your back and positioning the wheel horizontally: now the illusion that your body is rotating in the horizontal plane around your head becomes more intense.

When you are upright, the slight vection is accompanied by a very strong alteration of the vertical subjective. This alteration is made obvious in the following way: the subject is upright in front of the revolving disk whose center is level with the subject's eyes. At the center is a smaller disk 20 centimeters in diameter, on which a line is painted. This disk is stationary, but the subject can turn it manually using a button. This enables the subject to tilt the line at will. He is asked to place the line in such a way that it is parallel with the vertical he perceives. Recall that he sees only the revolving disk and has no view of the environment; he must thus rely on an internal representation of the perceived vertical, which is called subjective.[4] In general, subjects tilt the line 5 to 20 degrees. This readjustment of the perceived vertical is accompanied by tilting the head and the body in the same direction.

The visual stimulus thus simultaneously induces an illusion of movement, a perception of change in the direction of the vertical, and a postural readjustment. This association of perceptions and motor effects is, to my mind, one of the most convincing indications of the intricacy of perceptual and motor mechanisms as well as of the fundamental role of perception as the source of postural reorganization. Posture is controlled by perception and not by local reflexes. Vision does not act only as a proprioceptor; it influences the perception of the vertical which, in turn, induces postural changes, as proposed long ago by Gurfinkel in Moscow.

The influence of vision on the control of posture is decisive in infants of 2 to 3 years. They fall down when the walls of the box are displaced only a few centimeters.[5] If a seated baby is placed in front of a disk revolving around the axis of gaze, the baby leans to the side when the disk turns. This effect increases up to the age of 2 years, providing evidence of the fundamental role of vision in the development of postural control. It diminishes gradually, disappearing around age 16.

I think that this diminution of visual dependence is connected to the gradual installation of complex methods of postural control based on internal models that make vision less dependent on the sensory environment. But vision reassumes its role if the other sensors controlling posture are impaired. For example, patients presenting with accidental or surgical lesions of the vestibular system are dependent on vision for control of posture.[6] It is not always possible to detect the disturbance simply by asking them to close their eyes; sometimes they have to be subjected to a situation called stabilized vision. Illuminated boxes are placed over their heads so that vision is activated, but the vertical is not signaled. Using this same trick, we discovered that in the absence of gravity on board space stations, astronauts recover their visual dependence during the first moments of flight when they have to recalibrate mechanisms of postural control.[7]

In the late 1970s, to better study the postural readjustments induced by the movement of visual scenes, we equipped subjects with a portable mirror device,[8] derived from the one that enabled me to study vection (Figure 11.1).[9] These mirrors reflected visual scenes moving from front to back or from back to front, projected onto the ceiling. In this optic tunnel, the subject had the il-

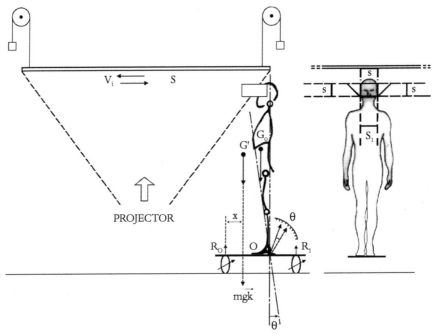

Figure 11.1. Device for studying the influence of visual movement on the control of posture. The subject is upright. Visual scenes in motion are projected onto a screen, *S*. The scrolling velocity of the scenes *(V_i)* is controlled by a computer. Mirrors create virtual screen images *(s)*. In this way, the subject is in a virtual tunnel and has the impression of moving forward or backward according to the direction of visual movement. The tilting of the body that results is measured by potentiometers located at his ankles.

lusion of moving forward or backward in a sense contrary to the motion of the scene. It was a primitive virtual reality machine. The illusion was accompanied by general tilting of the body in the same direction as the motion of the scene. Training of the body by vision and by changing the perception of the vertical occurs at very low frequencies—less than one oscillation per second—of visual movement and depends as much on the velocity of movement of the scene as on its spatial composition. Training is reinforced if the visual scene includes objects and contours.

The following anecdote shows how a negative result can lead to a discovery. One day, a subject proved to be totally insensitive to all visual stimulation. Her case was a real challenge, and I struggled to understand the reason for her resistance. I asked her what she did for a living, and she told me she was a flutist. I hypothesized that she had developed a visual inhibition so she could play in an orchestra and not be distracted by the movements of the other instrumentalists. To suppress this mechanism, I asked her to count backwards from 100.[10] And, as had other subjects, the flutist immediately responded to visual stimulation. It was then clear that for her, training to play the flute entailed selection of sensory information that made her in some sense blind or less sensitive to visual perturbations.[11] It may also be that her resistance to visual perturbation was a true gender difference. Experiments from our lab show that when subjected to visuo-vestibular conflict, women recalibrate their perception of body rotation less than men.[12]

This observation is instructive for several reasons: first, postural mechanisms depend on the general context of the action; second, people resolve the problem of sensorimotor integration in their own ways. Captives of statistics, researchers sometimes try too hard to define an average behavior, a typical subject, and standard means and deviations in normal populations. Psychophysiology of sensorimotor integration must now move in new directions, indicated in France by the differential psychology of Reuchlin.[13] Researchers need to take into account both general rules of behavior and individual solutions. All the observations that have been made concerning postural strategies in subjects with vestibular lesions, in astronauts during and after space flights, and so on lead to a challenge. The challenge consists of constructing a physiology of individual perceptuomotor strategies and of the common repertoire of behaviors that these strategies draw on. The functional flexibility that makes possible the choices and decisions characteristic of the brain of primates and man must be fully explored. I will come back to this point in the chapter on adaptation.

Toward a Projective Physiology

An emphasis on functional flexibility also suggests a top-down model of regulation of postural mechanisms of selection and anticipation as opposed to the accepted bottom-up model of reflexes. For me, a chance encounter opened new prospects.

One day, I was presenting the results of these experiments at an international physiology conference in Paris. I was standing in front of my poster. Opposite me a young American, Lewis Nashner, was also presenting a poster. He wore splendid cowboy boots, was abroad for the first time, and spoke not a word of French. His poster claimed that proprioception contributes to rapid postural reactions during a perturbation on a moving platform. However, when his subjects closed their eyes, he observed no difference in the amplitude of their fall and concluded that vision made no contribution. In contrast, my results indicated that vision played a very important role.

We had in common the fact that no one was the least interested in our results. So we began to talk. I felt compelled to criticize both his premise and his conclusions. In fact, I explained to him that when a person closes his eyes, the brain no longer has access to visual cues and thus reorganizes other cues that it uses instead. I argued that if vision were preserved by inhibiting visual cues about velocity, the brain, expecting this information, would make a mistake, and a major perturbation would result. I made him a friendly challenge and invited him to come and try the experiment, which, one year later, he did.

Together we constructed a novel device, placing the subject upright on a platform that could be abruptly displaced.[14] With respect to the paradigm of postural perturbation, our device was nothing new. The trick consisted in surrounding the subject's head with a box made of light polystyrene that slid from front to back. This box could be fixed in space or suddenly joined to the movement of the head by an electronic command: the perturbation was transmitted, the subject fell forward, and, right at that moment, an electronic device initiated the coupling between the subject's head and the box. In this way the visual world was stabilized in a transitory manner. The brain perceived no movement by vision. We had fooled it. Only the proprioceptors of the muscles and the joints, along with the vestibular sensors, signaled the movement.

The result was spectacular: the muscular reaction of the legs was cancelled for a short interval. Lacking visual information, the brain did not trigger a compensatory reaction. This experiment convinced me that when vision is effective, the brain waits for a cue from this preferred sensor, together with others, to initiate preprogrammed reactions of balancing. These signals require a specific configuration and combination to trigger the reaction. Vision is both a sensor of velocity and an analyzer that initiates responses when its stimulation by a movement is conjointly signaled by several other sensors.

The multisensory control of balance is thus not due to a simple string of responses to stimuli. It involves comparison of the state of the sensors with prediction. Now a good many contradictory findings became clear. All the authors before us who had claimed that vision played no role in rapid postural reactions had asked their subjects to close their eyes. Comparing the eyes-closed performance with the eyes-open performance revealed no differences. My interpretation was that when the subject closes his eyes, the brain actively reorganizes the configurations of sensors on which it bases its decisions and expectations. Hence the new idea that the brain chooses sensory inputs according to context.

This theory was reinforced by the so-called broom experiment proposed by Vidal, one of my collaborators. He positioned a broom handle horizontally before a subject. This broom handle constituted a stationary rod on a platform, enabling the subject to stop himself with his arms. During a postural perturbation, instead of activating the muscles of his legs, the subject reacted exclusively with his arms, which caused a complete reorganization of his postural reactions. This experiment was also conducted and published by Cordo and Nashner. They called this flexible functional state "set," or state of preparation. It remains to reconcile the concept with the definition given by Bernstein of the posture he called "readiness to move."

THE SET CONCEPT

The term *Einstellung* (set) was introduced around 1905 at Würzburg by Külpe, Ach, and Watt.[15] It was adopted first of all to describe the internal state of subjects awaiting instruction to concentrate their attention on this or that aspect of an object. For example, if a subject was asked to describe the number of objects present in a room, he had trouble remembering their color several moments later, which gave rise to the idea of a perceptual filter.

The set concept is tightly linked to the problem of relationships between posture and movement. Sherrington considered these to be two distinct mechanisms and proposed that posture accompanied movement like a shadow.[16] Later, movement was considered to be scarcely more than a succession of postures. This idea turns up in various recent formulations of the so-called point of equilibrium theory, which holds that movement is due to a gradual change in static balance between the forces exerted by the muscles located around a joint, called antagonist muscles (for example, in the arm, the biceps bends the forearm, and the triceps stretches it out).[17] According to Sherrington, this shift is induced by a central command that regulates the relation between the forces of the muscles, which produces the transition from one position of balance to another. For some, a movement intended to reach a target—a teleokinetic movement—can only occur when it is supported by an initial so-called ereismatic posture.[18] It was also suggested that there are two submechanisms of postural control, one that steadies the reference point of the movement, the other the distribution of muscle tone in the moving limbs.

The set concept was assigned a wide range of meanings, as various contexts for set in the literature attest: mental, neural, voluntary, unconscious, postural, organic, preparatory, task-oriented *(Aufgabe)*, situational, goal-oriented, temporary, permanent, reactive, perceptual, expectational, anticipatory, intentional, and so on. It has also been defined as a preparatory state for receiving a stimulus that has not yet arrived.[19] Discussions of the set concept were accompanied by numerous studies of events associated with impending movement.[20]

Concurrently with this research carried on in the West, a very original program of research was developing in the Soviet Union, which at the time was closed to scientific communication. The idea of a preparatory influence on perception and control of movement had already been suggested by Sechenov: "When a stimulus is expected, the activity of any other mechanism interferes with the phenomenon, and curbs and delays the reflex."[21]

In the 1960s, Gurfinkel and his colleague Alexejev in Moscow established the importance of anticipation, instruction, and a person's mental state in the control of posture. Indeed, if you say to someone, "Wait for the bus. It will be here in two hours," the gain of this reflex is much less than if you say to him, "Watch out! I'm going to kick you!" This observation might appear trivial. But the idea that rigid reflexes underlay the organization of motor function was so anchored in people's minds, when in fact they are modulated by a global plan of action, that it led to a genuine revolution.

How to Pick Up a Load without Falling Down

Posture thus both supports and prepares for movement. But the predictive brain has devised something even better. It maintains a repertoire of postures specific to each animal. In Chapter 7 I defined the concept of synergy. Balance is guaranteed by a repertoire of synergies that are organized into strategies. Let me explain. When the bus brakes abruptly, you can only bend your body two ways: either by rotating around your ankles and keeping your body rigid like a reverse pendulum (ankle synergy), or by flexing your trunk at the pelvis, leaving your legs rigid (trunk synergy).[22] A third strategy would be to take a step forward. These synergies involve sequential activation and inhibition of the muscles of the legs, hips, and trunk whose temporal organization is very precise. Thus time is as important in organizing a synergy as the distribution of activities.

The goal of some of these synergies is to avoid loss of balance caused by ordinary body movements. Indeed, any number of operations can lead to loss of balance: rising to your tiptoes, lifting your leg, bending over to pick something up, and so on. Try standing on a beam or an unstable part of the floor: simply lifting your arm is enough to shift the center of gravity forward and cause you to fall.[23] To avoid these disruptions, the brain automatically triggers a slight tilting of the body backwards prior to the main gesture of lifting the arm.

Now try standing on your left leg and lifting your right leg to the side as high as you can, as dancers do. The first time you do it, your torso and head have a tendency to move slightly to the right before your leg goes up. This slight adjustment enables the center of gravity to shift directly below your standing foot to maintain balance.[24] Training entails a reorganization of the gesture and the posture that accompanies it: you no longer incline to the right, and the movements of the body are redistributed around a perfectly vertical head and torso. With Pozzo (see Chapter 4), I showed that the head is used as a stabilized frame of reference for coordinating movements of the limbs.

Try one last exercise: Bend over to pick something up. Position yourself a few millimeters away from a wall without touching it with your back. You will notice that just *before* you bend over, your buttocks touch the wall. A movement of the body backward anticipates the bending forward. It displaces the center of pressure that your body exerts on the floor toward the back and in this way compensates, in advance, for the body's forward displacement. So balance is not maintained by detecting an error and correcting it; it is maintained by anticipating the consequences of action. This synergy is called the Babinski synergy.[25] It also comes into play when you lean backward. Babinski pointed out that it is absent in patients with lesions of the cerebellum. Thus the cerebellum contributes to this motor anticipation.

Finally, a very common but interesting phenomenon is the initiating movement of a step, which is also preceded by anticipatory postural activity. Suppose that you lift your foot to take a step forward. You cannot do it without lifting the 30 or 40 kilograms of body weight that are bearing on that foot. To relieve the weight and initiate the step, your center of gravity has to be shifted forward by a burst of activity in the flexor muscle of the ankle, the anterior tibialis. The brain accomplishes this anticipation by producing a shift in the center of gravity just before you lift your foot. This anticipatory motor synergy is disturbed in patients with Parkinson's, who have a dopamine deficit in the basal ganglia, the nuclei at the center of brain that are essential in organizing movements.[26] Initiating activity of the anterior tibialis is very weak in these patients, and they can no longer coordinate their movements. The basal ganglia are thus indispensable in anticipatory shifts of the center of gravity associated with the initiation of movement. Together with the supplemental motor area of the frontal cortex,[27] they play a fundamental role in coordinating posture and movement. These anticipatory postural synergies, which are

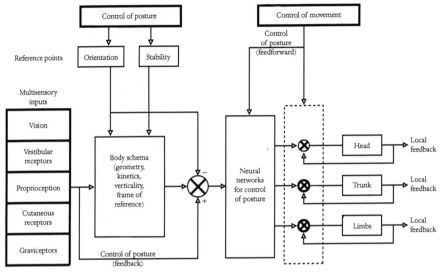

Figure 11.2. Coordination between posture and movement. At left, postural control is achieved based on frames of reference that indicate the desired orientation of the body or limbs and the degree of stability required. These data are sent to the parts of the brain that contain the body schema and are used to update it and to detect discrepancies between the assumed posture and the value required for the task. The sensory data not only trigger postural readjustments by the reflexes, they also contribute to changes in the body schema, which is the principal determinant. Making a move *(right)* results in anticipatory adjustment of posture (feedforward).

also found in animals,[28] are involved in manual tasks requiring coordination between the position of the arm and movement (for example, in lifting a load).[29]

The repertoire of natural motor synergies—all preceded by slight movements of anticipation that enable posture to be reorganized in advance, as it were, and not, as reflex theory predicted, to correct the difference between the ideal position and the desired position—is the best proof that instead of simply being the result of passive servomechanisms, posture is programmed into general planning for action (Figure 11.2).

This is also confirmed by the observation that when acrobats walk on their hands, using their arms rather than their legs to maintain their balance, the postural regulatory activities are immediately transferred to the muscles of their arms.[30] You might say that the control of balance obeys the law of motor equivalence. Motor equivalence refers to a simple and remarkable property of the brain that enables the same movement to be made using very different effectors. For example, I can write the letter *A* with my hand, or my foot, or my mouth. I can even draw a letter *A* while walking on the beach. This property fascinates neurobiologists specializing in motor function and is considered to be proof that the brain encodes a motor structure (morphokinesis) very generally, which then enables it to express the structure or to execute it using very different combinations of muscles. This property is also valid for maintaining balance. The transfer of postural programs from the legs to the arms demonstrates, once again, the global rather than local character of mechanisms of balance and posture.

Posture is thus not a passive state of reactions initiated by reflexes. It is a state of preparation to move based on internal simulation of predicted sequences of movement and general goals of action.

The Concept of Body Schema

What are the higher-level mechanisms of control of balance and posture? A clue to this mystery lies in a concept that is a little woolly but turns out to be surprisingly useful: that of the body schema. The neurologists Head and Holmes were among the first to suggest that the cerebral cortex contains what they called "a schema of the body" that directs posture and the coordination of movements.[31] According to them, the brain contains an internal model of the relative sizes of body segments, their relationships, and their positions. The body schema might also be the source of our perception of the body.[32] They situated the body schema in the parietal cortex, which, as Sherrington foresaw, assigned the brain a top-down rather than bottom-up control of posture.

This idea also shows up in the writings of Adrian, another of the great British physiologists: "In this way the organism carries in its head not only a map of external events but a small-scale model of external reality and of its own possible actions."[33] It is the word "possible" that is so unusual for the time: the body schema was not conceived as a *representation* of the body but as a *blueprint* of possible actions.

The Phantom Arm

Phantom arm syndrome is further proof of the reality of the body schema. People whose arm is severed either accidentally or surgically continue to sense their hand as if it were still there. They feel it move and experience aches and pains that existed before the amputation.

A situation similar to amputation is that of ischemia, that is, anesthesia of the arm. Subjects with this condition also perceive a phantom arm, whose position does not coincide necessarily with that of the real arm. Gurfinkel conducted the following experiment based on this condition.

Rest your right arm on the table and angle it about 45 degrees to your right. Close your eyes and imagine that your arm is anesthetized and that you perceive a phantom arm at 45 degrees to the left, that is, to the other side. Lay a pencil on the table in front of you and move it to the left: it represents your phantom arm (you will have to open your eyes briefly to arrange the pencil). Now bring your left hand, that is, the one that is not anesthetized, before you, resting your finger on the table, right in front of you. If you have followed the instructions correctly, your real right arm is to the right, the finger of your left hand is in the center, and the phantom arm (the pencil) is off to the left.

Gurfinkel asked his subjects to close their eyes and to displace their anesthetized arm toward the finger of the left hand; in your case, you will have to move your right arm toward the center. The subjects who perceived the phantom arm to the left made a move of the real arm toward the right; that is, this movement was dictated by the position of the phantom arm in space. This experiment shows that planning a movement is determined by the central representation of the position of the body in space and not by its real position as it is detected by proprioception.

The importance of the central representation of movement of the body and its domination by peripheral sensory factors is also illustrated by illusions of movement induced by the vibrations of muscles I described earlier.[34] For example, if the nape of the neck is vibrated, subjects perceive an illusory movement of the head. The postural reactions of the legs caused by this vibration of the nape of the neck are consistent with the illusion of movement of the head and not with the local motor reflexes caused by the vibration.

You can do another experiment that tests the same principle. Sit comfortably in an armchair. Turn your head to the right and keep it there, eyes closed. During the first few minutes, you will have the impression that your head is turned to the right, and an electromyographic recording of the activity of your legs would show a very slight asymmetry of muscle tone that reflects your postural asymmetry. After 8 or 10 minutes, if you have the patience to keep your head fixed to the right, you will experience the illusion that your head is gently returning to the center and, after another 10 minutes or so, you will have the distinct impression that your head has resumed its natural position, facing front. At the same time, the muscular activity of your legs will also regain symmetry. Here again, muscular activity follows the illusory perception of the position of the head. It is as if you have a phantom head that is looking straight ahead, even though your real head is turned to the right, and postural regulation is following the phantom head and not your real head.

Gurfinkel and his collaborators also used hypnosis to show the importance of the internal representation of the body. Hypnotized subjects were persuaded that their head was turned to one side, even though it was perfectly straight. At the same time that the illusory percept was established, muscular contractions were observed in the legs, of the kind that generally accompany a real rotation of the head. *The percept directed the postural activity.*

Similar results were obtained using vestibular stimulation. It is well known in neurology that pathways descending from the vestibular sensors and flowing through the vestibular nuclei ensure postural reactions that keep the body upright, stable, and in balance. They also initiate compensatory reflexes during tilting of the body. These are the so-called vestibulospinal reflexes. In principle, they are well-identified and depend on rigid neural networks that connect each semicircular canal with the postural muscles and those of the nape of the neck. It is thus reasonable to assume that in humans, stimulation of the labyrinth by application of a constant current (galvanic stimulation) placed close to the ear would induce highly stereotyped postural reactions. Yet these reactions also change completely based on the illusory position of the head.

It follows that the internal representation of the body in space is capable of altering the organization and direction of reflexes, which means that the whole definition of reflex itself must be reconsidered.[35]

INTERNAL MODELS AND BODY SCHEMA

Does the brain contain one or several body schemas? A long time ago, Russian physiologists of the Moscow school demonstrated that electrical pulse stimulation of the spinal cord of a frog, severed at the level of the neck, caused the frog's foot to move to very precise positions that varied depending on where the foot was positioned at the outset.[36] This suggests that a neural mechanism located in the spinal cord has available a model of the limb that enables automatic readjustment of the set of muscular synergies necessary to reach a goal. This internal model is an elementary form of body schema, for it includes both the geometry and dynamics of the limbs.

The second example is that of the cerebellum. It is currently being suggested that the cerebellum is an inverse model or a proactive model used by the brain to evaluate the consequences of movement and to create predicted trajectories even before they are executed.[37] In this sense, the cerebellum might thus contain unique elements of a body schema or a specific form of body schema.

A third example is found in the thalamus, whose multimodal neurons integrate sensory messages with motor activity.[38] In fact, the grouping that comprises the thalamus, cortex, and basal ganglia, which are interconnected in the form of circuits and traversed by sustained rhythmic activity (see Figure 1.3), might also constitute an internal representation of the body and its possible movements.

Finally, an even higher level is probably located in the parietal cortex, where the body schema as such is situated in the context of external space, whose properties are analyzed by vision. Head and others after him situated the body schema in the parietal cortex probably because it contains the only level of generality capable of directing motor behavior in an integrated way toward an external goal. It is most likely at this level that a *global* body schema is to be found, whereas in the medulla, the schema is *local.*

In all probability there is not one body schema but multiple body schemas, each adapted to a particular function, just as there are multiple representations of the body. This idea of multiple representations of the body was taken up again recently in connection with patients with impaired body recognition, or autotopoagnosia. This deficit is often associated with others such as aphasia, apraxia, disorders of manual grasping, and hemispatial neglect. However, autotopoagnosia is a specific impairment of body recognition. Some patients cannot indicate a part of their body verbally but can move it if the examiner touches it.[39] The same subjects have trouble referring to parts of an object external to their body. The impairment is perhaps not unique to body function but a general function of designating the parts of a whole. However, patients may have specific difficulties in indicating a "discrete image of the body."[40]

A recent observation also argues for multiple representations of the body.[41] A patient whose cerebral lesions were diffuse could neither localize the parts of her own body nor, moreover, those of other people, although she could name them. In other words, she could not localize them spatially, but she could identify them verbally. The authors of this observation suggest that there are at least four types of processes that enable one to represent one's body. The first processes *semantic* and *lexical* messages about the parts of the body; the second, *specific visuospatial* representations of a person's own body and also objects in the environment, that is, the location of parts in relation to each other (for instance, the nose is in the middle of the face, the ears are on each side of the head) and their boundaries; the third is a *body frame of reference* based on a body schema; and the final process involves *movements themselves,* which organize perception of the body.

Consequently, it is futile to seek a single biological reality underlying the concept of body schema. Gross reached similar conclusions concerning the frames of reference the brain uses to integrate sensory information (see Chapter 4). There are as many sensorimotor spaces as there are limbs. Given the extreme complexity of the construction of a single representation of the body, especially when it is in motion, how is this coherence assured? This is an important but still unanswered question.

Brain-Imaging the Body Schema

Recent findings obtained by brain imaging using PET elucidate the neural bases of the body schema.[42] Bonda and co-workers recently conducted the following experiment. Subjects were shown photographs of hands in a variety of postures in different orientations and were asked whether they were looking at the right hand or the left hand. This experimental protocol derives from cognitive psychology experiments on mental rotation, which showed that subjects respond more quickly the more natural a posture appears. If you try the experiment, for instance, by asking someone to show you his hands in various positions while hiding behind a screen, as in a puppet theater, you will be aware that you are making a mental rotation of your own hand to align it with the hand on display. Performing this task thus involves reference to the body schema.

The areas of the brain activated during this task are the following: the anterodorsal portion of the insula (which is not the retroinsular area, where vestibular activities occur), the superior parietal cortex, especially the left, at the point of the anterior parietal sulcus, and the motor cortex. Mesulam maintained that area 25, the insula, the cingulate cortex, the prefrontal cortex, and area 5 constitute a set of structures that contribute to the body schema.[43] Mishkin also suggested that the insula is involved in representations of the body and that there is cooperation between the superior parietal cortex and the anterior insula to produce what might be called a mental kinematics of the body.

I think that understanding the mechanisms of control of balance and coordination between posture and movement will in future require intense cooperation among mathematicians, physicists, neurobiologists, and neuropsychologists. The time when posture was conceptualized as nested reflexes is gone. The time when the idea of a body schema can be expressed in terms of precise neural functioning is not yet come.

Dog approaching another dog with hostile intentions.

Dog in a humble and affectionate frame of mind.

Figure 11.3. To the left is an aggressive dog; to the right, an affectionate dog waiting to be petted. Note that each posture both expresses an emotion and prepares for action: the posture of the aggressive dog is adapted to fighting, that of the submissive dog to being petted. This is also the position taken by a female in heat ready to receive the male, and it has been shown recently how this posture involves activation of a special synergy of the back muscles.

Relationships between posture and emotions also have to be taken into account. Darwin showed how different postural synergies are used to express feelings. The images of the aggressive dog and the affectionate dog in Figure 11.3 show how these expressive postures are also a preparation for action—in the first case, preparation to aggress, and in the second to be submissive and to be petted.

12

ADAPTATION

One of the most important properties of the nervous system is its ability to adapt. This term encompasses several mechanisms that complement one another.

The first mechanism is diminution of activity over time. For example, tactile receptors stop firing when pressure is maintained, which enables them to avoid overloading the nervous system with perpetual signals. When you are seated, you forget that your chair is supporting you, which is not a particularly useful bit of information while you concentrate on reading this book! Similarly, the membrane of motor neurons is endowed with adaptive properties that under the control of numerous neuromediators act to diminish their discharge in reaction to a sustained stimulus. The visual system is also equipped with mechanisms that suppress the transmission of stimuli when they are prolonged, and so on.

A second mechanism regulates the sensitivity of the sensory receptors. For example, in Chapter 2, I mentioned that the γ motor neurons control the activity of the neuromuscular spindles, thus modulating the intensity of sensory messages from the muscles and regulating the influence of proprioception on the motor commands for walking and balance. Another example is mechanisms of accommodation, which regulate the optic properties of the eye at a distance from objects, just as other mechanisms regulate the amplitude of the vestibulo-ocular reflex according to the distance from the object of regard. If I want to look at my nose, the reflex has to be cancelled; if I wish to keep my gaze fixed on a tree in the distance, my eye must make a displacement equal to and in the opposite direction from that of my head.

A third mechanism involves changing strategies. For example, when I am on a bus, I can respond to a slight braking with a simple contraction of my leg muscles; but if the braking is abrupt, I respond by moving my foot forward. If the target I am following with my eyes is moving too rapidly for ocular pursuit, I follow it with saccades. When an astronaut no longer has gravity to rely on as a reference, he employs his own body as a frame of reference and uses vision to anchor his moves in the space station, and so on.

The major physiological theories on adaptation all take into account this diversity of solutions and considered phenomena of plasticity at the cellular level, as well as higher-level mechanisms like functional substitution that involve reorganization of behavior as a whole.

I will now examine a few examples to give an idea of the state of the art of research on this subject and to show how the brain's predictive functions come into play in adaptation. Though this chapter barely touches on this immense domain, it will at least contribute some additional information. The brain possesses mechanisms enabling it to choose solutions that anticipate the results of action, and that *activity* is a necessary tool for testing these assumptions. Adaptation to disturbances of our senses, to sensorimotor conflicts, or to lesions triggers strategies that bring to light the underlying processes, which alter local mechanisms through global direction of the organism.

Adaptation and Substitution

The example that I propose is a familiar one. If you wear glasses, you will have noticed that when you first put on a pair that magnify slightly, you need some time—perhaps several days—to get used to this new vision. Hunters know that it is practically impossible to walk while looking through binoculars, even if they only magnify twofold. And if you are a lover of opera and theater, it does not take you long to figure out that if you do not want to lose sight of the actors while looking at them through opera glasses, you have to keep your head still. Why? Very simple—because the amplitude of the vestibulo-ocular reflex (see Chapters 2 and 4) that stabilizes the image of the world on the retina is, normally, exactly equal to that of the movement of the head. When an error is introduced by glasses or binoculars, the brain has to automatically adjust the reflex to the new speed of visual motion. It detects the error, then implements neural mechanisms that adapt the reflex to this new situation. Two extreme conditions allow experimenters to probe the limits of this mechanism.

The first consists in placing a subject, human or animal, on a revolving chair inside an illuminated cylinder that moves with the subject. Under these circumstances, the visual world is stable, and the reflex must be cancelled. In the other experiment, subjects are asked to wear prism goggles called Dove prisms that reverse the direction of visual movement. When the head turns to the right, the visual world, instead of turning to the left as it would under natural conditions, turns in the same direction as the head. The brain not only has to adjust the amplitude of the movement of the eye but also to reverse the direction of movement. The entire anatomy of the vestibular neurons whose remarkable arrangement I described in Chapters 2 and 4 is useless in this case. The creation of millions of years of evolution has to be undone. How does the brain do it?

At the beginning of this century, psychologists and ethologists showed that chickens could quite easily adapt and peck at their grain or earthworms while wearing prism goggles. In humans, wearing prisms induces profound transformations that are adaptively reversible. Gonshor and Melvill-Jones conducted a critical experiment that revealed this surprising adaptability of the central nervous system.[1] They had a subject wear prism goggles for three weeks and tested his vestibulo-ocular reflex every day. The tests were conducted in the dark, without allowing the subject any visual cues regarding the actual movement of his head. After five or six days, the reflex was reversed. In other words, an adaptive mechanism had totally altered the functioning of this remarkable wiring.

A good dozen laboratories all over the world have been working for 30 years to understand the mechanism of this adaptation, which has become a model for the neurobiology of sensorimotor plasticity. An experimental fact very soon helped to focus the theories: a lesion of a small area of the cerebellum—the flocculus—prevented this adaptation in animals (see Figure 2.4). Why the flocculus? The answer came from anatomy and neurophysiology: projections of the vestibular receptors reach the Purkinje cells of the flocculus via the mossy fibers of the cerebellum and are combined with projections from the accessory optic system described in Chapter 3. The flocculus is thus the center for a visuo-vestibular convergence. What critical event happens at this level to determine adaptation? What happens is that the flocculus projects to the neurons of the vestibular nuclei, which are involved in the reflex.[2] This is an inhibitory projection, so it is able to modulate the amplitude of the reflex. You can experiment with this modulation yourself. Hold a finger in front of you and oscillate your head horizontally from one side to the other, moving your finger with your head so that your finger follows your head's movement exactly. Under these conditions, you will notice that your gaze is fixed in the socket. There is no longer a vestibulo-ocular reflex. Patients with lesions of the cerebellum lose this visual inhibition of the vestibulo-ocular reflex.

In other words, the flocculus is part of a reflex control loop modulated by stimuli that are both visual and vestibular. Ito subsequently suggested the so-called flocculus hypothesis.[3] He proposed that a synaptic mechanism detects the difference between visual and vestibular inputs at the level of the Purkinje cells and, based on this error signal, alters their influence on the reflex.

Ito later refined his hypothesis based on a mechanism he called "long-term depression," caused by the coincidence of two different synaptic events (those associated with the climbing fibers and the mossy fibers, which are the two main groups of neurons that convey visual and vestibular information to the cerebellum) that initiate this alteration. Very fine molecular mechanisms underlie this adaptive capacity, which Ito used as a general model of motor training. But the model was controversial. Some saw motor training as a more complex function, not localized to a single synapse or a single long-term suppressive mechanism. Others situated the critical synapse for adaptive changes not at the level of the cerebellum but at the level of the brainstem, where specialized neurons ensure its regulation: according to this conceptualization, the cerebellum is also implicated in alterations, but as controller.[4] Adaptation thus operates on at least two levels. Whatever their differences, these theories assume that adaptation is the product of neurons situated within the networks that control the reflex.

But there is another way of thinking about adaptation that leads to a completely different theory: with Melvill-Jones, I proposed that instead of reversing the vestibulo-ocular reflex, the brain replaces it with a reverse pseudo-reflex using another element of the oculomotor repertoire, such as saccades or pursuit.

To prove this theory, we repeated the original experiment together, asking subjects to wear Dove prism goggles. After putting the goggles on, the subjects succeeded within several hours, and sometimes immediately, in producing an adaptive movement composed of a sequence of saccades that gradually merged to approximate a vestibular reflex.[5] The brain had suppressed the vestibulo-ocular reflex via the mechanism described above involving the flocculus, instead using the saccadic system to construct a new adaptive movement.

We hypothesized that this pseudo-reflex was guided not by visual targets but by an internal simulation of the movement of targets; that is, by a percept that directs adaptation from top to bottom just as the body schema controls posture. A high-level mechanism, calling into play the cortical structures involved in the representation of space and motor imagery, appeared to be responsible for this adaptation. We reasoned that the physical presence of visual targets was not necessary to achieve adaptation, and that simply imagining the target might be enough. Consequently, we asked subjects to sit in a revolving chair that oscillated and to imagine that they were focusing on a target attached to their nose. The conflict they experienced had to do with the fact that their vestibular system was telling their brain that it was turning, thus causing a vestibulo-ocular reflex that inhibited the very intention of fixing their regard on the target attached to their nose. After several hours of this exercise, we observed that the vestibulo-ocular reflex had changed. A simple internal mental effort—subjects keeping their gaze focused in front of them during a rotation of the head in the dark—was enough to cause adaptation.[6] Here, as for postural control, adaptation is caused by a percept of movement worked out in the areas of the cerebral cortex involved in the representation of space and not by low-level reflexes. We have recently shown that there is a great difference between men and women in their adaptation to a visuo-vestibular conflict. We have exposed both normal and anxious subjects (the anxiety level was judged by standard psychiatric tests) to a conflict by turning their bodies by a certain amount and turning the visual world in a virtual reality helmet by a different amount. We tested their vestibular perceptions before and after the exposure to the conflict and observed that men adapt their vestibular perception more than do women under these circumstances (or women remain more independent). Also, anxious men adapt even more. We are performing these studies on anxiety because I believe that part of the mechanism underlying spatial anxiety is related to the difficulty of building a coherent perception of the relation between the body and space and difficulties in coping with incongruous sensory inputs.[7]

The brain thus employs various mechanisms to adapt. Some are low-level and promote neural plasticity; others are high-level and, especially in humans, rely on remarkable capacities for functional substitution, based on internal simulation of the signals to be controlled. In other words, the brain is not a servo system for regulating variables; it is a composer that orchestrates the instruments at its disposal. The hierarchy of nerve centers developed over the course of evolution enables the brain to find new solutions when adaptation is required. Low-level mechanisms then stabilize the new solutions. In fact, brain imaging has revealed that during motor training, the cerebral cortex is used at the beginning of the training and gradually becomes silent; the activity is then transferred to the basal nuclei and to the cerebellum, which leaves the cortex free to confront new problems and find new solutions to them.

For all these reasons, inter- and intraindividual differences must regain the central status they have lost in the behavioral sciences and neurosciences. If the brain does have this ability to choose solutions, we must vary our efforts to find one or several solutions, rather than focusing on typical behavior, as we have done. A diversity of strategies better clarifies the mechanisms at work than does an average profile. However, it must also be recognized that the number of possible solutions is not infinite. It is constrained by the genetic legacy of sensorimotor subsystems. To what extent it is possible to escape simple selection to achieve genuine invention is a fine subject, but it is a matter for another discussion.

The Rheumatologist and the Ophthalmologist

Once upon a time, in Lisbon, there lived a rheumatologist whose patients were suffering from back pains. He noticed that they had ocular disturbances too. In the same city, an ophthalmologist was also attending to patients with back pains. These two doctors shared a fondness for wine, so much so that they frequented the same wine-tasting establishments specializing in fine Portuguese wines. During one oenological meeting, they were discussing their overlap in presenting syndromes and agreed to exchange patients. Over the years, they came to the conclusion that some of their patients were suffering from the same illness. Indeed, if these patients were asked to close their eyes and to position their arms horizontally, they placed their arms at different heights; if they were asked to close their eyes and to position their feet one next to the other, exactly parallel, the doctors noted that the patients maintained a considerable gap between their feet; if the patients were asked to walk in a straight line, the doctors observed that the patients staggered. Ophthalmological tests revealed significant asymmetries in the perception these patients had of the position of objects in three-dimensional space. Generally, they had a distorted perception of their own body and its spatial localization.

Da Cunha, the rheumatologist, identified what he called a "syndrome of disorientation" and hypothesized that this syndrome was caused by a distortion of the body schema, which might explain the constellation of postural oculomotor symptoms, and possibly even gastric or cephalic symptoms. He achieved spectacular remissions by a combination of having patients wear prism goggles, under ophthalmological supervision, and of motor reeducation aimed at making subjects aware of the position of their own bodies; for example, by making them move in front of a mirror. Electroencephalography confirmed his theory, revealing asymmetric activity of the parietal cortex in certain patients. The asymmetry disappeared with reeducation.

Such a theory would have been inadmissible even a few years ago, and these doctors would have been charged with quackery, so rigid was the conception of the brain. The possibility that the cortical structures involved in sensorimotor function are accessible to training was pretty much unthinkable. Recently, however, neuronal recordings in the monkey and in humans using brain-imaging techniques have shown that neurons from the somatosensory areas of the cerebral cortex can reorganize the maps where these parts of the body are represented.[8]

A similar discovery was made recently by Rousié, a physician from Lille who is attached to the maxillofacial surgery clinic of a university known for expertise in oculography, anatomy (especially of the vestibular system), and neurology. This physician noticed that many patients who came to consult for facial asymmetries also presented symptoms that appeared to indicate a more general etiology than simple malformation of the bones of the face or jaw. Together with several specialists, we designed a complete battery of postural, vestibular, ophthalmological, and oculomotor tests, and proposed that a basic vestibular asymmetry was the source of all the disturbances. Dr. Rousié ordered radiological examinations, which did indeed show a functional asymmetry of the vestibular receptors in these patients.[9] These vestibular asymmetries caused a perturbation of the body schema that was compensated for by other postural or oculomotor asymmetries. The local symptoms are influenced by this global perturbation of a patient's perception of his body.[10]

The Role of Activity in Compensating for and Preventing Disorientation

It is well known that following a unilateral operation on a vestibular organ, patients are often afflicted, albeit transiently, by postural disturbances, abnormal ocular movements, vertigo, and general spatial disorientation. When they are walking along the street, they are often confused, and do not know where they are. These symptoms generally disappear after several months in a process called compensation. The nature of these compensations is still not well understood, and they are as likely to involve low-level local mechanisms affecting the structures of the brainstem as higher-level mechanisms of cortical or limbic structures associated with spatial orientation. It may be that the disturbances caused by these vestibular asymmetries do not just affect the vestibular nuclei but also affect the structures of the cortex and the hippocampus that reconstruct movements of the body in space. In certain patients, these disturbances never completely disappear.

How do we reeducate these patients? Several practitioners tried to subject them to optokinetic stimulation while standing upright, projecting points of light into a room in a continuous stream. Surprisingly, exposure to visual scenes in motion diminished the symptoms in a fair number of cases. Indeed, the result raises many questions, for it only takes a few intermittent and brief exposures to decrease a postural susceptibility to a visual stimulus. Whereas the posture of patients subjected to this stimulus for the first time is highly perturbed, after a month they can stand without difficulty. Moreover, they no longer feel disoriented walking around in a natural environment. We measured eye movements caused by the visual movement and discovered that they revert to symmetry after reeducation.[11]

I think that no assumption of local action on the neural mechanisms of each subsystem resulting from this brief stimulation can explain this absolutely spectacular therapeutic effect. A global theory is needed. When subjects are standing in a room in which the visual world is moving in a complex way (if the points of light are distributed over all the walls and moving in various directions, for example), they are completely disoriented. They are in an unnatural situation that obliges their brains to find a solution to the conflict between vision, which tells them that their bodies are falling in several directions simultaneously, and the vestibular system and proprioception, which tell them that their bodies are motionless.

One solution is to reduce retinal slip; that is, to force the optokinetic nystagmus in a way that the eye constantly follows the visual movement, with no resulting visual perception of bodily motion. In this case, because the image of the world is perfectly stabilized on the retina, the body only has to worry about the information supplied by all the other sensory receptors to regulate posture. This is probably the strategy used by the brain, since the amplitude of the nystagmus is controlled and, at the end of training, the eye manages to follow the scene very well. This specific training of the optokinetic nystagmus might also plausibly be used by subjects when they move in a total visual environment, such as in the street or in natural surroundings, and would thus help in stabilizing the image of the world on their retinas. At issue is an effect on one of the subsystems responsible for postural control and spatial orientation. A second strategy might be to de-emphasize the importance attached to vision and narrow the focus and concentration of the subjects, as the flutist I mentioned earlier did (see Chapter 11).

Owing to the work of Igarashi and Lacour, it is well known that motor activity is fundamental in compensating for disturbances associated with vestibular processes. It is probably what enables the nervous system to find ways of adapting and to discover new solutions. A supporting argument was provided by a recent space flight. An astronaut flying for the first time was particularly sensitive to motion sickness in space. He was determined not to be sick in space and trained his brain—for example, using the revolving armchair that obliges astronauts to tilt their heads, a reliably nausea-inducing stimulus—to identify the conflict but to disregard it. Carrying out this cognitive distancing from an elementary mechanism is a very interesting process of adaptation, and one that has not been much investigated. It is believed that it is a product of mental manipulations in space. Several astronauts have stressed that after several days in flight, they could quickly change what they perceived to be the floor and the ceiling. This flexibility, this ability to decide frames of reference seems to accompany the disappearance of symptoms of motion sickness in space. One astronaut who does not get sick told us that she "decided to believe what [she] saw" and not to believe what the disoriented part of her body was telling her. This is why it is necessary to construct what I call a cognitive theory of adaptation to sensory conflicts.

The intervention of high-level cognitive mechanisms in adaptation to sensory conflicts is really a problem of frames of reference. As I discussed earlier, alterations in the body schema can cause serious disturbances. Similarly, an abrupt change in frame of reference can lead to difficulties. During a recent space flight, an astronaut reported to us that she was losing track of everything she set down in the space station, even though she was able to handle objects without difficulty. The trouble she had remembering where things were supports the idea of a dissociation between the brain's use of allocentric encoding of the identification of objects and of their place in the environment and an egocentric encoding of handling. Several studies in microgravity show that in the absence of a gravitational reference point, astronauts tend to revert to an egocentric frame of reference. All manipulations and perceptual actions (to use Janet's terminology) in this frame of reference are thus intact, even when those that require allocentric encoding have deteriorated. To recall where an object is in the space station requires constructing an allocentric representation of the space station; but to recall it for the purpose of picking it up, the two systems of encoding have to be integrated.

The ability to move from one frame of reference to another and to combine these frames of reference are probably among the most important aspects of adaptation to microgravity as well the basic elements of reeducation following lesions of the vestibular system.

13

THE DISORIENTED BRAIN: ILLUSIONS ARE SOLUTIONS

> Here we put our finger on the mistake of those who maintain that perception
> springs from what is properly called the sensory vibration, and not from a sort
> of question addressed to motor activity.
>
> —H. Bergson

In this chapter, I maintain that perceptual illusions are solutions devised by the
brain to deal with sensory messages that are ambiguous, or that contradict either each other or the internal assumptions that the brain makes about the external world. An illusion is generally held to be a sensory error, as in this text
by Sully, written in 1883:

> We see, then, that in spite of obvious differences in the form, the pro-
> cess in all kinds of immediate cognition is fundamentally identical. It is
> essentially a bringing together of elements, whether similar or dissimi-
> lar and associated by a link of contiguity, and a viewing of these as con-
> nected parts of a whole; it is a process of synthesis. And illusion, in all
> its forms, is bad grouping or carelessly performed synthesis. This holds
> good even for the simplest kinds of error in which a presented element
> is wrongly classed; and it holds good for those more conspicuous errors
> of perception, memory, expectation, and compound belief, in which
> representations connect themselves in an order not perfectly answering
> to the objective order.[1]

I suggest that illusion is not an error or a bad solution, but rather the best
possible hypothesis.

Illusion: The Best Possible Hypothesis

What do I mean by "hypothesis"? In the Introduction, I put forward the idea that the brain is a generator of hypotheses. This idea is not exactly new. Indeed, Lashley wrote: "There are many indications that . . . in the discrimination box, responses to position, to alternation, or to cues from the experimenter's movements usually precede the reaction to light and *represent attempted solutions* that are within the rat's customary range of activity."[2] Krechevsky had the following things to say, also on the subject of rats:

> To the animal any new situation is not a confused, meaningless conglomeration of sensory impressions to which he makes confused, meaningless uncoordinated and unrelated responses. The animal is not altogether a victim of his immediate environment in the sense that each specific reaction is the result of a specific, momentarily-acting stimulus. He brings to each new situation a whole history of experiences. These experiences the animal is ready to apply . . . Such responses, "false solutions," "early systematic attempts," etc., we have dubbed with the dubious name of "hypotheses." The use of the term "hypothesis" to describe such behavior is a confession of failure. We have been unsuccessful in finding any one term which might adequately describe such behavior yet not carry with it the stigma of being fantastically anthropomorphic . . . "Hypothesis" carries with it all the mentalistic implications that have become associated with such words as "reasoning," "consciousness" and many others. "Hypothesis" as we wish to use it in this instance, however, need not and does not do so.[3]

Krechevsky continues, specifying that what he means by "hypothesis" is a situation in which the animal exhibits multifaceted behavior: it is systematic, goal-oriented, to a degree abstract, and not entirely dependent on the immediate environment, either to be undertaken or to be carried out. Finally, he adds: "A hypothesis is a person's interpretation of data and not a phenomenon that derives from the data themselves." Hypothesis is thus inference, to use Helmholtz's terminology.

Here are a few examples. The published literature includes an immense quantity of articles and reports about visual illusions, but most of these are static. In line with the theme of this book, I will focus on illusions connected to spatial orientation and the sense of movement.

Some visual illusions are connected to vestibular perceptions of bodily orientation.

The Moon Illusion

On a night when the moon is full, it seems larger at the horizon than at its zenith. This is a perceptual illusion. If you are not convinced, wait for a time when the moon is near the horizon and try the following simple experiment, which requires no special equipment. If you are at the beach, it is very easy to do; if you are in the city, it is better to go up to the roof of the building where you live. In either case, the exercise is the same: Turn your back to the moon and look at it from between your legs by bending forward and looking at it upside down. You will be amazed to see that it will seem to have gotten smaller. I will leave you to your amazement. What might explain this illusion? It is possible that the otolithic receptors play an essential role in this perceptual asymmetry. Indeed, the apparent size of the moon, such as it is perceived, is not a datum linked exclusively to its objective size on the retina, which can change, for example, according to the density of the atmosphere. It is commonly known that the diameter of the moon is slightly greater on foggy days, when it is nearer to the horizon, than when it is at its zenith. It is likely that a basic perceptual asymmetry related to our perception of top and bottom has something to do with this illusion. But it has also been proposed that the illusion is connected to the non-Euclidean character of distant visual space.

So the position of our body in space can cause changes in the apparent size of objects as the result of visual and gravitational factors. In this case, the illusion is not a solution, it is a sensory error. As Poincaré said, representative space is not identical to geometrical space.

Dynamic Illusions

Another illusion—this one a little more difficult to demonstrate—concerns the change in the apparent position of a point of light in space. To do the experiment, you will have to stand on an enclosed rotating circular platform or a centrifuge and look outward, turning your back to the center of rotation. Under these conditions, the vestibular receptors are subjected to a centrifugal acceleration that goes from your ears to your nose. Now, if the lights are turned off and you are told to follow a little point of light that moves along with you in the compartment, you will suddenly have the impression that when the platform or centrifuge starts up, the light is moving up or down according to

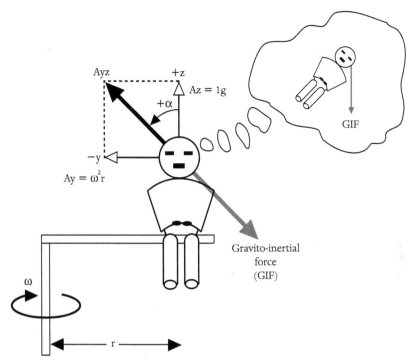

Figure 13.1. Illusion of body tilt on a centrifuge. When the centrifuge turns with an angular velocity ω, the acceleration of gravity Az (directed upward and leading to a downward gravitational force) and acceleration due to rotation Ay (directed toward the center and creating a centrifugal force) combine to produce a gravito-inertial acceleration Ayz tilted at angle α. This acceleration causes a resultant gravito-inertial force that the brain interprets as the true vertical.

the direction of the acceleration. If you are shown an illuminated horizontal bar, you will see it go up or go down as well. As I mentioned in Chapter 3, this effect of linear acceleration on visual perception is probably connected to the existence of vestibular projections on the visual cortex. The stability of the perceived world thus depends on the vestibular sensors and their detection of gravity. On the centrifuge, the apparent direction of gravity is changed, because the gravitational vector, which is vertical, is added to the centrifugal force: the sum of the forces of acceleration is a vector that is tilted in relation to the axis of the body. Subjects who are seated facing the centrifuge, as in Figure 13.1, have the impression that the vertical is tilted. If the subjects are in the dark, they may have the illusion that their bodies are tilted, because they assume that the vertical really is vertical. The illusion is both a solution and a problem.

Another way of experiencing an illusion whose source is otolithic is to stand on a turntable whose axis of rotation is tilted between 20 and 30 degrees.[4] The table is then rotated at a constant velocity. After 20 to 30 seconds, stimulation of the semicircular canals ceases, leaving only stimulation of the otoliths by the gravitational vector, whose projection to these receptor elements turns with the rotation. You would then perceive yourself propelled by a strange conical movement. The perceived movement corresponds precisely to what physiological examination of the otolithic receptors predicts. The brain thus constructs a perception in accordance with the aggregate of sensory inputs.

The same would be true if you were to sit on a revolving chair, like a barber's chair, in total darkness except for a tiny visual target in front of you, attached to your head. When the chair is halted, you will see the target turn to the left or the right in the horizontal plane. This illusion is due to the fact that after the stool stops moving, the semicircular canals remain activated, and an illusory movement is perceived by the vestibular system. I have shown that the vestibular signals also activate visual and multisensory structures such as the parietoinsular cortex. But neither the other proprioceptors (muscular, tactile) nor the other sensors signal movement of the body. When faced with this activation of multisensory cortical structures, the brain chooses—decides—that the perceived movement is due to the visual world moving.

In reality, then, these illusions are solutions that the brain has found to interpret a relationship between a perceived change in the vertical subjective and maintenance of a visual stimulus.

The Cemetery Illusion and Other Illusions Connected to Aerial Acrobatics

I have just described two illusions related to changes in the apparent direction of gravity under conditions of motionlessness (the illusion of the moon) and acceleration. Under other conditions (for example, airplane flight), accelerations combine in a much more complex way, and this leads to illusions that can have very serious consequences. In Chapter 2 I mentioned that on takeoff, the pilot of an airplane can have a false impression of bodily orientation, connected to ambiguous messages from the otoliths (Figure 13.2). In the case of more complex configurations, a whole catalog of illusions has been described—especially by the American military—that have led, and still do, to dangerous accidents.

For example, the inverse illusion occurs when a plane climbs rapidly and then gradually regains the horizontal. You can feel minor effects from it in a

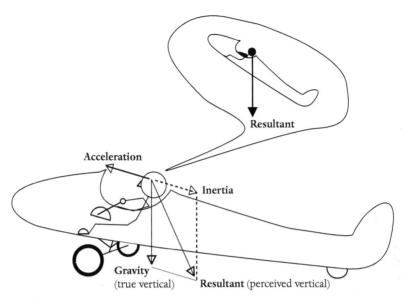

Figure 13.2. The ambiguity of otolithic messages can cause accidents. On take-off, an airplane simultaneously accelerates and inclines. When takeoff occurs in the dark or in fog, the pilot has only vestibular and tactile information to rely on to detect the pitch of the airplane. The otoliths are stimulated by two accel-erations: gravity and the forward motion of the airplane. The two component forces (gravitational and inertial) combine in a resultant force that is perceived by the brain as gravity (in the absence of any other visual cue to indicate up or down). Thus, the pilot believes that his airplane is inclining very sharply. He compensates for this illusory pitching, and the airplane crashes at the end of the runway. Fortunately, modern instruments provide pilots with a measure of the true pitch of the plane.

jet plane when it has finished its climb and suddenly levels off. Owing to the plane's circular or parabolic trajectory, centrifugal force combines with gravity to produce a force oriented upwards. Under these conditions, the pilot, who had the impression during the climb that he was sitting down, his bottom pointing toward earth, suddenly has the illusion that he is upside down. This illusion is also experienced in space flights: when gravity disappears, it is as if there were an acceleration in a direction contrary to normal gravity. The brain interprets the absence of gravity to be the result of acceleration in the oppo-site direction, and the illusion of inversion felt by the cosmonauts is very often associated with serious symptoms of motion sickness.

Another illusion, called the cemetery illusion because it has been the cause of numerous catastrophes, may affect pilots in airplanes performing acrobatic maneuvers descending in a helix. After several helical turns, the pilot loses the sensation that he is turning. Indeed, if the angular acceleration is constant, the sensations from the semicircular canals abate. When the pilot stops his rotation, he receives a vestibular stimulus due to the combined effects of angular deceleration and gravity that gives him the impression that he is turning in the opposite direction, which leads to fatal accidents. This illusion, which occurs in exceptional conditions, also occurs under more normal conditions when, for example, the pilot makes a simple circle in a plane flying horizontally for a fairly long while. After a time, the messages from the canals disappear, and at the instant the plane resumes a straight-line trajectory, the acceleration caused by straightening out gives the pilot the mistaken impression that he is inclining in the direction opposite to that of the circular trajectory. He inclines the airplane further to correct it, which is obviously very dangerous.

Illusions of Movement of the Limbs

KOHNSTAMM'S ILLUSION

An illusion that is very interesting because it causes observable movement was described for the first time by a physiologist, Kohnstamm. The illusion is evidence of a sort of motor memory, and it seized the imagination of neurophysiologists. Try the following experiment: Stand upright next to a wall, in profile, the back of your hand against the wall and the palm of your hand against your trousers (or skirt). You can also do the same experiment without standing next to a wall, but leaning the back of your hand on the edge of the table. Now, applying as much force as you can, push the back of your hand for 2 or 3 minutes against the wall or the table, as if you wanted to push it away from your body. What you are doing is applying a sustained tonic force involving the extensor muscles of your arm. At the end of this time, move away from the wall or the table, and relax. The experiment can be done entirely with your eyes open, but it is better to do it with your eyes closed. You will feel your arm lift all by itself, as if it were pulled by an invisible thread, and remain elevated for an instant. This illusion tells us something about the basic mechanisms of postural control. It could in fact be called involuntary postcontraction.[5]

The experiment can also be done with the feet, by sitting in a chair and pressing very hard against a wall or a table. The illusion only results when the muscle making the contraction is a proximal muscle—one that belongs to the postural system. Kohnstamm's phenomenon cannot be induced using the distal muscles of the hands or the feet. For instance, if you hang a 3-kilogram weight from your foot and let the weight drop abruptly, the involuntary contraction appears in the quadriceps (thigh muscle). Similarly, if you hang a 3-kilogram weight from your hand for 5 minutes, the involuntary contraction phenomenon is induced in the biceps muscle, not in the hand. Thus it is a motor reaction of the postural muscles. Why is this reaction the result of an illusion? Why is it the solution to a problem of the nervous system?

The interpretation of this phenomenon is the following: During the phase when your hand is pressed against the wall, you are exerting a constant force on it. The brain adapts to this situation in which motionlessness is accompanied by a constant muscular effort. When your hand pulls away from the wall, the brain continues to apply this force, because the brain reads the force as motionlessness, up to the point where the stretch receptors in the muscles signal clear movement.

In other words, while force is being exerted, the brain learns that motionlessness is accompanied by motor activity. Once the obstacle is removed, it still believes this to be true.

DRAWING WITH VIBRATIONS

At the beginning of this book (Chapter 2), I touched on illusions of bodily self-motion induced by the vibration of tendons, and I considered the interpretation proposed by several physiologists: activation of muscle receptors causes a sensation of movement that simulates an actual movement. A recent series of experiments went beyond this simple interpretation, showing that the configural vibration of the muscles of the hand can cause design illusions with complex geometric shapes. Little vibrators were applied to pairs of muscles that normally control hand gestures. For example, take a pencil and draw a circle, or a square, or a triangle: the activation sequences of the muscles of the fingers are very different depending on whether you draw one or the other. Vibrating the same muscles in order of activation stimulates the sensory receptors that follow different sequences corresponding to various geometric shapes.[6] Remarkably, a subject thus vibrated, who has not moved his hand, will have the illusion that he is drawing the corresponding shapes. Subjects are genuinely convinced that they have moved their fingers and are very surprised by the sensation, because they did not intend to move them. As in the case of vection, this nonintentional illusion of shape generation is generally felt as a very agreeable experience.

These results can be interpreted in the following way: the parameters of the illusory sensation can be expressed at any moment by a vector whose origin coincides with the extremity of the limb and whose direction, in the case of a curvilinear shape, is tangential to the trajectory of the drawing (a turning vector). The radius of curvature of the illusory drawing is determined by the velocity with which the frequency of the vibration applied to the two covibrated muscles increases or decreases. For example, when the change in frequency is stable and the same vibrational profile is applied to all successive pairs of muscle groups of the wrist, the sensation is that of a circle. When the change in frequency varies from one pair of muscles to another, the sensation is that of the shape of an ellipse, and so on. Another important aspect of this illusion is the fact that subjects recognize real geometrical shapes (triangles, ellipses, circles); in other words, shapes in the Gestalt sense and not simply outlines of movement.

But, as the authors of the experiment write,

> it is also worth mentioning that, in addition to perceiving illusory movements in response to proprioceptive stimulation, the subjects were able to recognize and to describe accurately the motor forms elicited by the various vibratory patterns used. This means that the various instantaneous kinesthetic sequences must have been memorized and linked up into identifiable geometrical figures belonging to a repertoire of possible geometrical shapes. It therefore emerges that muscle proprioception is able to generate spatiotemporal inputs that may subserve complex cognitive operations such as those involved in the memorizing and recognition of motor forms, including symbolic actions performed with the human hand such as writing or drawing.[7]

These results are important for us because they confirm the revival of the motor theory of perception. They reinforce the concept of a prespecification of sensory inputs during control of movement. Finally, they validate the idea that illusions are solutions resulting from an endogenous repertoire of motor or perceptual forms with which configurations of sensory inputs are compared.

Space and Motion Sickness

Everyone has had the misfortune of experiencing the disagreeable symptoms of motion sickness in a car, on a boat, (infrequently) on a train, and even at the movies. What mother has not told her child at least once, "Stop jumping around like that! You're making me queasy"? Motion sickness is thus connected to movement, whether visual or corporeal. Astronauts are susceptible to it, and for most the only relief is medication.

The mechanism of this peculiar bodily reaction is still not known. One interesting hypothesis that so far has not been rejected is called sensory conflict. It explains motion sickness as a conflict between the sensory messages supplied by the visual and vestibular systems (and possibly the graviceptors of the abdomen) about movements of the head. More generally, this symptom is produced by an incongruity between the messages relayed by the sensors of orientation and those of movement. For example, the visual system of a child sitting in the back seat of a car or a sailor below deck detects no movement, whereas the vestibular and proprioceptive systems do detect movement. Similarly, in the case of complex movements of the head that generate so-called Coriolis accelerations, the brain cannot establish coherence between accelerations detected by the vestibular system and those indicated by vision. In space, it is the failure of the otoliths to detect tilting of the head owing to the absence of gravity versus the ability of vision and the other sensors to detect it that is the cause of sickness, although cardiovascular reorganization may also contribute.

Thus, difficulty in reconstructing a coherent perception of movement and orientation is responsible for central nervous system activity leading to successive stages of motion sickness, from drowsiness to nausea. We still do not know why this incoherence manifests vegetatively as it does. It has long been known that dogs whose flocculus (the portion of the cerebellum that integrates visual and vestibular information) has been removed do not get sick. So it is possible that low-level mechanisms are responsible for motion sickness. But I would like to propose another hypothesis.

The hippocampus contributes to the memory of configurations of sensory information and others connected to action, and at this level visuovestibular convergences are probably treated on a global level. For example, in the monkey the neurons of the hippocampus discharge during movements of translation and rotation, but only when the animal is in a certain region of the room, or is heading toward a door.[8] It is thus possible that motion sickness is caused by much higher level mechanisms involving the entirety of the internal system of simulation of global movement of the body in the space constructed by the hippocampus, the prefrontal cortex, and the parietal cortex. Problems in establishing coherence might lead to vegetative effects whose origin is cortical.[9] This hypothesis explains why training plays such an important role in preventing motion sickness in space and, among other things, why it can be controlled by exercises that mentally manipulate orientation, as I mentioned at the end of the previous chapter. It is also possible that the two mechanisms exist concurrently, which would explain why this malady is so resistant to study and why it is so difficult to devise countermeasures.

If you wish to avoid motion sickness, do not read on winding roads or when the driver brakes abruptly when you are riding in a car. Focus on the distant landscape so your visual system has the same reference as your vestibular system. And when you are on a boat, go up to the deck and look at the horizon: this provides your vision with the exact same reference as the vestibular system. Once on the deck, make active movements to remain stable, for these motor commands will add their messages to the other sensory information. And if you are still feeling queasy, you may console yourself by recalling that Admiral Nelson was seasick when at sea for his entire life!

A Few Other Illusions

THE SAILORS' ILLUSION

There is another illusion that sailors feel after several days at sea. When they return to port, they have the impression that the ground is undulating in the same way as the sea. This illusion reflects a remarkably flexible solution that explains a lot about how the nervous system adapts. When the illusion occurs, it means that populations of neurons are animated by a rhythmic modulation in the centers that detect movement (the vestibular nuclei, vestibular cortex, and so on). How is this memory of movement constructed? It isn't known for certain, but it is assumed that at sea the brain establishes a dynamic modulation in phasic opposition (antiphase) to the undulations of the water. Thus, when the head is accelerated, this endogenous wave causes the activity in the centers of movement to diminish. The result is a sensation of stability. And, in fact, when you are on the water for several days, you eventually forget the motion of the boat. The same phenomenon is produced when you are walking: after some time, you no longer have the impression that the visual world is unfolding so rapidly. In contrast, when you stop after an hour of walking, you may perceive an apparent motion of the environment, a little like the famous illusion of the waterfall: when you look at a waterfall for some minutes and then you look round about, you perceive an apparent movement of the mountain. A dynamic memory of movement establishes a dynamic neural profile that stabilizes the perception and that, I think, enables the brain to free itself of this perpetual stimulation to be able to think about something else. Illusions that result from exposure reveal the solutions the brain has devised to adapt.

Vertigo and Agoraphobia: Illusions without Solutions

Vertigo is an illusion that has no solution. Fear of heights results when proprioception of the legs signals small oscillations to the brain but vision detects no movement, owing to the remoteness of the landscape. It is also due, in certain cases, to difficulty in processing visual information. Why are some people unable to resolve this conflict? No one knows, but I think that my hypothesis of a high-level mechanism involving the structures of the cortex or of the limbic system responsible for perception of the body and the environment is one worth pursuing.

The same is true for agoraphobia and the so-called panic attacks that come over people when they suddenly find themselves in wide-open spaces. A body of research indicates a possible disturbance of vestibular origin in these patients, and we are investigating these mechanisms in the laboratory. But the idea of a breakdown of the representation of space and difficulty in constructing coherence among multiple corporeal frames of reference and sensory messages also bears looking into. Panic attacks are a consequence of an inability to find a solution, even in the form of an illusion!

Illusions and Hallucinations

Illusion is a solution to an incongruity, to the loss of perceptual coherence. Hallucination, which is a creation of the brain, is a different story entirely.[10] Hallucination is not the result of sensations that the brain fails to integrate into a coherent perception but of the sudden combination of endogenous memories of perceptions. In some sense, hallucination is a waking dream, the autonomous functioning of internal circuits that normally work to simulate the consequences of action. Somewhere in the brain a purely internal activation initiates perceptual hypotheses. Like the painter of Lascaux for whom shadows suggested the images of the animals he hunted and that he projected onto the cave walls, here a pathological internal activity triggers the hallucination. Hallucinations are such that they are projected onto the world, exemplifying a fundamental property of the brain. Whereas illusion provides the solution to conflict, hallucination constitutes clear evidence for the central theme of this book: the anticipatory, projective function of the brain imposes its assumptions and memories on the world, and it reconstructs movement based on the slightest hint of change.

Dreaming is also essential evidence. Shepard, speculating on the nature of dreams, wrote: "Thus, although J. Gibson (1970) held that perceiving is an entirely different kind of activity from thinking, imagining, dreaming, or hallucinating, I like to caricature perception as *externally guided hallucination*, and dreaming and hallucination as *internally simulated perception*. Imagery and some forms of thinking could also be described as internally simulated perceptions, but at more abstract levels of simulation."[11]

14

ARCHITECTS HAVE FORGOTTEN THE PLEASURE OF MOVEMENT

Perception is inseparable from action. It predicts the future. It was organized over the course of evolution according to the natural properties of the physical world and of biological mechanisms. Not so long ago, our natural world was replaced by an artificial world of cities, trains, and airplanes. Architects are responsible for constructing an environment that meets our brains' expectations. During three thousand years of architecture, humans have experimented with many varieties of habitats. I think that most contemporary architects have betrayed these centuries of research. What they have built and continue to build is a tragedy for our brain, its emotions, and the pleasure it takes in movement.

Evolution made the brain particularly sensitive to the basic elements that constitute perceived nature, our milieu, our *Umwelt,* as Uexküll says. The first of these basic elements is regularity. Regularities can be geometric, like crystals or leaves—for instance, the way a leaf stems from a tree branch. They can be rhythmic, like the undulation of branches in the wind, waves on the sea, or sounds that form regularities in time. The second of these basic elements is chance, which disrupts regularity according to its own laws. We like this game of regularity disrupted by chance.

The third element is movement. Sometimes it demands attention, like the movement of the sea etched in its pebbles and rocks, and reflected in the shape of boats; the movement of the wind, which shapes plants, trees, houses, and the skin, fur, and skeletons of animals who had to adapt to resist it, even animals so small as insects; the flicker of fire, light and elusive, which suggests shapes whose very transience delights us. Not only is there movement in the elements and in animals, there is also human movement: the gestures of harvesting, which dictate the height of trees (and perhaps even the choice of certain fruits); and the gestures of manual capture, which determine the shape of the objects we choose to grasp, throw, and catch.

These three basic elements—regularity, chance, and movement—are what make nature appear as it does. They are also the elements that constitute perception. Among all the people who build artificial objects—engineers, couturiers, and so on—it seems that only architects have forgotten movement. Designers of cars, airplanes, and bicycles have necessarily integrated movement into the architecture of the things they create: a Boeing 747, a high-speed train, a beautiful car, a fast motorboat, or a sailboat. Even vacuum cleaners have a pleasing shape! Their rounded forms reflect both their status as moving objects and our concern that they not bump into the corners of tables.

But in the last twenty years architects have forgotten movement. They build gigantic structures that resemble filing cabinets, and with a few exceptions, like Niemeyer, André Bruyère, or Ricardo Porro, use only straight lines.

By ignoring movement, they offer nothing but flat and square houses to explore. The monotony of their boxes, built on piles and covered with what looks like bathroom tiling, discourages lingering glances. These buildings deny our senses the pleasure of shape and movement that curves and mass would provide.

For painters, use of a vertical and horizontal frame for their canvases provides a point of reference, probably required by the selectivity of visual neurons for these directions as well as the influence of the gravitational vertical and the vestibular system. With distinctly different effect, architects trap our gaze in bleak prisons of lines that instead of fluidly intermingling, intersect at the one angle guaranteed to produce an impression of a collision, an accident, a painful shock: the right angle.

I realized this the day a young architect told me that she had to construct the grid for her project. This rectangular network is a yoke on the imagination. Of course, had it not been for Thales of Miletus, who invented the grid, there would be no modern cities. But that is not the point. It has become an architectural end in itself, as well as the end of architecture.

I actually like the rectangle of a window or a painting. But, oh, right angle, you are neither in a wave on the ocean nor in the wind, nor in the knowing coincidence of a sigh and a smile; you are neither in the petal nor in the leaf. You express the triumph of the most ordinary, lazy, petty geometric mindset over the sense of finesse, which might save me from despair.

Perception, which is simulated action, needs to find natural or artificial objects in the environment that imply action. So our brains take pleasure in playing, in guessing the real and the false, in lying, in laughing and crying, in capturing and fleeing, in predicting the future—in a word, in living. The architects of the Grande Bibliothèque, of the Opéra-Bastille—magnificent inside, but from the outside, like a big bathtub—of Beaubourg, now in the process of being refurbished, tried above all to arrange people and things in an orderly way, but they never thought about the pleasure of those who would visit or inhabit these spaces. Thus they condemned us to boredom. I accuse them of the crime of melancholy, of leading millions of people to despair, of crimes against the biological brain—its flexibility, its desire for movement and variation.

My generation was also to blame: we denuded the apartments we bought of gilt, friezes, and the stucco decorating the walls, and now the ceilings of the lower middle class dream of the splendors of Versailles. We admired Bauhaus and read the theories of its masters, Kandinsky and Gropius, simplistic though they were, who said that architecture, clothing, and objects should be combinations of three prototypic forms: circle, square, and triangle. Perhaps they were right after all. The recent work of Sakata, a Japanese neurophysiologist and disciple of Merleau-Ponty, demonstrates that neurons in the parietal cortex are activated in a specific way by shapes such as a cylinder. These findings might confirm the Bauhaus theory of prototypic shapes on which perception is based and on which artificial objects or houses must also be based.

But the brain does not merely store a static combinatorial system of shapes; it extracts these shapes from movement. The brain can extrapolate the three-dimensional shape of objects based on the movement of a cluster of points on a flat screen. Movement gives rise to structure.

Today, some architects have rediscovered the old compass in their desk drawer and are trying to create an illusion of movement by using semicircles, a shape that is still acceptable in calculating profit and in generating the Pert charts that are used to monitor the progress of a building project. But this is merely a mockery of the golden rules that intelligence and refinement have taught over the centuries, according to which curves are pleasing to the human body and mind. There is no more boring shape than a circle unless it takes the form of a crêpe Suzette, or an ellipse unless it is seen in the context of other shapes.

In the 1950s, Jean Prouvé, an architect who was ahead of his time, preached the advantages of building with glass and iron. His slides showed the hopeless entanglement of cities, a confusion of wires, poles, houses, and apartment blocks made from many materials. He advocated a return to a purity pleasing to the spirit, to a uniformity and lightness that offered new possibilities owing to the flexibility of these materials and to the prospect of letting light in. He was, I think, betrayed by money grubbers and victims of priorities and mass projects, who saw in these materials only the possibility of using them to make giant Lego constructions.

The architect Nouvel had no better option than to build the cultural center of the Boulevard Raspail out of glass. The only thing his generation knows how to do is to reflect the architecture of Paris. Look at the Grande Bibliothèque from the opposite bank. You will not detect any difference between its boxes, which purport to look like books, and the low-rent apartment buildings of the XIIIth arrondissement that form the backdrop. The real shape of a book is what it contains, and what beauty there is in the geometry of the letters!

This mirroring in most buildings is a loud and honest admission of the failure of modern architecture. In it, you see everything around, the sky as well as the city of Haussmann.[1] After three thousand years of art, architects have invented a nonarchitecture that reflects the image of previous centuries.

Nevertheless, this architecture chills me, because it interjects a screen between me and the city, which prevents me from touching it, from palpating it with my eye. These buildings do not invite my desire, but rather form a wall off of which my vision bounces. They no longer portray the weight of the snow sliding off a roof, the breath of the wind that buildings must resist, the trickling of water, the slow movement of the sun, or the gesture of opening a window.

Yet further proof of this disdain for subtlety is the disappearance of wrought iron. Lift your eyes in Paris! See how your gaze delights in following the infinitely varied swirls of Parisian balconies. These balconies of wrought iron play an essential double role: they please the eye and break up the regularity of the facades by giving them a rhythm. When you are inside the house, they provide a transparent screen that divides your view of your neighbor into endlessly different fragments depending on the room you are in. They frame faces perceived on the street in a thousand new and pleasing ways. Their shapes play with the leaves of the sycamores and chestnut trees, the movement of the wind sometimes impelling, sometimes impeding. They prevent the dispiritedness that is a by-product of living in the same neighborhood forever. These forms and spaces suggest movement; they symbolize the possibility of escape.

Now look at what architects have given us in recent years. Prison bars! The appalling repetition that the brain so despises, shape without variation, the curse of preemptory predictability. The brain is a biological forecasting machine. It follows that its pleasure consists in taking gambles. And it can only gamble on a reality in motion, ever changing. Shape, even motionless shape, is an opportunity for mind shifts, for imaginative changes of direction, which criminal architects would prevent us from enjoying.

I recently visited a renovated building that had a staircase going up to the second floor. It was extended from the building's original lower staircase. I compared the two banisters. The more recent one was square, as rigid as an insensitive person, sad and gloomy, going nowhere; the staircase that belonged to the old building had that marvelous curvature you see in Parisian banisters, the legacy of a thousand years of craftsmanship. It showed the way without hesitation, its subtle helix contrasting with the drumbeat of the steps that sectioned space and promised a difficult, jerky climb. The old banister was shape and movement; already I felt in the palm of my hand the rounded curves that embraced me as I got my spinal locomotor generator into gear. The banister was refinement itself, the lover of the craftsman who had created her. She was unlike any other, she was calm regularity, she was movement. I loved her.

What can I say about the grillwork on doors? The best example of failure is that of several schools in Paris that shall remain nameless. For the hundreds of children who cross its threshold each day, the schoolyard gate is the transition from family to school, between the city and this neutral and protected place that the philosopher Alain was so fond of. The gate is the symbol of a touching, eminently contradictory moment of transition between two worlds: the place where parents wait, and the passage through which children take flight at the close of school. It is a symbol both of enclosure and of liberation from family constraints and the teacher's clutches. This object must serve as a border and a crossing, a versatile window on the city and a peephole for catching a glimpse of the sanctuary of school. The gate must be pleasing to cling to, to resist, to rest one's eyes on, to distract one's gaze while waiting, or to examine for symbols of a changing world. What have architects made of it? Prison bars. Total renunciation of the unexpected gesture, the desired meaning.

It took me a long time to understand why the architect André Bruyère was so interested in doors, but now it is obvious to me. The door is one of the most important elements in architecture, experienced as a front for emotions, fading actions, and the construction of memories.

I understand about priorities. I understand the requirements of economies of scale and of standardization. I understand that the straight line is the shortest path from one point to another, and that the curve is a luxury that poverty cannot afford. Nor do I deny the importance of repetition. It is a component of our body. It is pleasing to the brain.

But look at a face. It engages us because it is organized both with a rigorous geometry and a marvelous suppleness of form. Nothing about it is ornamental; everything about it is functional; yet nothing about it is fixed. Every expression is accompanied by some action, by the movement that shapes it. But these movements are subtle, like the line of my lips when I imagine gently crushing a morsel of chocolate cake in a gourmand's kiss. The shapes are free and unconstrained; they are at the service of my imagination, of my desire to act, and neither master nor jailer.

Now it is time to reflect upon the distance covered and try to sum up the ideas presented in this book. To be honest, these ideas are more a work in progress than a finished theory. Let me suggest some thoughts to begin the discussion.

The Computer Metaphor: A New Dualism

The brain is neither a computer nor an artificial intelligence machine. It is an original biological structure, the product of the slow and astonishing work of evolution and, in the case of the human brain, of history and culture. The brain-computer metaphor is useful as a guide to thinking about the brain, but it is very dangerous. Indeed, it implicitly assumes a separation between complex cognitive processes (so-called computational processes similar to software) and a neural substrate whose molecular biology and neurobiology disclose its extraordinary capabilities (still largely unexplored). For many years this metaphor bolstered the functionalist theory, which was recently revived by one of its most ardent defenders, the philosopher Hilary Putnam. It is possible to see in it a modern form of Bergson's mentalism, which I cannot agree with when he says, in substance, that the mind is to the brain what the coat is to the hook that holds it.

Listen to Bergson:

> Anyone who approaches, without preconceived ideas and on the firm ground of facts, the classical problem of the relations of soul and body, will soon see this problem as centering upon the subject of memory, and, even more particularly, upon the memory of words: it is from this

quarter, undoubtedly, that will come the light which will illumine the obscurer parts of the problem . . . Speaking generally, the physical [sic] state seems to us to be, in most cases, immensely wider than the cerebral state. I mean that the brain state indicates only a very small part of the mental state, that part which is capable of translating itself into movements of locomotion. Take a complex thought which unrolls itself in a chain of abstract reasoning. This thought is accompanied by images, that are at least nascent. And these images themselves are not pictured in consciousness without some foreshadowing, in the form of a sketch or a tendency, of the movements by which these images would be acted or played in space—would, that is to say, impress particular attitudes upon the body, and set free all that they implicitly contain of spatial movement. Now, of all the thought which is unrolling, this, in our view, is what the cerebral state indicates at every moment. He who could penetrate into the interior of a brain and see what happens there, would probably obtain full details of these sketched-out, or prepared, movements; there is no proof that he would learn anything else.[1]

In other words, Bergson accepts the idea that the brain is used in the internal simulation of movement, but he limits its contribution to this mental structuring and refuses to grant that it plays any role whatsoever in the highest cognitive functions that are the exclusive domain of the mind. Moreover, in the same passage he says that the whole of psychophysiology provides the brain with a level of comprehension no greater than that of an observer watching the comings and goings of actors on stage to understand the play. Bergson is confident, however, that the brain is a tool for simulating. He writes: "I should say that the brain is an organ of pantomime, and of pantomime only. Its part is to play the life of the mind, and to play also the external situations to which the mind must adapt itself. The work of the brain is to the whole of conscious life what the movements of the conductor's baton are to the orchestral symphony."[2] The computer metaphor also assumes an implicit dissociation between the hardware (the neurons of the brain) and software (the symphony).

The Sense of the Senses

There are more than five senses. Consider, for example, the vestibular sensors, the proprioceptors of the muscles and joints, and, for certain species, other senses such as echolocation and magnetic sense. But to give any meaning to perception, what is needed most is a complete revision of the word "sense." I suggest returning to a classification of the senses that corresponds to perceptual function. So to the sense of taste and smell, touch, vision, and hearing, add, as the vernacular already does, the sense of movement, space, balance, effort, self, decision, responsibility, initiative, and so on. This idea of the senses shows the way, determined by the subject, toward a goal. To varying degrees, separately and together, each of the senses calls upon the receptors, and even certain properties of each receptor, that it needs. The brain filters the messages supplied by the senses according to its own plans. The mechanisms of this selection have yet to be deciphered: we understand only a few forms of selectivity.

In other words, we must completely change the way we study the senses. We must begin by considering the goal of the organism, so that we may understand how the brain is informed by the receptors, how it regulates their sensitivity, combines messages from them, and estimates their value, according to an internal simulation of the expected consequences of action.

The relationships between perception and action require study that takes into account several key features of the brain and their complementarity, which no one yet has been able to characterize. According to so-called centralist theories, action is determined by internal plans of action (my projective formulation of a schema) whose mechanisms are endogenous and are based on a repertoire of innate and acquired behaviors characteristic of each species and each individual. Functions are not confined to a single structure of the brain but are the result of cooperation among specific structures. This cooperation constitutes pathways for neural activity that is characteristic for each structure. Many brain functions are not just the product of pathways along which processing proceeds sequentially, but of closed circuits. Within these circuits, structures that are activated sequentially, or in a synchronous way through oscillations, are often connected by what Edelman calls "reentrant loops." These functional loops interact at several levels and thus influence each other in multiple ways, and this interconnectivity is subtended by assemblies of very specific neurons. In other words, it is wrong to think that in the brain everything is connected with everything else.

The subsystems that define a repertoire of sensorimotor behaviors are subject to variability. The brain has mechanisms for exploiting this variability and combining elements of the repertoire to form strategies that are specially adapted to its current purposes. I think that functional flexibility, recently discovered at the level of the cortex, is only a subsidiary manifestation of this fundamental property. The brain can choose a variety of solutions to the same problem. This ability to choose is at the root of illusions, which are solutions to problems of perceptions that are ambiguous or incompatible with environmental cues. In contrast, hallucinations, like dreams, are creations that the environment cannot confirm.

The Problem of Coherence

The central issue is the neural basis of coherence. Indeed, all the findings of neuropsychology and of modern neurophysiology confirm the multiplicity of cooperating subsystems, the role of segregation, the modularity of cerebral operations, and their interrelatedness. There are many frames of reference in which the body is represented; the body schema may be nothing more than the synthesis of multiple schemas through as yet unknown mechanisms. Each sense breaks down perceptible reality into elements that are subsequently reconstructed and connected. The concept of coherence is thus central. The theory of temporal encoding asserts that temporal synchronization constitutes de facto coherence through simultaneity.

I think that major illnesses like autism, schizophrenia, and agoraphobia share disturbances of coherence that take diverse forms according to the principal features of each illness.

A Memory for Predicting the Future

Memory is essentially a tool for predicting the future. In this sense, the concept of working memory and its association with the prefrontal cortex is important, but it accounts for only a portion of memory. Researchers began with the idea that memory served the brain to recall bygone experiences of youth in old age. Then it was discovered that this memory could in fact be broken down into several categories associated with various parts of the brain, and ultimately one of these categories is working memory. I think that it is critical to start with the idea that memory is a mechanism that enables us to predict the future. The simulating brain uses memories to implement mental processes of prediction. So the hippocampus—the highest level for encoding messages in the nervous system, a supramodal level that takes into account the set of relationships between behavior and the messages encoded in each specialized system—together with the prefrontal and parietal cortex might play a central role in memory as a guide to future action: it reconstructs multisensory episodes from partial elements, as in the example of Proust's madeleine. But many other centers participate in the memory of movement. The parietal cortex is probably involved in long-term motor memory and in the mental simulation of movements;[3] the cerebellum, along with the basal ganglia, in manual dexterity, and so on.

Perception and Emotion

We are far from understanding the relationships between perception and emotion. Yet we have just entered a century that Malraux said would be either "religious or not be at all." If recent expressions of religion betraying only intolerance, fanaticism, and hatred are any indication, this century, if it is religious, will be a century of unprecedented violence. Citizens in every country have an enormous responsibility. To be able to wander about with a bomb strapped to one's back as a young Palestinian fanatic did in March 1995, to murder thousands of Muslims as the Serbs did, or a million Cambodians as did Pol Pot, to murder a playmate in his backyard, like two French youths who were caught up in a game of fantasy role playing, to kill hundreds of one's own followers—to commit such acts the mind must be so constrained by the perception that fascinates it that it suppresses any alternatives. The functioning of the brain is corrupted, as it is with hypnosis. Perception and action should be interesting not only to educators, ergonomists, or physicians, but to all those who feel a responsibility to fight the hatred of others. My generation believed that the basis of selfishness was economic. We were not completely wrong, and the present return to the ruthless struggle between the strong and the weak and the disappearance of feelings of solidarity can only result in sorrow and suffering.

But there is more. The human brain operates with powerful mechanisms of territorialism, of distrust toward those who are different, of selfishness that precludes the open hand of generosity. The most bloodthirsty dictators are supported not only by economic forces but by the forces of obscurantism, which derive their hateful influence from archetypal images that are stiff, petty, and grotesque, and which blot out consideration for the other and justify his destruction. In this book I have maintained that the brain projects its internal perceptions onto the world, that it constructs its perceptions with reference to actions, much the same way that Lascaux man saw bison on the walls of his cave. Unfortunately, this leaves the brain with a propensity for retreating into pre-established schemas that it then projects onto the world and onto others. Skilled alchemists of the mind can thus manufacture ready-made perceptions or caricatures, like those seen in the antisemitic exhibitions organized under the Vichy regime, or those presently being forced on people around the world to feed religious wars. The same process is probably at work in sects, where gurus painstakingly brainwash their followers, distorting the way others are perceived, so that perception of them is tainted by unconditional adoration of the master. And what is the point of playing this role, if not to impose a schema for perception that defines the relationships the guru compels a person to maintain with others? All these various social structures have a common mechanism for restricting the operation of the brain within a set of rigid interpretations, using methods somewhat akin to hypnosis. This perversion of the faculty—the projective nature of perception—must be resisted. It is a scourge graver than the gravest illness, for it does not merely destroy the person; it leads its victims to destroy others.

Anger and hatred often lead us back to these perceptual prototypes, which we suddenly project onto every object or every person in sight. When we do, anticipation becomes a prison for perception and a trap for action; the serene path of reason is abandoned for that of the emotions. Tolerance requires a generous and benevolent appreciation of differences. Along with the pleasure of discovering the mechanisms of perception and action, and the obligation to enhance well-being, human beings also have a duty to understand the perversions of the mind, so that this new century may be one of tolerance. Failing that, we risk the worst.

NOTES

INTRODUCTION

1. J. von Uexküll, *Umwelt und Innenwelt der Tieren und Menschen* (Berlin: Springer, 1934).

2. G. Châtelet, *Les Enjeux du mobile—Mathématiques, physique, philosophie* (Paris: Le Seuil, 1993), p. 39.

3. Johann Wolfgang von Goethe, *Faust,* trans. Randall Jarrell (New York: Farrar, Straus, and Giroux, 1959), p. 61.

4. Immanuel Kant, *Critique of Pure Reason,* trans. Norman Kemp Smith (New York: St. Martin's Press, 1961), p. 47.

5. H. Bergson, *The Creative Mind,* trans. Mabelle L. Andison (New York: Philosophical Library, 1946), pp. 243–245. I have recently published an anthology of one hundred years of the "Leçons inaugurales" [inaugural lectures] of the professors at the Collège de France that indicate this central role of physiology in cognitive sciences in France. See A. Berthoz, *Leçons sur le corps, le cerveau et l'esprit* (Paris: Odile Jacob, 1999).

6. T. Ribot, *The Psychology of the Emotions,* Contemporary Science Series (London: W. Scott, 1897), p. 2.

7. J.-P. Changeux, *L'Homme neuronal* (Paris: Fayard, 1983).

1   PERCEPTION IS SIMULATED ACTION

1. P. Viviani, "Motor-Perceptual Interactions: The Evolution of an Idea," in M. Imbert, P. Bertelson, R. Kempson, D. Osherson, H. Schnelle, N. Streitz, A. Thomassen, and P. Viviani, eds., *Cognitive Science in Europe: Issues and Trends* (Heidelberg: Springer, 1987), pp. 11–39.

2. R. H. Lotze, *Medizinische Psychologie oder Physiologie der Seele* (Leipzig: Weidemann, 1852).

3. H. von Helmholtz, *Treatise on Physiological Optics,* trans. J. P. C. Southall (New York: Dover, 1962).

4. William James, *The Principles of Psychology* (New York: Holt, 1890), vol. 2, pp. 582–585.

5. P. Janet, *Les Débuts de l'intelligence* (Paris: Flammarion, 1935), p. 31.

6. Ibid., p. 37. Janet talks about "kinetic melodies" to designate reflex action that is in no way a unique response but a succession of movements in a certain order.

7. Ibid., p. 43.

8. Ibid.

9. Ibid., p. 54. Bergson said, "To recognize a common object is mainly to know how to use it." H. Bergson, *Matter and Memory,* trans. Nancy Margaret Paul and W. Scott Palmer (New York: Zone Books, 1988), p. 93; and further on, "To follow an arithmetical addition is to do it over again for ourselves," p. 116.

10. Or, literally, "Vision is palpation by gaze" [trans.]. M. Merleau-Ponty, *Le Visible et l'Invisible* (Paris: Gallimard, 1964), p. 177.

11. P. K. Anokhin, *Biology and Neurophysiology of Conditioned Reflexes and Their Role in Adaptive Behaviour* (Oxford: Pergamon, 1974).

12. *Human Motor Actions: Bernstein Reassessed,* ed. H. T. A. Whiting, Advances in Psychology (Amsterdam: North-Holland, 1984), vol. 17, p. 363.

13. α rhythm is an electrical wave that can be recorded on the cranial surface when a human subject is at rest; this rhythm ceases in states of attentive alertness and is replaced by other rhythms induced by other oscillating activities.

14. R. Llinás, "Possible Role of Tremor in the Organisation of the Nervous System," in L. J. Findley, R. Capildeo, eds., *Movement Disorders, Tremor* (New York, Oxford University Press, 1984), pp. 475–477; "The Intrinsic Electrophysiological Properties of Mammalian Neurons: Insights into Central Nervous System Function," *Science,* 242 (1988): 1654–1664; and "The Noncontinuous Nature of Movement Execution," in D. R. Humphrey and H. J. Freund, eds., *Motor Control: Concepts and Issues* (New York: Wiley, 1991), pp. 223–242.

15. E. N. Sokolov and O. S. Vinogradova, *Neuronal Mechanisms of the Orienting Reflex* (Hillsdale, N.J.: Erlbaum, 1975).

16. J. J. Bouyer, M. F. Montaron, J. M. Vahnée, M. Albert, and A. Rougeul, "Anatomical Localization of Beta Rhythms in Cats," *Neuroscience,* 22 (1987): 863–869; M. Chatila, C. Milleret, P. Buser, and A. Rougeul, "A 10 Hz 'Alpha-Like' Rhythm in the Visual Cortex of the Waking Cat," *Electroencephalography and Clinical Neurophysiology,* 83 (1992): 217–222; C. M. Gray, P. König, A. K. Engel, and W. Singer, "Oscillatory Responses in Cat Visual Cortex Exhibit Inter-columner Synchronization Which Reflects Global Stimulus Patterns," *Nature,* 338 (1989): 334–337; M. Joliot, U. Ribary, and R. Llinás, "Human Oscillatory Brain Activity Near 40 Hz Coexists with Cognitive Temporal Binding," *Proceedings of the National Academy of Sciences,* 91 (1994): 11748–11751.

17. R. A. Schmidt, "A Schema Theory of Discrete Motor Skill Learning," *Psychological Review,* 82 (1975): 225–260.

18. U. Neisser, *Cognition and Reality* (San Francisco: Freeman, 1976).

19. D. G. MacKay, *The Organization of Perception and Action* (New York: Springer, 1987), p. 7.

20. Ibid., pp. 1–2.

21. G. Rizzolatti, M. Gartilucci, R. M. Camarda, V. Gallex, G. Luppino, M. Matelli, and L. Fogassi, "Neurons Related to Reaching-Grasping Arm Movements in the Rostral Part of Area 6 (Area 6a)," *Experimental Brain Research,* 82 (1990): 337–350. Rizzolatti and his co-workers also discovered neurons that discharge in relation to other motor actions like precision gripping. Other teams, such as Sakata's [H. Sakata, M. Taira, A. Murata, and S. Mine, "Neural Mechanisms of Visual Guidance of Hand Action in the Parietal Cortex of the Monkey," *Cerebral Cortex,* 5 (1995): 429–438] in Japan found neurons in the parietal cortex that encode simple motor behaviors. This article describes a conceptual schema for guiding movements of the hand.

22. J. Bouveresse, *Langage, perception et réalité* (Nîmes: Jacqueline Chambon, 1995), p. 54.

23. Ibid., p. 50.

24. M. Merleau-Ponty, *Résumés de cours au Collège de France* (Paris: Gallimard, 1968).

25. These are transistors whose properties are a model for control of membrane potential by ionic currents and that can be used to construct so-called neuromorphic networks.

26. M. Jeannerod, "The Representing Brain: Neural Correlates of Motor Intention and Imagery," *Behavioral and Brain Sciences,* 17 (1994): 187–202.

27. M. Wexler, S. Kosslyn, and A. Berthoz, "Motor Processes in Mental Rotation," *Cognition*, 68 (1998): 77–94.

2 THE SENSE OF MOVEMENT

1. Aristotle, *De Sensu and De Memoria*, trans. G. R. T. Ross (New York: Arno, 1973), p. 55.
2. The first experiments using vibrations in humans were conducted by Hagbarth in Sweden around 1966, then by P. B. C. Matthews in the cat around 1970. Many groups subsequently used vibrations to study mechanisms of perception and control of movement, for example, Gurfinkel, in Moscow; Jeannerod, in Lyon; Roll, in Marseille; and our laboratory. Vibrations have also been used in experiments in space.
3. The illusion of movement is thus a perceptual solution that is chosen by the brain and leads to motor consequences. For more on this subject, see Chapter 13.
4. G. S. Sherrington, "Observations on the Sensual Role of the Proprioceptive Nerve Supply of Extrinsic Ocular Muscles," *Brain*, 41 (1918): 332–343.
5. F. Emonet-Dénand, Y. Laporte, P. B. C. Matthews, and J. Petit, "On the Subdivision of Static and Dynamic Fusimotor Actions on the Primary Ending of the Cat Muscle Spindle," *Journal of Physiology*, 261 (1977): 827–861; Y. Laporte, F. Emonet-Dénant, and L. Jami, "The Skeletofusimotor or β Innervation of Mammalian Muscle Spindles," *Trends in Neuroscience*, 4 (1981): 97–99; J. Celichowski, F. Emonet-Dénand, Y. Laporte, and J. Petit, "Distribution of Static γ Axons in Cat Peroneus Tertius Spindles Determined by Exclusively Physiological Criteria," *Journal of Neurophysiology*, 71 (1994): 722–732.
6. J. P. Roll, J. C. Gilhodes, R. Roll, and J.-L. Velay, "Contribution of Skeletal and Extraocular Proprioception to Kinaesthetic Representation," in M. Jeannerod, ed., *Attention and Performance XIII* (Hillsdale, N.J.: Erlbaum, 1990): 549–566; J. P. Roll, R. Roll, and J.-L. Velay, "Proprioception as a Link between Body and Extra-personal Space," in J. Paillard, ed., *Brain and Space* (New York: Oxford University Press, 1991), pp. 112–132.
7. C. Perret and A. Berthoz, "Evidence on Static and Dynamic Fusimotor Actions on the Spindle Response to Sinusoidal Stretch during Locomotor Activity in the Cat," *Experimental Brain Research*, 18 (1973): 178–188.
8. A. Berthoz and S. Metral, "A Torque-Producing Stimulator for the Study of Muscular Response to Variable Forces," *Proceedings of the International Symposium on Biomechanics III* (Basel: Karger, 1973), pp. 158–164.
9. R. J. Nudo, G. W. Milliken, W. M. Jenkins, and M. M. Merzenich, "Use-Dependent Alterations of Movement Representations in Primary Motor Cortex of Adult Squirrel Monkeys," *Journal of Neuroscience*, 16 (1996): 785–807.
10. H. Flor, T. Elbert, S. Knecht, C. Wienbruch, C. Pantev, N. Birbaumer, W. Larbig, and E. Taub, "Phantom-Limb Pain as a Perceptual Correlate of Cortical Reorganization following Arm Amputation," *Nature*, 375 (1995): 482–484.
11. L. Jami, "Golgi Tendon Organs in Mammalian Skeletal Muscle: Functional Properties on Central Actions," *Physiological Reviews*, 72 (1992): 623–666.
12. J.-J. Slotine and W. Li, *Applied Nonlinear Control* (Englewood Cliffs, N.J.: Prentice Hall, 1991).
13. S. Hanneton, A. Berthoz, J. Droulez, and J.-J. Slotine, "Does the Brain Use Sliding Variables for the Control of Movements?" *Biological Cybernetics*, 77 (1997): 381–393.

14. J. Decéty, M. Jeannerod, and C. Prablanc, "The Timing of Mentally Represented Actions," *Behavioral and Brain Research*, 34 (1989): 35–42.

15. J. Decéty, M. Jeannerod, M. Germain, and J. Pastène, "Vegetative Response during Imagined Movement Is Proportional to Mental Effort," *Behavioral and Brain Research*, 24 (1991): 1–5.

16. Scarpa worked at the University of Pavia.

17. P. Flourens, *Recherches expérimentales sur les propriétés et les fonctions du système nerveux dans les animaux vertébrés* (Paris: Crèvot, 1824).

18. M. G. Jones and K. E. Spells, "A Theoretical and Comparative Study of the Functional Dependence of the Semi-circular Canal upon Its Physical Dimensions," *Proceedings of the Royal Society of London. Series B: Biological Sciences*, 157 (1963): 403–419.

19. A. Sans and E. Scarfone, "Afferent Calyces and Type I Hair Cells during Development. A New Morphological Hypothesis," *Annals of the New York Academy of Sciences*, 781 (1996): 1–12.

20. The frontal plane is the vertical plane that passes through the two ears; the sagittal plane is the vertical plane that passes through the nose and the back of the skull.

21. H. Poincaré, *The Value of Science*, trans. George Bruce Halsted (New York: Science Press, 1907), p. 73.

22. Ibid.

23. Ibid., p. 74.

24. H. Poincaré, *Science and Hypothesis* (New York: Dover, 1952).

25. Ibid., p. 50.

26. Ibid., p. 51–52.

27. Ibid., p. 57.

28. Poincaré, *The Value of Science*, p. 47.

29. Poincaré, *Science and Hypothesis*, p. 58.

30. Ibid., pp. 58–59.

31. A. De Kleijn and T. Magnus, "Ueber die Funktion der Otolithen," *Pflugers Archiv*, 186 (1921): 6–81.

32. G. G. J. Rademaker, *Réactions labyrinthiques et équilibre* (Paris: Masson, 1935).

33. R. Lorente de Nó, "Vestibulo-ocular Reflex Arc," *Archives of Neurology and Psychiatry*, 30 (1933): 245–291.

34. J. Szentagothai, "The Elementary Vestibulo-ocular Reflex Arc," *Journal of Neurophysiology*, 13 (1950): 395–407.

35. A. Berthoz, R. Baker, and W. Precht, "Labyrinthine Control of Inferior Oblique Motoneurons," *Experimental Brain Research*, 18 (1973): 225–241; R. Baker and A. Berthoz, "Organization of Vestibular Nystagmus in the Oblique Oculomotor System," *Journal of Neurophysiology*, 37 (1974): 195–217, and *Control of Gaze by Brain Stem Neurons* (Amsterdam: Elsevier, 1977).

36. R. Baker and A. Berthoz, "Is the Prepositus Hypoglossi Nucleus the Source of Another Vestibular Ocular Pathway?" *Brain Research*, 86 (1975): 121–127; R. Baker, M. Gresty, and A. Berthoz, "Neuronal Activity in the Prepositus Hypoglossi Nucleus Correlated with Vertical and Horizontal Eye Movements in the Cat," *Brain Research*, 101 (1976): 366–371.

37. A. J. Pellionisz and R. Llinás, "Tensorial Approach to the Geometry of Brain Function. Cerebellar Coordination via a Metric Tensor," *Neuroscience*, 5 (1980): 1761–1770.

38. B. W. Peterson, J. Baker, and A. J. Pellionisz, "Comparison of Spatial Transformation in Vestibulo-ocular and Vestibulo-spinal Reflexes," *Proceedings of the Symposium on the Representation of Three-Dimensional Space in the Vestibular, Oculomotor, and Visual Systems* (The Barany Society: Bologna, 1988). See also the chapter by Pellionisz in A. Berthoz and G. Melvill-Jones, *Adaptative Mechanisms in Gaze Control* (Amsterdam: Elsevier, 1985).

39. A. J. Pellionisz and R. Llinás, "Brain Modeling by Tensor Network Theory and Computer Simulation. The Cerebellum: Distributed Processor for Predictive Coordination," *Neuroscience*, 4 (1979): 232–348, "Tensorial Approach," and "Space-Time Representation in the Brain. The Cerebellum as a Predictive Space-Time Metric Tensor," *Neuroscience*, 7 (1982): 2949–2970.

40. A. J. Pellionisz, "Coordination: A Vector-Matrix Description of Transformations of Overcomplete CNS Coordinates and a Tensorial Solution using the Moore-Penrose Generalized Inverse," *Journal of Theoretical Biology*, 101 (1984): 353–375.

41. M. Arbib and S. Amari, "Sensorimotor Transformations in the Brain (With a Critique of the Tensor Theory of Cerebellum)," *Journal of Theoretical Biology*, 112 (1985): 123–155.

42. J. L. Barbur, A. J. Harlow, and L. Weiskrantz, "Spatial and Temporal Response Properties of Residual Vision in a Case of Hemianopia," *Philosophical Transactions of the Royal Society of London. Series B: Biological Sciences*, 343 (1994): 366–371.

43. E. Mach, *Grundlinien der Lehre von den Bewegungsempfindungen* (Amsterdam: E. J. Bonset, 1967).

44. D. N. Lee and E. Aronson, "Visual Proprioceptive Control of Standing in Human Infants," *Perception and Psychophysics*, 15 (1974): 529–532; D. N. Lee and J. R. Lishman, "Visual Proprioceptive Control of Stance," *Journal of Human Movement Studies*, 1 (1975): 87–95.

45. J. Dichgans and T. Brandt, "Visual-Vestibular Interaction and Motion and Perception," in J. Dichgans and E. Bizzi, eds., *Cerebral Control of Eye Movements and Motion Perception* (New York: Karger, 1972), pp. 327–338, and "Visual-Vestibular Interactions: Effects on Self-Motion Perception and Postural Control," in H. Heibowitz and H. L. Teuber, eds., *Handbook of Sensory Physiology* (Berlin: Springer, 1978), vol. 5, pp. 755–804; T. Brandt, J. Dichgans, and E. Koenig, "Differential Effects of Central versus Peripheral Vision on Egocentric and Exocentric Motion Perception," *Experimental Brain Research*, 16 (1973): 746–491.

46. A. Berthoz, B. Pavard, and L. Young, "Perception of Linear Horizontal Self-Motion Induced by Peripheral Vision (Linear Vection)," *Experimental Brain Research*, 23 (1974): 471–489; B. Pavard and A. Berthoz, "Perception du mouvement et orientation spatial (Revue bibliographique)," *Le Travail humain*, 2 (1976): 207–226, and "Linear Acceleration Modifies the Perceived Velocity of a Moving Scene," *Perception and Psychophysics*, 6 (1977): 529–540.

3 BUILDING COHERENCE

1. A. R. Damasio, "The Brain Binds Entities and Events by Multiregional Activation from Convergence Zones," *Neural Computation,* 1 (1989): 123–132.
2. Aristotle, *Psychologie. Opuscules* (Paris: Dumont, 1847), p. 17.
3. D. Ferrier, *The Functions of the Brain* (New York: G. P. Putnam's Sons, 1876), p. 60.
4. The interactions between visual and auditory perception are described in B. E. Stein and M. A. Meredith, *The Merging of the Senses* (Cambridge, Mass.: MIT Press, 1993), and are currently the subject of renewed interest. But they have been little studied in humans.
5. I. M. Sechenov, *Selected Works* (Leningrad: State Publishing House for Biological and Medical Literature, 1935), p. 305.
6. R. Held and A. Hein, "Movement-Produced Stimulation in the Development of Visually Guided Behavior," *Journal of Comparative Physiological Psychology,* 56 (1963): 872–876.
7. J. J. Gibson, *The Senses Considered as Perceptual Systems* (Boston: Houghton Mifflin, 1966).
8. M. T. Turvey and P. N. Kugler, "An Ecological Approach to Perception and Action," in *Human Motor Actions. Bernstein Reassessed* (Amsterdam: Elsevier, 1984).
9. Gibson, *Senses Considered,* and "The Theory of Affordances," in R. E. Shaw and J. Bransford, eds., *Perceiving, Acting, and Knowing* (Hillsdale, N.J.: Erlbaum, 1977).
10. R. E. Wurtz, C. Duffy, and J. P. Roy, "Motion Processing for Guiding Self-Motion," in T. Ono, L. R. Squire, M. Raichle, D. I. Perrett, and M. Fukuda, eds., *Brain Mechanisms of Perception and Memory: From Neuron to Behavior* (Oxford: Oxford University Press, 1993), pp. 141–182; C. Salzman and W. T. Newsome, "Neural Mechanisms for Forming a Perceptual Decision," *Science,* 265 (1994): 231–237; D. C. Bradley, M. Maxwell, R. A. Andersen, M. S. Banks, and K. V. Shenoy, "Mechanisms of Heading Perception in Primate Visual Cortex," *Science,* 273 (1996): 1544–1547.
11. M. N. Shadlen and W. T. Newsome, "Motion Perception: Seeing and Deciding," *Proceedings of the National Academy of Sciences,* 93 (1996): 628–633.
12. G. A. Orban, P. Dupont, B. De Bruyn, R. Vogels, R. Vanderberghe, and L. Mortelmans, "A Motion Area in Human Visual Cortex," *Proceedings of the National Academy of Sciences,* 92 (1995): 993–997.
13. Compare descriptions in O. J. Grüsser and T. Landis, *Visual Agnosias and Related Disorders,* vol. 12, Vision and Visual Dysfunction (Basingstoke, UK: Macmillan, 1991).
14. J. Simpson, "The Accessory Optic Systems," *Annual Review of Neuroscience,* 7 (1984): 13–14; K. P. Hoffman, "Responses of Single Neurons in the Pretectum of Monkeys to Visual Stimuli in Three-Dimensional Space," in B. Cohen and V. Henn, eds., *Representation of Three-Dimensional Space in the Vestibular, Oculomotor, and Visual Systems* (New York: New York Academy of Sciences, 1988), pp. 1–261.
15. J. I. Simpson and W. Graf, "The Selection of Reference Frames by Nature and Its Investigators," in Berthoz and Melvill-Jones, *Adaptive Mechanisms in Gaze Control,* pp. 3–20.
16. J. Dichgans, C. L. Schmidt, and W. Graf, "Visual Input Improves the Speedometer Function of the Vestibular Nuclei in the Goldfish," *Experimental Brain Research,* 18 (1973): 319–322; V. Henn, L. R. Young, and C. Finley, "Vestibular Nucleus Units in

Alert Monkeys Are Also Influenced by Missing Visual Field," *Brain Research*, 71 (1974): 144–149.

17. Baker and Berthoz, "Organization of Vestibular Nystagmus"; A. Berthoz, K. Yoshida, and P. P. Vidal, "Horizontal Eye Movement Signals in Second-Order Vestibular Nuclei Neurons in the Alert Cat," *Annals of the New York Academy of Sciences*, 374 (1981): 144–156; A. Berthoz, J. Droulez, P. P. Vidal, and K. Yoshida, "Neural Correlates of Horizontal Vestibulo-ocular Reflex Cancellation during Rapid Eye Movements in the Cat," *Journal of Physiology*, 419 (1989): 717–751.

18. M. Magnin, M. Jeannerod, and P. T. S. Putkonen, "Vestibular and Saccadic Influences on Dorsal and Ventral Nuclei of the Lateral Geniculate Body," *Experimental Brain Research*, 21 (1974): 1–18.

19. K. Yoshida, R. A. McCrea, A. Berthoz, and P. P. Vidal, "Interneurones Inhibiteur de la saccade oculaire horizontale étudiée chez le chat éveillé à l'aide d'injections intra-axonique de peroxydase," *Comptes rendus hebdomadaires de l'Académie des sciences (Paris, série III)*, 290 (1980): 636–638, "Eye Movement Related Activity of Identified Second-Order Vestibular Neurons in the Cat," in A. Fuchs and W. Becker, eds., *Progress in Oculomotor Research* (Amsterdam: Elsevier, 1981), pp. 371–378, and "Properties of Immediate Premotor Inhibitory Burst Neurons Controlling Horizontal Rapid Eye Movements in the Cat," in Fuchs and Becker, *Progress in Oculomotor Research*, pp. 71–81; R. A. McCrea, K. Yoshida, C. Evinger, and A. Berthoz, "The Location, Axonal Arborization, and Termination Sites of Eye-Movement-Related Secondary Vestibular Neurons Demonstrated by Intra-axonal HRP Injection in the Alert Cat," in Fuchs and Becker, *Progress in Oculomotor Research*, pp. 379–386; Berthoz et al., "Neural Correlates," pp. 717–751.

20. E. von Holst and H. Mittelstaedt, "Das Reafferenzprinzip. Wechselwirkungen zwischen Zentralnervensystem und Peripherie," *Naturwissenschaften*, 37 (1950): 464–476.

21. W. Penfield and E. Boldrey, "Somatic Motor and Sensory Representation in the Cerebral Cortex of Man as Studied by Electrical Stimulation," *Brain*, 60 (1937): 389–443; W. Penfield and T. Rasmussen, *The Cerebral Cortex of Man. A Clinical Study of Localization of Function* (New York: Macmillan, 1957); W. Penfield, "Vestibular Sensation and the Cerebral Cortex," *Annales d'otorhinolaryngologie*, 66 (1957): 691–698.

22. B. H. Smith, "Vestibular Disturbances in Epilepsy," *Neurology*, 10 (1960): 465–469.

23. J. Silberpfennig, "Contributions to the Problem of Eye Movements. III. Disturbance of Ocular Movements with Pseudo-hemianopsia in Frontal Lobe Tumors," *Confinia Neurologica*, 4 (1941): 1–13.

24. L. Friberg, T. S. Olsen, P. Roland, O. B. Paulson, and N. A. Lassen, "Focal Increase of Blood Flow in the Cerebral Cortex of Man during Vestibular Stimulation," *Brain*, 108 (1985): 609–623; P. Tuohimaa, E. Aantaa, K. Toukoniitty, and P. Mäkelä, "Studies of Vestibular Cortical Areas with Short Living Oxygen 15 Isotopes," *Otorhinolaryngologica*, 45 (1983): 315–321; G. Bottini, R. Sterzi, E. Paulesu, G. Vallar, S. Cappa, F. Erminio, R. E. Passingham, C. D. Frith, and R. S. J. J. Frackowiack, "Identification of the Central Vestibular Projections in Man: A Positron Emission Tomography Activation Study," *Experimental Brain Research*, 99 (1994): 164–169; E. Lobel, J. F. Kleine, D. Le Bihan, A. Leroy-Willig, and A. Berthoz, "Functional MRI of Galvanic Vestibular Stimulation," *Journal of Neurophysiology*, 80 (1998): 2699–2709. Caloric stimulation, a technique com-

mon in clinical otorhinolaryngology, is used to activate the vestibular receptors during brain-imaging tests. Hot or cold water is injected into the ear, and this sudden and local change in temperature induces an activity in the vestibular receptors as if they had been stimulated by a rotation of the head. The mechanism of this activity is still not well known. Barany, a Nobel laureate who devised many vestibular tests, suggested that the receptors are excited by convection currents induced in the endolymph by a temperature gradient. But an experiment conducted on the U.S. space shuttle, to which our laboratory contributed, contradicted this hypothesis. In fact, in space—in microgravity—convection disappears, and yet the injection of cold air into the ears of the astronauts produced an ocular movement (nystagmus) characteristic of activation of the semicircular canals. Our results do not, however, call into question the utility of the test.

25. T. Brandt and M. Dieterich, "Skew Deviation with Ocular Torsion: A Vestibular Brainstem Sign of Topographic Diagnostic Value," *Annals of Neurology,* 33 (1993): 528–534; M. Dieterich and T. Brandt, "Thalamic Infarctions: Differential Effects on Vestibular Function in the Roll Plane (35 Patients)," *Neurology,* 43 (1993): 1732–1740.

26. W. O. Guldin and O. J. Grüsser, "Single Unit Responses in the Vestibular Cortex of Squirrel Monkeys," *Neuroscience Abstracts,* 13 (1987): 1224. For a summary, see A. Berthoz and P. P. Vidal, *Noyaux vestibulaires et vertiges* (Paris: Arnette, 1993).

27. O. J. Grüsser, "Cortical Representations of Head Movement in Space and Some Psychophysical Considerations," in A. Berthoz, P. P. Vidal, and W. Graf, eds., *The Head-Neck Sensory-Motor System* (New York: Oxford University Press, 1991), pp. 497–509; O. J. Grüsser, M. Pause, and U. Schreiter, "Localisation and Responses of Neurones in the Parieto-insular Vestibular Cortex of Awake Monkeys *(Macaca fascicularis)," Journal of Physiology,* 430 (1990): 537–557, and "Vestibular Neurons in the Parieto-Insular Cortex of Monkeys *(Macaca fascicularis):* Visual and Neck Receptor Responses," *Journal of Physiology,* 430 (1990): 559–583; S. Akbarian, O. J. Grüsser, and W. O. Guldin, "Corticofugal Projections to the Vestibular Nuclei in Squirrel Monkeys: Further Evidence of Multiple Cortical Vestibular Fields," *Journal of Comparative Neurology,* 332 (1993): 7270–7281.

28. Grüsser and Landis, "Visual Agnosias." In the monkey, the six cortical areas implicated in vestibular activity are areas 7a, 6, 3a, 2v, T3, and PIVC. Bottini et al. ("Central Vestibular Projections in Man") found equivalent areas in humans. In addition, E. Vitte, A. Sémont, and A. Berthoz ["Repeated Optokinetic Stimulation in Conditions of Active Standing Facilitates the Recovery from Vestibular Deficits," *Experimental Brain Research,* 102 (1994): 141–148] discovered activity in the hippocampus.

29. K. Kawano, M. Sasaki, and M. Yamashita, "Vestibular Input to Visual Tracking Neurons in the Posterior Parietal Association Cortex of the Monkey," *Neuroscience Letters,* 17 (1980): 55–60; H. Sakata, H. Shibutani, and K. Kawano, "Functional Properties of Visual Tracking Neurons in Posterior Parietal Association Cortex of the Monkey," *Journal of Neurophysiology,* 49 (1983): 1364–1380; P. Thier and R. G. Erickson, "Responses of Visual-Tracking Neurons from Cortical Area MST-I to Visual, Eye, and Head Motion," *European Journal of Neuroscience,* 4 (1992): 539–553. For a review of the cortical influences of the vestibular system, see A. Berthoz, "How Does the Cerebral Cortex Process and Utilize Vestibular Signals?" in R. W. Baloh and G. M. Halmagyi, eds., *Disorders of the Vestibular System* (Oxford: Oxford University Press, 1996), pp. 113–125. Demon-

stration of vestibular input on hippocampal formation was provided by E. Vitte, C. Derosier, Y. Caritu, A. Berthoz, D. Hasboun, and D. Soulié, "Activation of the Hippocampal Formation by Vestibular Stimulation: A Functional Magnetic Resonance Study," *Experimental Brain Research*, 112 (1996): 523–526 and V. V. Gavrilov, S. I. Wiener, and A. Berthoz, "Whole Body Rotations Enhance Hippocampal Theta Rhythm Slow Activity in Awake Rats Passively Transportated on a Mobile Robot," *Annals of the New York Academy of Sciences*, 781 (1996): 385–398.

30. P. Buisseret and M. Imbert, "Visual Cortical Cells: Their Developmental Properties in Normal and Dark Reared Kittens," *Journal of Physiology*, 255 (1976): 511–525; P. Buisseret and E. Gary-Bobo, "Development of Visual Cortical Orientation Specificity after Dark Rearing: Role of Extra-ocular Proprioception," *Neuroscience Letters*, 13 (1979): 259–263.

31. X. M. Sauvan and E. Peterhans, "Neural Integration of Visual Information and Direction of Gravity in the Prestriate Cortex of the Alert Monkey," in T. Mergner and F. Hlavackà, eds., *Multisensory Control of Posture* (New York: Plenum, 1995), pp. 43–50.

32. J. S. Taube, R. U. Muller, and J. B. Ranck, Jr., "Head-Direction Cells Recorded from the Postsubiculum in Freely Moving Rats. I. Description and Quantitative Analysis," *Journal of Neuroscience*, 10 (1990): 420–435, and "Head Direction Cells Recorded from the Postsubiculum in Freely Moving Rats. II. Effects of Environmental Manipulations," *Journal of Neuroscience*, 10 (1990): 436–447. L. L. Chen, L. Lin, E. J. Green, C. A. Barnes, and B. McNaughton, "Head Direction Cells in the Rat Posterior Cortex. I. Anatomical Distribution and Behavioural Modulation," *Experimental Brain Research*, 101 (1994): 8–23; L. L. Chen, L. Lin, C. A. Barnes, and B. McNaughton, "Head Direction Cells in the Rat Posterior Cortex. II. Contributions of Visual and Ideothetic Information to the Directional Firing," *Experimental Brain Research*, 101 (1994): 24–34.

33. A. B. Rubens, "Caloric Stimulation and Unilateral Visual Neglect," *Neurology*, 35 (1985): 1019–1024; S. Cappa, R. Sterzi, G. Vallar, and E. Bisiach, "Remission of Hemineglect and Anosognosia during Vestibular Stimulation," *Neuropsychologia*, 25 (1987): 775–782; G. Vallar, R. Sterzi, G. Bottini, S. Cappa, and M. L. Rusconi, "Temporary Remission of Left Hemianesthesia after Vestibular Stimulation. A Sensory Neglect Phenomenon," *Cortex*, 26 (1990): 123–131.

34. Many theories have been proposed to explain spatial neglect. For reviews see M. Jeannerod, *Neurophysiological and Neuropsychological Aspects of Spatial Neglect* (Amsterdam: Elsevier, 1987), and G. Rizzolatti and V. Galeze, "Mechanisms and Theories of Spatial Neglect," in F. Boller and J. Gravman, eds., *Handbook of Neuropsychology* (Amsterdam: Elsevier, 1988), pp. 289–313.

35. G. Vallar, R. Sterzi, G. Bottini, S. Cappa, and M. L. Rusconi, "Temporary Remission of Left Hemianesthesia after Vestibular Stimulation. A Sensory Neglect Phenomenon," *Cortex*, 26 (1990): 123–131.

36. Ibid.

37. G. Bottini, E. Paulesu, R. Sterzi, E. Warburton, R. J. Wise, G. Vallar, R. S. Frackowiak, and C. D. Frith, "Modulation of Conscious Experience by Peripheral Sensory Stimuli," *Nature*, 376 (1995): 778–781.

38. Ibid., p. 781.

39. J. Gertsmann, "Problems of Imperception of Disease and of Impaired Body Territories

with Organic Lesions. Relation to Body Scheme and Its Disorders," *Archives of Neurology and Psychiatry*, 48 (1942): 890–913.

40. E. Bisiach, M. L. Rusconi, and G. Vallar, "Remission of Somatoparaphrenic Delusion through Vestibular Stimulation," *Neuropsychologia*, 29 (1991): 1029–1031.

41. H. O. Karnath, K. Christ, and W. Hartje, "Decrease of Contralateral Neglect by Neck Muscle Vibration and Spatial Orientation of Trunk Midline," *Brain*, 116 (1993): 383–396; Vallar et al., "Improvement of Left Visuo-spatial Hemineglect," pp. 73–82.

42. K. Lorenz, *L'Envers du miroir* (Paris: Flammarion, 1975); J. E. Cutting and P. M. Vishton ["Perceiving Layout and Knowing Distances: The Integration, Relative Potency, and Contextual Use of Different Information about Depth," in *Perception of Space and Motion* (San Diego: Academic, 1995), pp. 69–117] provide a nice clarification of this question.

43. Poincaré, *Science and Hypothesis*, p. 53.

44. Y. Trotter, S. Celebrini, B. Stricane, S. Thorpe, and M. Imbert, "Neural Processing of Stereopsis as a Function of Viewing Distance in Primate Visual Cortical Area V1," *Journal of Neurophysiology*, 76 (1996): 2872–2885.

45. Cited in Bouveresse, *Langage, perception et réalité*, p. 37.

46. Stein and Meredith, *Merging of the Senses*, p. 47.

47. "Sensoritopy" refers to the spatial organization of neurons in a structure of the central nervous system that corresponds to the spatial distribution of the receptors at the periphery.

48. M. Graziano and C. Gross, "Coding of Visual Space by Premotor Neurons," *Science*, 266 (1994): 1054–1057, and "The Representation of Extrapersonal Space: A Possible Role for Bimodal-Tactile Neurons," in M. Gazzaniga, ed., *The Cognitive Neurosciences* (Cambridge, Mass.: MIT Press, 1994): 1021–1034.

49. Stein and Meredith, *Merging of the Senses*, p. 137.

50. Merleau-Ponty, *Le Visible et l'invisible*, p. 107.

51. J.-P. Sartre, *The Psychology of Imagination* (Westport, Conn.: Greenwood Press, 1978), p. 115.

52. J. De Ajuriaguerra, *Résumés de cours. Collège de France* (Paris, 1976).

53. P. Bach-y-Rita, "Vision Substitution by Tactile Image Projection," *Nature*, 221 (1989): 963–964.

54. F. Uhl, P. Franzen, G. Lindinger, W. Lang, and L. Deecke, "On the Functionality of the Visually Deprived Occipital Cortex in Early Blind Persons," *Neuropsychologia*, 33 (1991): 256–259.

55. Graziano and Gross, "Coding of Visual Space," 1054–1057, and "The Representation of Extrapersonal Space," 1021–1034. Similar neurons have also been found for the mouth (Rizzolatti, 1987); moreover, J.-R. Duhamel, C. L. Colby, and M. E. Goldberg ["Congruent Representations of Visual and Somatosensory Space in Single Neurons of Monkey Ventral Intraparietal Cortex (area VIP)," in Paillard, *Brain and Space*, pp. 223–236] proposed that, in the parietal cortex, these neurons participate in a "supramodal analysis of the environment" by establishing a connection between one representation centered on the body and another centered on the retina (retinotopic).

56. Merleau-Ponty, *La Nature*.

57. E. Husserl, *Idées directrices pour une phénoménologie et une philosophie phénoménologique*

pure, vol. 2, *Recherches phénoménologiques pour la constitution* (Paris: Presses Universitaires de France, 1982), p. 207.

58. A. J. Mistlin and D. I. Perrett, "Visual and Somatosensory Processing in the Macaque Temporal Cortex: The Role of 'Expectation,'" *Experimental Brain Research*, 82 (1990): 437–450.

59. J. K. Hietanen and D. I. Perrett, "A Role of Expectation in Visual and Tactile Processing within the Temporal Cortex," in Ono et al., *Brain Mechanisms*, pp. 83–103; quotation from p. 89.

60. A. Prochazka, "Sensorimotor Gain Control: A Basic Strategy of Motor Systems?" *Progress in Neurobiology*, 33 (1989): 281–307; quotation from p. 301.

61. J. MacIntyre, E. Gurfinkel, M. Lipshits, J. Droulez, and V. Gurfinkel, "Measurement of Human Force Control During Constrained Arm Motion using a Force-Activated Joystick," *Journal of Neurophysiology*, 17 (1995): 1201–1222.

62. Proust, Marcel, *Remembrance of Things Past*, trans. C. K. Scott Moncrieff and Terence Kilmartin (London: Chatto and Windus, 1981), p. 64.

63. J. Droulez and C. Darlot, "The Geometric and Dynamic Implications of the Coherence Constraints in Three-Dimensional Sensorimotor Coordinates," in M. Jeannerod, ed., *Attention and Performance XIII* (Hillsdale, N.J.: Erlbaum, 1989), pp. 495–526.

64. J. Droulez and V. Cornilleau-Pérès, "Application of the Coherence Scheme to the Multisensory Fusion Problem," in A. Berthoz, ed., *Multisensory Control of Movement* (Oxford: Oxford University Press, 1993), pp. 485–508; quotation from p. 490.

65. M. Abeles, *Local Cortical Circuits* (Berlin: Springer, 1982); C. von der Malsburg, "Nervous Structures with Dynamical Links," *Berichte der Bunsengesellschaft für physikalische Chemie*, 89 (1985): 703–710; E. Bienenstock and C. von der Malsburg, "Statistical Coding and Short Term Synaptic Plasticity: A Scheme for Knowledge Representation in the Brain," in E. Bienenstock, F. Fogelman, G. Weisbuch, *Disordered Systems and Biological Organization* (Berlin: Springer, 1986): 247–272; Gray et al., "Oscillatory Responses"; W. Singer, "Search for Coherence: A Basic Principle of Cortical Self-Organization," *Concepts in Neuroscience*, 1 (1990): 1–26. For further references, see G. Buzsáki, Z. Horváth, R. Urioste, J. Hetke, and K. Wise, "High Frequency Network Oscillations in the Hippocampus," *Science*, 256 (1992): 1025–1027.

66. Frith, *Autism: Explaining the Enigma*, Cognitive Development (Oxford: Basil Blackwell, 1989), pp. 103–104.

67. Ibid., p. 110.

68. Ibid., p. 166.

4 FRAMES OF REFERENCE

1. N. Eshkol and A. Wachmann, *Movement Notation* (London: Weidenfeld and Nicolson, 1958).

2. Grüsser and Landis, *Visual Agnosias and Related Disorders;* Cutting and Vishton, "Perceiving Layout and Knowing Distances."

3. J. Hyvarinen, *The Parietal Cortex of Monkey and Man* (Berlin: Springer, 1982).

4. V. H. Mountcastle, J. C. Lynch, A. Georgopoulos, H. Sakata, and C. Acuna, "Posterior Parietal Association Cortex of the Monkey: Command Functions for Operations within Extrapersonal Space," *Journal of Neurophysiology*, 38 (1975): 871–908.

5. G. Vallar, E. Lobel, G. Galati, A. Berthoz, L. Pizzamiglio, and D. Le Bihan, "A Fronto-Parietal System for Computing the Egocentric Spatial Frame of Reference in Humans," *Experimental Brain Research*, 124 (1999): 281–286.

6. J. Paillard, "Les déterminants moteurs de l'organisation spatial," *Cahiers de psychologie*, 14 (1971): 261–316; "Le corps et ses langages d'espace," in E. Jeddi, ed., *Le Corps en psychiatrie* (Paris: Masson, 1982), pp. 53–69; "Posture and Locomotion: Old Problems and New Concepts. Foreword," in B. Amblard, A. Berthoz, and F. Clarac, eds., *Posture and Gait: Development, Adaptation, and Modulation* (Amsterdam: Elsevier, 1988), pp. 5–12.

7. E. Muybridge, *The Human Figure in Motion* (New York: Dover, 1957).

8. A. Berthoz and T. Pozzo, "Intermittent Head Stabilisation during Postural and Locomotory Tasks in Humans," in Amblard et al., *Posture and Gait*, pp. 189–198; T. Pozzo, A. Berthoz, and L. Lefort, "Head Stabilisation during Various Locomotor Tasks in Humans. I. Normal Subjects," *Experimental Brain Research*, 82 (1990): 97–106; M. J. Dai, I. Curthoys, and G. M. Halmagyi, "A Model of Otolith Stimulation," *Biological Cybernetics*, 60 (1989): 185–194.

9. K. D. Walton, D. Lieberman, A. Llinás, M. Begin, and R. Llinás, "Identification of a Critical Period for Motor Development in Neonatal Rats," *Neuroscience*, 51 (1992), 763–767.

10. P. Buisseret and M. Imbert, "Visual Cortical Cells."

11. H. Aubert, "Über eine scheinebare Drehung von Objecten bei Neigung des Kopfes nach rechts oder links," *Virchows Archiv*, 20 (1967): 381–393.

12. J. Müller, "Über das Aubert'sche Phänomen," *Zeitschrift für Psychologie der Sinnesorgane*, 49 (1916): 109–249.

13. G. Clément, V. S. Gurfinkel, F. Lestienne, M. Lipshits, and K. Popov, "Adaptation of Postural Control to Weightlessness," *Experimental Brain Research*, 57 (1984): 61–72; G. Clément and F. Lestienne, "Adaptive Modifications of Postural Attitude in Conditions of Weightlessness," *Experimental Brain Research*, 72 (1988): 381–389.

14. H. Mittelstaedt, "The Role of Otoliths in the Perception of the Orientation of Self and World to the Vertical," *Zoologische Jahrbücher der Physiologie*, 95 (1991): 419–425; S. Glasauer and H. Mittelstaedt, "Determinants of Orientation in Microgravity," *Acta Astronautica*, 27 (1992): 1–9; W. Haustein and H. Mittelstaedt, "Evaluation of Retinal Orientation and Gaze Direction in the Perception of the Vertical," *Vision Research*, 30 (1990): 255–262; H. Mittelstaedt and S. Glasauer, "Illusions of Verticality in Weightlessness," *Clinical Investigations*, 71 (1993): 732–739.

15. H. Mittelstaedt, "Evidence of Somatic Graviception from New and Classical Investigations," *Acta Otolaryngologica*, 115 (suppl. 520) (1995): 186–187, and "Somatic Graviception," *Biological Psychology*, 42 (1996): 53–74.

16. Ferrier, *Functions of the Brain*, pp. 63–64.

17. V. S. Gurfinkel, "The Mechanisms of Postural Regulation in Man," in T. Turpaev, ed., *Physiology and General Biology Reviews* (Chur, Switzerland: Harwood Academic Publishers, 1994).

18. Rademaker, *Réactions labyrinthiques et équilibre*.

19. L. Young, M. Shelhamer, and S. Modestino, "M.I.T./Canadian Vestibular Experiments in Weightlessness," *Experimental Brain Research*, 64 (1986): 299–307.

20. See Chapter 2.

21. J. R. Lackner and M. S. Levine, "Changes in Apparent Body Orientation and Sensory Localization Induced by Vibration of Postural Muscles: Vibratory Myesthetic Illusions," *Aviation, Space, and Environmental Medicine,* 50 (1979): 346–354; J. R. Lackner, "Some Contributions of Touch, Pressure, and Kinesthesis to Human Spatial Orientation and Oculomotor Control," *Acta Astronautica,* 8 (1981): 825–830, "Some Proprioceptive Influences on the Perceptual Representation of Body Shape and Orientation," *Brain,* 111 (1981): 281–297, and "Orientation and Movement in Unusual Force Environments," *Psychological Science,"* 4 (1993): 134–142.

22. J. F. Soechting and B. Ross, "Psychophysical Determination of Coordinate Representation of Human Arm Orientation," *Neuroscience,* 13 (1984): 595–604; J. F. Soechting and M. Flanders, "Sensorimotor Representations for Pointing to Targets in Three-Dimensional Space," *Journal of Neurophysiology,* 62 (1989): 582–594.

23. D. I. Perrett, E. T. Rolls, and W. Caan, "Visual Neurons Responsive to Faces in the Monkey Temporal Cortex," *Experimental Brain Research,* 47 (1982): 342; D. Perrett, A. J. Mistlin, A. J. Chitty, P. A. J. Smith, D. D. Potter, R. Broennimann, and M. H. Harries, "Specialised Face Processing and Hemispheric Asymmetry in Man and Monkey: Evidence from Single Unit and Reaction Time Studies," *Behavioural Brain Research,* 29 (1988): 245–258.

24. D. I. Perrett, J. K. Hietanen, M. W. Oram, and P. J. Benson, "Organisation and Function of Cells Responsive to Faces in the Temporal Cortex," *Philosophical Transactions of the Royal Society. Series B: Biological Sciences,* 335 (1992): 23–30.

25. S. M. Kosslyn, C. F. Chabris, C. J. Marsolek, and O. Koenig, "Categorical versus Coordinate Spatial Representations: Computational Analysis and Computer Simulations," *Journal of Experimental Psychology: Human Perception and Performance,* 18 (1992): 562–577.

26. M. Arbib, T. Iberall, and G. Bingham, "Opposition Space as a Structuring Concept for the Analysis of Skilled Hand Movements," *Experimental Brain Research,* 15 (1986): 158–173.

27. M. A. Arbib, "Interaction of Multiple Representations of Space in the Brain," in Paillard, *Brain and Space,* pp. 380–403; quotation from p. 380.

28. M. Jeannerod, "A Neurophysiological Model for the Directional Coding of Reaching Movements," in Paillard, *Brain and Space,* pp. 49–69; Goodale et al. [M. A. Goodale, A. D. Milner, L. S. Jakobson, and D. P. Carey, "Neurological Dissociation between Perceiving Objects and Grasping Them," *Nature,* 349 (1991): 154–156] showed that patients with cervical lesions dissociate between perception and scaling of grasp.

29. Arbib, "Multiple Representations of Space," p. 385.

30. J. Droulez and A. Berthoz, "Servo-Controlled (Conservative) versus Topological (Projective) Modes of Sensory Motor Control," in W. Bles and T. Brandt, eds., *Disorders of Posture and Gait* (Amsterdam: Elsevier, 1988), pp. 83–97.

31. A. Berthoz, "Reference Frames for the Perception and Control of Movement," in Paillard, *Brain and Space,* pp. 82–111. The ability to switch frames of reference—to change perspective—is surely one of the distinctive features of the human brain.

32. M. F. Levin and A. G. Feldman, "The Role of Stretch Reflex Threshold Regulation in Normal and Impaired Motor Control," *Brain Research,* 657 (1994): 23–30; A. G. Feldman and M. F. Levin, "The Origin and Use of Positional Frames of Reference

in Motor Control," *Behavioral and Brain Sciences*, 18 (1995): 723–744; E. Bizzi, N. Accornero, W. Chapple, and N. Hogan, "Posture Control and Trajectory Formation during Arm Movement," *Journal of Neuroscience*, 4 (1984): 2738–2744.

33. Feldman and Levin, "Origin and Use of Positional Frames of Reference," p. 727.

5 A Memory for Predicting

1. See Chapter 1. Ferrier wrote: "In calling up an idea, or when engaged in the attentive consideration of some idea or ideas, we are in reality throwing into action, but in an inhibited or suppressed manner, the movements with which the sensory factors of ideation are associated in organic cohesion" (Ferrier, *Functions of the Brain*, p. 285).

2. This line of argument is close to that of Poincaré, cited in Chapter 2.

3. Paillard, *Brain and Space*, p. 163.

4. C. Darwin, "Origin of Certain Instincts," *Nature*, 179 (1887): 417–418; quotation from p. 418.

5. T. Gladwin, *East Is a Big Bird* (Cambridge, Mass.: Harvard University Press, 1970).

6. J. S. Barlow, "Inertial Navigation as a Basis for Animal Navigation," *Journal of Theoretical Biology*, 6 (1964): 76–117.

7. J. S. Beritoff, *Neural Mechanisms of Higher Vertebrate Behavior* (New York: Brown, 1965).

8. S. Miller, M. Potegal, and L. Abraham, "Vestibular Involvement in a Passive Transport and Return Task," *Physiological Psychology*, 11 (1983): 1–10.

9. H. Mittelstaedt and M. L. Mittelstaedt, "Mechanismen der Orientierung ohne richtende Aussensreise," *Fortschritte der Zoologie*, 21 (1973): 46–58, and "Homing by Path Integration in a Mammal," *Naturwissenschaften*, 67 (1992): 566–567.

10. A. S. Etienne, R. Mowrer, and F. Saucy, "Limitations in the Assessment of Path-Dependent Integration," *Behaviour*, 106 (1988): 81–111; B. L. Matthews, J. H. Ryu, and C. Bockaneck, "Vestibular Contribution to Spatial Orientation," *Acta Otolaryngologica*, 468 (1989): 149–154.

11. Recent experiments conducted in our laboratory show that patients with vestibular lesions have trouble finding their way in the dark.

12. Etienne et al., "Limitations in Assessment." For a recent review of experiments and projects concerning the study of the effect of gravity on sensory motor systems, see A. Berthoz and A. Guell, "Space Neuroscience Research," *Brain Research Reviews*, 28 (1998): 1–234. This review was written following a planning symposium for the international space station being built by the United States. (See also Chapter 9, note 4.)

13. J. Bloomberg, G. Melvill-Jones, B. Segal, S. McFarlane, and J. Soul, "Vestibular Contingent Voluntary Saccades Based on Cognitive Estimates of Remembered Vestibular Information," *Advances in Oto-Rhino-Laryngology*, 40 (1988): 71–75.

14. I. Israël, S. Rivaud, P. Pierrot-Deseilligny, and A. Berthoz, "Depayed VOR: An Assessment of Vestibular Memory for Self-Motion," in J. Reguin and J. Stelmach, eds., *Tutorials in Motor Neuroscience* (Dordrecht: Kluwer, 1991), pp. 599–607; I. Israël, S. Rivaud, A. Berthoz, and C. Pierrot-Deseilligny, "Cortical Control of Vestibular Memory-Guided Saccades," *Annals of the New York Academy of Sciences*, 656 (1992): 472–484.

15. T. Metcalfe and M. Gresty, "Self-controlled Reorienting Movements in Response to Ro-

tational Displacements in Normal Subjects and Patients with Labyrinthine Diseases," in D. L. Tomko, B. Cohen, and F. E. Guedry, eds., *Sensing Motion* (New York: Annals of the New York Academy of Sciences, 1992).

16. I. Israël, "Memory-Guided Saccades: What Is Memorized?" *Experimental Brain Research*, 90 (1992): 221–224; Israël et al., "Cortical Control," pp. 472–484; C. Pierrot-Deseilligny, I. Israël, A. Berthoz, S. Rivaud, and B. Gaymard, "Role of the Different Frontal Lobe Areas in the Control of the Horizontal Component of Memory-Guided Saccades in Man," *Experimental Brain Research*, 95 (1993): 166–171.

17. A. Berthoz, A. Grantyn, and J. Droulez, "Some Collicular Neurons Code Saccadic Eye Velocity," *Neuroscience Letters*, 72 (1987): 289–294; I. Israël and A. Berthoz, "Contribution of the Otoliths to the Calculation of Linear Displacement," *Journal of Neurophysiology*, 62 (1989): 247–263; I. Israël, N. Chapuis, S. Glasauer, O. Charade, and A. Berthoz, "Estimation of Passive Linear Whole Body Displacement in Humans," *Journal of Neurophysiology*, 70 (1993): 1270–1273.

18. A. Berthoz, I. Israël, P. Georges-François, R. Grasso, and T. Tsuzuku, "Spatial Memory of Body Linear Displacement: What Is Being Stored?" *Science*, 269 (1995): 95–98. I. Israël, R. Grasso, P. Georges-François, T. Tzuzuku, and A. Berthoz, "Spatial Memory and Path Integration Studied by Self-Driven Linear Displacement," *Journal of Neurophysiology*, 77 (1999): 3180–3192. We have recently found [Y. P. Ivanenko, R. Grasso, I. Israël, and A. Berthoz, "The Contribution of Otoliths and Semicircular Canals to the Perception of Two-Dimensional Passive Whole-Body Motion in Humans," *Journal of Physiology*, 502 (1997): 223–233], using a combination of translations and rotations (semicircular path) during passive transport in robots, that in humans there may be a dissociation between the processing of information relative to translations and to rotations, as found in the rodents (see note 12 above).

19. Poincaré, *The Value of Science*, p. 74.

20. J. Droulez and A. Berthoz, "The Dynamic Memory Model and the Final Ooculomotor and Cephalomotor Integrator," in H. Shimazu and Y. Shinoda, eds., *The Oculomotor System* (Tokyo: Japan Scientific Societies Press, 1990), pp. 1–19; A. Berthoz, "Hippocampal and Parietal Contribution to Topokinetic and Topographic Memory," in N. Burgess, K. J. Jeffery, and J. O'Keefe, eds., *The Hippocampal and Parietal Foundations of Spatial Cognition* (Oxford: Oxford University Press, 1999), pp. 381–399.

21. M. F. Land and D. N. Lee, "Where We Look When We Steer," *Nature*, 369 (1994): 742–744. It is, in fact, easy to show that the curvature of the road is connected to direction t of the point tangent to the trajectory of the car using the following expression: $C = 1/([d \cos t] - 1/d)$ where the curvature (C) is the inverse of the radius of the bend of the road and (d) is the lateral distance of the driver in relation to the bend. We recently described a similar gaze anticipation during locomotion [R. Grasso, S. Slasauer, U. Takei, and A. Berthoz, "The Predictive Brain: Anticipatory Control of Head Direction for the Steering of Locomotion," *NeuroReport*, 7 (1996): 1170–1174].

22. B. Milner, "Visual Recognition and Recall after Right Temporal Lobe Excision in Man," *Neuropsychologia*, 6 (1968): 191–209.

23. The hippocampus is a structure that is currently the subject of much interest owing to its role in memory and especially the acquisition of memories based on events.

Among the many recent reviews on its very complex role, see especially T. Ono, B. L. McNaughton, S. Molotchnidoff, E. T. Rolls, and H. Nishijo, *Perception, Memory, and Emotion: Frontiers in Neuroscience* (Cambridge: Pergamon, 1996); for a review on relations of the hippocampus with the prefrontal and parietal cortex, see L. G. Ungerleider, "Functional Brain Imaging Studies of Cortical Mechanisms for Memory," *Science*, 270 (1995): 769–775.

24. D. Marr, "Simple Memory: A Theory for Archicortex," *Proceedings of the Royal Society of London. B: Biological Sciences*, 262 (1971): 23–81.

25. M. Mishkin, "A Memory System in the Monkey," *Philosophical Transactions of the Royal Society. Series B: Biological Sciences*, 98 (1982): 85–95.

26. J. O'Keefe and L. Nadel, *The Hippocampus as a Cognitive Map* (Oxford: Oxford University Press, 1978).

27. R. Tamura, T. Ono, M. Fukuda, and K. Nakamura, "Recognition of Egocentric and Allocentric Visual and Auditory Space by Neurons in the Hippocampus of Monkeys," *Neuroscience Letters*, 109 (1990): 293–298, and "Spatial Responsiveness of Monkey Hippocampal Neurons to Various Visual and Auditory Stimuli," *Hippocampus*, 2 (1992): 307–322; E. T. Rolls and S. M. O'Mara, "Neurophysiological and Theoretical Analysis of How the Hippocampus Functions in Memory," in Ono, *Brain Mechanisms of Perception*.

28. H. Eichenbaum, S. I. Wiener, M. L. Shapiro, and N. J. Cohen, "The Organization of Spatial Coding in the Hippocampus: A Study of Neural Ensemble Activity," *Journal of Neuroscience*, 9 (1989): 2764–2775; S. I. Wiener, C. A. Paul, and H. Eichenbaum, "Spatial and Behavioral Correlates of Hippocampal Neuronal Activity," *Journal of Neuroscience*, 9 (1989): 2737–2763.

29. S. M. O'Mara, E. Rolls, A. Berthoz, and R. P. Kesner, "Neurons Responding to Whole Body Motion in the Primate Hippocampus," *Journal of Neuroscience*, 14 (1994): 6511–6523; S. I. Wiener, V. A. Korshunov, R. Garcia, and A. Berthoz, "Inertial, Substratal, and Landmark Cue Control of Hippocampal CA1 Place Cell Activity," *European Journal of Neuroscience*, 7 (1995): 2206–2219. For a recent review of the the various models concerning the neural basis of spatial processing during navigation, see O. Trullier, S. Wiener, A. Berthoz, and J. A. Meyer, "Biologically Based Artificial Navigation Systems: Reviews and Prospects," *Progress in Neurobiology*, 51 (1997): 483–544.

30. A. Treves and E. T. Rolls, "What Determines the Capacity of Autoassociative Memories in the Brain?" *Network*, 2 (1991): 371–397, and "Computational Constraints Suggest the Need for Two Distinct Input Systems to the Hippocampal CA3 Network," *Hippocampus*, 2 (1992): 189–199; E. T. Rolls and S. M. O'Mara, "Neurophysiological and Theoretical Analysis of How the Hippocampus Functions in Memory," in Ono et al., *Brain Mechanisms of Perception*, pp. 276–297. Rolls and Treves [E. Rolls and A. Treves, *Neural Networks and Brain Function* (Oxford: Oxford University Press, 1998)] have recently reviewed the computational hypothesis concerning the hippocampus in the general frame of brain networks.

31. Y. Myashita, "Where Visual Perception Meets Memory," *Annual Review of Neuroscience*, 16 (1993): 245–263; K. Tanaka, "Inferotemporal Cortex and Object Vision," *Annual Review of Neuroscience*, 19 (1996): 109–139.

32. R. N. Shepard, "Ecological Constraints on Internal Representation: Resonant Kinematics of Perceiving, Imagining, Thinking, and Dreaming," *Psychological Review,* 91 (1984): 417–447; quotation from p. 433.

33. J. O'Keefe and M. Recce, "Phase Relationship between Hippocampal Place Units and the EEG Theta Rhythm," *Hippocampus,* 3 (1993): 317–330.

34. J. E. Lisman and M. A. P. Idiart, "Storage of 7 ± 2 Short-Term Memories in Oscillatory Subcycles," *Science,* 267 (1995): 1512–1515. High-frequency (200 Hz) oscillations in the hippocampus have also been demonstrated by G. Buzsáki et al., "High Frequency Network Oscillations."

35. A. Michotte, *Causalité, permanence et réalité phénoménales* (Paris: Béatrice-Nauwelaerts, 1962).

36. *The Essential Writings of Merleau-Ponty,* ed. Alden L. Fisher (New York: Harcourt, Brace, 1969), pp. 257–258.

6 Natural Movement

1. The term "rule" is used here instead of "law" to stress the fact that it is not a question only of the laws of mechanics but also of constraints connected to intrinsic processes of the nervous system.

2. Ariane Mnouchkine maintains that Oriental theater is the only theater (from a playbill).

3. *Poems of Pierre de Ronsard,* trans. and ed. Nicholas Kilmer (Berkeley: University of California Press, 1979), p. 201.

4. Slotine and Li, *Applied Nonlinear Control.*

5. C. Darwin, *The Expression of the Emotions in Man and Animals* (New York: Philosophical Library, 1955).

6. E. J. Marey, *Le Mouvement* (Paris: Masson, 1894); *La Machine animale* (Paris: Revue EPS, 1993).

7. P. Viviani and C. Terzuolo, "Trajectory Determines Movement Dynamics," *Neuroscience,* 72 (1982): 431–437.

8. N. A. Bernstein, "Some Emergent Problems of the Regulation of Motor Acts," *Questions of Psychology,* no. 6 (1957), reprinted in *Human Motor Actions: Bernstein Reassessed,* ed. H. T. A. Whiting, Advances in Psychology (Amsterdam: North-Holland, 1984), vol. 17, pp. 354–355.

9. C. De Waele, W. Graf, A. Berthoz, and F. Clarac, "Vestibular Control of Skeletal Geometry," in A. Berthoz and F. Clarac, eds., *Posture and Gait* (Amsterdam: Elsevier, 1988), pp. 423–432; Berthoz et al., *Head-Neck Sensory-Motor System.*

10. F. Lacquaniti, J. Soechting, and C. Terzuolo, "Path Constraints on Point to Point Arm Movements in Three-Dimensional Space," *Neuroscience,* 17 (1986): 313–324.

11. Soechting and Ross, "Psychophysical Determination of Coordinate Representation"; Soechting and Flanders, "Sensorimotor Representations."

12. C. C. A. M. Gielen and E. J. Van Zuylen, "Coordination of Arm Muscles during Flexion and Supination: Applications of the Tensor Analysis Approach," *Neuroscience,* 17 (1986): 527–539.

13. P. Viviani and T. Flash, "Minimum-Jerk, Two-Thirds Power Law, and Isochrony. Converging Approaches to Movement Planning," *Journal of Experimental Psychology (Human Perception)*, 21 (1995): 32–53; quotation from p. 34.

14. P. Viviani and R. Schneider, "A Developmental Study of the Relation between Geometry and Kinematics of Drawing Movements," *Journal of Experimental Psychology (Human Perception)*, 17 (1991): 198–218.

15. Viviani and Flash, "Minimum Jerk," pp. 32–53.

16. Ibid., p. 35.

17. Listing's law is explained in several chapters in A. Berthoz, ed., *Multisensory Control of Movement* (Oxford: Oxford University Press, 1993).

18. F. C. Donders, "Beitrag zur Lehre von den Bewegungen des menschlichen Auges," in *Hollandischen Beitragen zu den Anatomischen und Physiologischen Wissenschaften* (Amsterdam, 1847), vol. 1, pp. 104–145.

19. P. Viviani, G. Baud-Bovy, and M. Redolfi, "Perceiving and Tracking Kinaesthetic Stimuli: Further Evidence of Motor-Perceptual Interactions," *Journal of Experimental Psychology*, 23 (1997): 1232–1252.

20. S. Yasui and L. R. Young, "Perceived Visual Motion as Effective Stimulus to Pursuit Eye Movement System," *Science*, 190 (1975): 906–908.

21. Hence the following formula, taken from Viviani and Flash, "Minimum Jerk," p. 35:

$$CF = 1/2 \int_{t1}^{t2} [(d^3x / dt^3)^2 + (d^3y / dt^3)^2] dt$$

According to this expression, movement would be maximally smooth when the value of the function is at a minimum. Calculus shows that only one pair of equations produces this minimum value for a given set of boundary conditions; that is, the conditions that define the beginning and the end of movement. The trajectory, specified by the horizontal and vertical components of a movement, is therefore expressed as a fifth-order polynomial function of time. Such functions determine the value of various parameters all along the trajectory, and the model provides a way to define the temporal structure of movement. This is how Edelman and Flash demonstrated in 1987 that a simulation of natural motor activity—for example, writing—could be obtained using this principle of minimum jerk.

7 SYNERGIES AND STRATEGIES

1. V. E. Belen'kii, V. S. Gurfinkel, and Y. I. Pal'tsev, "On Elements of Voluntary Movement Control," *Biofizika*, 12 (1967): 135–141; J. Paillard, "L'intégration sensori-motrice et idéo-motrice," in M. Richelle, J. Requin, and M. Robert, eds., *Traité de psychologie expérimentale* (Paris: Presses Universitaires de France, 1994), pp. 925–961.

2. We need only retain a few very brief ideas from this field of study.

3. A. Berthoz, J. Droulez, P. P. Vidal, and K. Yoshida. "Neural Correlates of Horizontal Vestibulo-ocular Reflex Cancellation during Rapid Eye Movements in the Cat," *Journal of Physiology* 419 (1989): 717–751.

4. Y. Shinoda, T. Ohgaki, T. Futami, and Y. Sugiushi, "Vestibular Projections to the Spinal Cord: The Morphology of Single Vestibulo-Spinal Axons," in O. Pompeiano and J.

Allum, eds., *Vestibulospinal Control of Posture and Movement,* Progress in Brain Research (Amsterdam: Elsevier, 1988), pp. 17–27.

5. E. von Holst, "Relations between the Central Nervous System and the Peripheral Organs," *Journal of Animal Behaviour,* 2 (1954): 89–94; Von Holst and Mittelstaedt, "Das Reafferenzprinzip."

6. W. Graf and R. Baker, "Adaptive Changes in the Vestibulo-ocular Reflex of the Flatfish Are Achieved by Reorganization of Central Nervous Pathways," *Science,* 221 (1983): 777–779.

7. E. V. Evarts, "Relation of Pyramidal Tract Activity to Force Exerted during Voluntary Movements," *Journal of Neurophysiology,* 31 (1968): 14–27, and "Role of Motor Cortex in Voluntary Movements in Primates," in V. B. Brooks, ed., *Handbook of Physiology,* sect. 1, vol. 2, *Motor Control* (Bethesda, Md.: American Physiological Society, 1981), pp. 1083–1120.

8. A. P. Georgopoulos, J. F. Kalaska, R. Caminiti, and J. T. Massey, "On the Relations between the Direction of Two-Dimensional Arm Movements and Cell Discharge in Primate Motor Cortex," *Journal of Neuroscience,* 2 (1982): 1527–1537; A. P. Georgopoulos, A. B. Schwartz, and R. E. Kettner, "Neuronal Population Coding of Movement Direction," *Science,* 233 (1986): 1416–1429; J. F. Kalaska, R. Caminiti, and A. P. Georgopoulos, "Cortical Mechanisms Related to the Direction of Two-Dimensional Arm Movements: Relations in Parietal Area 5 and Comparison with Motor Cortex," *Experimental Brain Research,* 51 (1983): 247–260; A. P. Georgopoulos, "Current Issues in Directional Motor Control," *Trends in Neurosciences,* 18 (1995): 506–510.

9. A. P. Georgopoulos, M. D. Crutcher, and A. B. Schwartz, "Cognitive Spatial Motor Processes. III. Motor Cortical Prediction of Movement Direction during an Instructed Delay Period," *Experimental Brain Research,* 75 (1989): 183–194; A. P. Georgopoulos, J. T. Lurito, M. Petrides, A. B. Schwartz, and J. T. Massey, "Mental Rotation of the Neuronal Population Vector," *Science,* 243 (1989): 234–236.

10. L. Rispal-Padel, F. Cicirata, and C. Pons, "Cerebellar Nuclear Topography of Simple and Synergistic Movements in the Alert Baboon *(Papio papio),*" *Experimental Brain Research,* 47 (1982): 365–380.

8 CAPTURE

1. N. A. Bernstein, *The Coordination and Regulation of Movement* (New York: Pergamon Press, 1967).

2. J. P. Ewert, "Neural Mechanisms of Prey-Catching and Avoidance Behavior in the Toad *(Bufo bufo* L.)," *Brain Behaviour and Evolution,* 3 (1970): 36–56, and "Neuroethology of Releasing Mechanisms: Prey-Catching in Toads," *Behavioral and Brain Sciences,* 10 (1987): 337–405.

3. Schmidt, "A Schema Theory," *Psychological Review,* 32 (1975): 225–260.

4. D. N. Lee, "A Theory of Visual Control of Braking Based on Information about Time-to-Collision," *Perception,* 5 (1976): 437–459.

5. D. N. Lee and P. E. Reddish, "Plummeting Gannets: A Paradigm of Ecological Optics," *Nature,* 293 (1985): 293–294; D. N. Lee and D. S. Young, "Visual Timing in Interceptive Actions," in D. J. Ingle, M. Jeannerod, and D. N. Lee, eds., *Brain Mechanisms and Spatial Vision* (Dordrecht: Martinus Nijhoff, 1996), pp. 1–30.

6. For reviews, see R. J. Bootsma and C. E. Peper, "Predictive Visual Information Sources for the Regulation of Action with Special Emphasis on Catching and Hitting," in L. Proteau and D. Elliot, eds., *Vision and Motor Control* (Amsterdam: Elsevier, 1992), pp. 285–314; L. Peper, R. J. Bootsma, D. R. Mestre, and F. C. Bakker, "Catching Balls: How to Get the Hand to the Right Place at the Right Time," *Journal of Experimental Psychology (Human Perception and Performance)*, 20 (1994): 591–612; Lee and Young, "Visual Timing," pp. 1–36.

7. Peper et al., "Catching Balls," p. 610.

8. Lee calculated two values: first, the value for the trajectory itself (the relationship between tangential velocity and the proximity to the vehicle, or time-to-collision); second, the visual angle subtended by the trajectory divided by the rate of change of the angle. Lee notes a connection between these two variables. In other words, he discovered a new relationship between the geometry and kinematics of movement.

9. J. A. S. Kelso, "Phase Transitions and Critical Behaviour in Human Bimanual Coordination," *American Journal of Physiology*, 14 (1984): R1000–1004; J. A. S. Kelso, J. D. Delcolle, and G. Schöner, "Action-Perception as a Pattern Formation Process," in M. Jeannerod, ed., *Attention and Performance* (Hillsdale, N.J.: Erlbaum, 1990), pp. 139–169; J. J. Buchanan and J. A. S. Kelso, "Posturally Induced Transitions in Rhythmic Multijoint Limb Movements," *Experimental Brain Research*, 94 (1993): 131–142.

10. G. Schöner, "A Dynamic Theory of Coordination of Discrete Movements," *Biological Cybernetiks*, 63 (1990): 257–270.

11. F. Lacquaniti and C. Maioli, "The Role of Preparation in Tuning Anticipatory and Reflex Responses during Catching," *Journal of Neuroscience*, 9 (1989): 134–148.

12. F. Lacquaniti, M. Carrozzo, and N. Borghese, "The Role of Vision in Tuning Anticipatory Motor Responses of the Limbs," in Berthoz, *Multisensory Control of Movement*, pp. 379–390; quotation from p. 390.

13. Like our sensory receptors, measuring instruments are imperfect. The information they provide has a high signal-to-noise ratio. To increase precision in the measurement of a physical variable, one simple solution is to make several measurements and to calculate their mean. The mean is an example of a filter: it eliminates much of the noise inherent in measurements, provided of course that each measurement is made independently and that there is no systematic bias. To calculate this mean, you can wait until you have all your measurements, then calculate their sum and divide it by the number of measurements. But this approach has several disadvantages. All previous measurements have to be memorized—there is no way to estimate the result before the final calculation—and the entire operation has to begin from scratch with each new measurement. A better method is one called iterative. This approach estimates an initial value that is adjusted with subsequent measurements by calculating a weighted mean of the preceding estimated value and the new measurement. Such an iterative filter continuously provides the best possible estimate of the value of a physical variable, especially if the noise is Gaussian. Kalman filters extend this idea of iterative filters to problems of optimal estimates of dynamic systems whose conditional variables evolve over time according to known laws. Each step of their calculation entails two phases that rely on two predictions. Based on the preceding estimate of the condi-

tional variables, a law of evolution is applied to predict a new state without taking into account the new measurements; this first estimate is used to predict the information supplied by the receptors with the help of an internal model of their functioning. The spread between the predicted information and the actual measurements facilitates optimal correction of the first estimate.

14. From Merleau-Ponty's discussion of von Uexküll's work in his book *La Nature*, pp. 220, 227. J. von Uexküll, *Umwelt und Innenwelt der Tieren und Menschen* (Berlin: Springer, 1934); *Streifzüge durch die Umwelten von Tieren und Menschen* (Berlin: Springer, 1965).

15. Merleau-Ponty, *La Nature*.

16. R. N. Shepard, "Ecological Constraints," p. 422.

17. P. Viviani and N. Stucchi, "Biological Movements Look Uniform: Evidence for Motor-Perceptual Interactions," *Journal of Experimental Psychology (Human Perception)*, 18 (1992): 603–623.

18. M. Kawato, K. Furukawa, and R. Suzuki, "A Hierarchical Neural Network Model for Control and Learning of Voluntary Movements," *Biological Cybernetiks*, 57 (1987): 169–185; H. Gomi and M. Kawato, "Adaptive Feedback Control Models of the Vestibulo-cerebellum and Spinocerebellum," *Biological Cybernetics*, 68 (1992): 105–114.

19. Walton et al., "Identification of a Critical Period."

20. Kelso, "Phase Transitions"; Kelso et al., "Action-Perception"; Schöner, "A Dynamic Theory"; Buchanan and Kelso, "Posturally Induced Transitions."

9   The Look that Investigates the World

1. Lorenz, *Behind the Mirror*, p. 56.

2. The history of the concepts and theories about gaze and vision are superbly reviewed in Grüsser and Landis, *Visual Agnosias and Related Disorders*, from which the events reported here have been borrowed.

3. Von Holst and Mittelstaedt, "Das Reafferenzprinzip."

4. Grasso et al., "The Predictive Brain." We have recently obtained evidence that there may be a dissociation between the storage in memory of distance and direction during the steering of locomotion. The length of the path and the direction may be controlled by two distinct mechanisms. This suggestion fits with the finding of "head direction cells," which code the static direction of the head in space, as discussed in Chapter 5. If verified, this dissociation between the representations of distance and direction, which has also been suggested for the control of arm movements, would be another proof of the brain's segregation of kinematic variables. R. Grasso, P. Prévost, Y. Ivanenko, and A. Berthoz, "Eye-Head Coordination for the Steering of Locomotion in Humans: An Anticipatory Synergy," *Neuroscience Letters*, 253 (1998): 115–118; R. Grasso, C. Assaiante, P. Prévost, and A. Berthoz, "Development of Anticipatory Orienting Strategies during Locomotor Tasks in Children," *Neuroscience and Biobehavioral Review*, 22 (1998): 533–539; A. Berthoz, A. Amorim, S. Glasauer, R. Grasso, Y. Takei, and I. Viaud-Delmon, "Dissociation between Distance and Direction during Locomotor Navigation," in R. G. Golledge, ed., *Wayfinding Behaviour* (Baltimore: John Hopkins University Press, 1999), pp. 328–348.

5. De Ajuriaguerra, *Résumés des cours.*

6. Heliodorus of Emesa, *The Aethiopica,* vol. 5. The Athenian Society's Publications (Athens: privately printed for the Athenian Society, 1897), p. 158.

7. P. Bourdieu, *Le Sens pratique* (Paris: Editions de Minuit, 1980), pp. 118–119.

8. Darwin, *Expression of the Emotions.*

9. Frith, *Autism,* p. 4.

10. I. P. Pavlov, *Les Réflexes conditionnés* (Paris: Presses Universitaires de France, 1927).

11. Sokolov and Vinogradova, *Neuronal Mechanisms.*

12. Ibid., p. 234.

13. Chatila et al., "A 10 Hz 'Alpha-Like' Rhythm"; Bouyer et al., "Anatomical Localization of Beta Rhythms in Cats."

10 VISUAL EXPLORATION

1. This word is the English translation of the name of a machine invented by the scientist Cosinus in Georges Columb [Christophe pseud.], *L'Idée fixe du savant Cosinus* (Paris: Colin, 1939).

2. D. G. MacKay, *The Organization of Perception and Action* (New York: Springer, 1987). For more on this topic see Bergson's observations in *Matter and Memory.*

3. J. C. Eccles, M. Ito, and J. Szentogothai, *The Cerebellum as a Neuronal Machine* (Berlin: Springer, 1967); R. Llinás and C. Sotelo, *The Cerebellum Revisited* (New York: Springer, 1992).

4. J. Decety and D. H. Ingvar, "Brain Structures Participating in Mental Simulation of Motor Behaviour: A Neuropsychological Interpretation," *Acta Physiologica Scandinavica,* 73 (1990): 13–34.

5. H. Korn and D. S. Faber, "Organisation and Cellular Mechanisms Underlying Chemical Inhibition in a Vertebrate Neuron," in J.-P. Changeux, ed., *Molecular and Cellular Interactions Underlying Higher Brain Functions,* Progress in Brain Research (Amsterdam: North Holland, 1983); D. S. Faber, W. S. Young, P. Legendre, and H. Korn, "Intrinsic Quantal Variability Due to Stochastic Properties of Receptor-Transmitter Interactions," *Science,* 258 (1992): 1494–1498; R. Miles, K. Toth, A. Gulyas, N. Hajos, and T. F. Freund, "Differences between Somatic and Dendritic Inhibition in the Hippocampus," *Neuron,* 16 (1996): 814–823. These researchers showed that inhibiting junctions have several ways of acting on their neuronal target: controlling the base potential of these cells, opposing the effect of excitatory inputs when these are not sufficiently intense (that is, "significant"), adjusting the rate at which these cells emit signals, and finally in synchronizing the activity of clusters of neighboring cells, which makes them more "effective." Moreover, inhibitory junctions evidence training and memory properties until recently unsuspected.

6. J. Piaget, *The Origin of Intelligence in the Child* (London: Routledge and Kegan Paul, 1953).

7. Von Holst and Mittelstaedt, "Das Reafferenzprinzip"; B. Bridgeman, "A Review of the Role of Efference Copy in Sensory and Oculomotor Control Systems," *Annals of Biomedical Engineering,* 23 (1995): 409–422.

8. Israël, "Memory-Guided Saccades."

9. I. Israël, S. Rivaud, P. Pierrot-Desilligny, and A. Berthoz, "Delayed VOR: An Assess-

ment of Vestibular Memory for Self Motion," in J. Requin and J. Stelmach, eds., *Tutorials in Motor Neuroscience* (The Netherlands: Kluwer, 1991), pp. 599–607; I. Israël, S. Rivaud, B. Gaymard, A. Berthoz, and C. Pierrot-Deseilligny, "Cortical Control of Vestibular-Guided Saccades in Man," *Brain,* 118 (1995): 1169–1183.

10. A. Berthoz, "Neural Basis of Decision in Perception and in the Control of Movement," in A. R. Damasio, et al., eds., *Neurobiology of Decision-Making* (Berlin: Springer, 1996), pp. 83–100; A. Berthoz and L. Petit, "Les mouvements du regard: une affaire de saccades. Un modèle pour l'étude des circuits de la décision et de l'imagination motrice," *La Recherche,* 289 (1996): 58–65.

11. Berthoz, "Neural Basis of Decision," pp. 83–100.

12. See A. Berthoz, "Coopération et substitution entre le système saccadique et les réflexes d'origine vestibulaires: faut-il réviser la notion de réflexe?" *Revue Neurologique,* 145 (1989): 513–526, for a detailed description of saccadic mechanisms in the brainstem. See also Berthoz and Petit, "Les mouvements du regard," pp. 58–65, for a detailed description of this schematic diagram.

13. I. S. Curthoys, C. H. Markham, and N. Furuya, "Direct Projection of Pause Neurons to Nystagmus-Related Excitatory Burst Neurons in the Cat Pontine Reticular Formation," *Experimental Neurology,* 83 (1984): 414–422.

14. D. P. Munoz and R. H. Wurtz, "Fixation Cells in Monkey Superior Colliculus. I. Characteristics of Cell Discharge," *Journal of Neurophysiology,* 70 (1993): 559–575; M. A. Segraves, "Effects of Frontal Eye Field Stimulation upon Omnipause and Burst Neurons in the Monkey Paramedian Pontine Reticular Formation," *Society of Neuroscience Abstracts,* 18 (1992): 296.10.

15. Berthoz et al., "Saccadic Eye Velocity."

16. A. Grantyn and R. Grantyn, "Axonal Patterns and Sites of Termination of Cat Superior Colliculus Neurons Projecting in the Tecto-bulbo-spinal Tract," *Experimental Brain Research,* 46 (1982): 243–256; A. Grantyn, R. Grantyn, V. Robine, and A. Berthoz, "Electroanatomy of Tectal Efferent Connections Related to Eye Movements in the Horizontal Plane," *Experimental Brain Research,* 37 (1979): 149–172; R. E. Wurtz and M. E. Goldberg, "Activity of Superior Colliculus in Behaving Monkey. III. Cells Discharging before Eye Movements," *Journal of Neurophysiology,* 35 (1972): 575–586; A. Grantyn and A. Berthoz, "The Role of the Tecto-reticulo-spinal System in Control of Head Movement," in G. W. Peterson and F. Richmond, eds., *Control of Head Movement* (Oxford: Oxford University Press, 1987), pp. 224–244; E. Oliver, A. Grantyn, M. Chat, and A. Berthoz, "The Control of Slow Orienting Eye Movements by Tectoreticulospinal Neurons in the Cat: Behavior, Discharge Patterns, and Underlying Connections," *Experimental Brain Research,* 93 (1993): 435–449; A. B. Moschovakis, A. B. Karabelas, and S. Highstein, "Structure Function Relationship in the Primate Superior Colliculus. I. Morphological Classification of Efferent Neurons," *Journal of Neurophysiology,* 60 (1988): 232–262, and "Structure Function Relationship in the Primate Superior Colliculus. II. Morphological Identity of Presaccadic Neurons," *Journal of Neurophysiology,* 60 (1988): 263–302.

17. A. B. Karabelas and A. K. Moschovakis, "Nigral Inhibitory Termination on Efferent Neurons of the Superior Colliculus. An Intracellular Horseradish Peroxidase Study in the Cat," *Journal of Cognitive Neuroscience,* 239 (1985): 309–329; J. M. Deniau and G.

Chevalier, "Disinhibition as a Basic Process in the Expression of Striatal Functions. The Striatonigral Influence on the Thalamocortical Cells of the Ventromedial Thalamic Nucleus," *Brain Research,* 334 (1981): 227–233.

18. C. Umilta, C. Mucignat, L. Riggio, C. Barbieri, and G. Rizzolatti, "Programming Shifts of Spatial Attention," *European Journal of Cognitive Psychology,* 6 (1994): 23–41; G. Rizzolatti, L. Riggio, and B. M. Sheliga, "Space and Selective Attention," in C. Umilta and M. Moscovitch, eds., *Attention and Performance XV* (Hillsdale, N.J.: Erlbaum, 1994), pp. 232–265.

19. O. Hikosaka and R. H. Wurtz, "Visual and Oculomotor Functions of Monkey Substantia Nigra Pars Reticulata. III. Memory-Contingent Visual and Saccade Responses," *Journal of Neurophysiology,* 49 (1983): 1268–1284.

20. O. Hikosaka, M. Sakamoto, and N. Miyashita, "Effects of Caudate Nucleus Stimulation on Substantia Nigra Cell Activity in Monkey," *Experimental Brain Research,* 95 (1993): 457–472.

21. O. Hikosaka, "Role of the Basal Ganglia in Motor Learning: A Hypothesis," in Ono et al., *Brain Mechanisms,* pp. 497–513.

22. J. Droulez and A. Berthoz, "The Concept of Dynamic Memory in Sensorimotor Control," in Humphrey and Freund, *Motor Control,* pp. 137–161.

23. Kalaska et al., "Cortical Mechanisms"; M. Jeannerod, "The Posterior Parietal Cortex as a Spatial Generator," in D. I. Ingle, J. Jeannerod, and D. Lee, eds., *Brain Mechanisms and Spatial Vision* (Dordrecht: Martinus Hijhoff, 1985), pp. 279–298; Andersen, "Visual and Oculomotor Functions"; Y. Burnod, *An Adaptative Network, the Cerebral Cortex* (Paris: Masson, 1988). For a review, see P. Thier and O. Karnath, *Parietal Lobe Contribution to Orientation in 3D Space* (Berlin: Springer, 1997).

24. R. A. Andersen and V. B. Mountcastle, "The Influence of the Angle of Gaze upon the Excitability of the Light Sensitive Neurons of the Posterior Parietal Cortex," *Journal of Neuroscience,* 3 (1983): 532–548.

25. R. A. Anderson, G. K. K. Essick, and R. M. Siegel, "Neurons of Area 7 Activated by Both Visual Stimuli and Oculomotor Behavior," *Experimental Brain Research,* 67 (1987): 316–322; R. A. Andersen, "Visual and Oculomotor Functions of the Posterior Parietal Cortex," *Annual Review of Neuroscience,* 12 (1989): 377–403.

26. H. Sakata and M. Kusunoki, "Organization of Space Perception: Neural Representation of Three-Dimensional Space in the Posterior Parietal Cortex," *Current Opinion in Neurobiology,* 2 (1992): 170–174; Andersen, "Visual and Oculomotor Functions," pp. 377–403; Andersen et al., "Neurons of Area 7," pp. 316–322; R. A. Andersen, R. M. Bracewell, S. Barash, J. W. Gnadt, and L. Fogassi, "Eye Position Effects on Visual, Memory, and Saccade-Related Activity in Area LIP and 7a of the Macaque," *Journal of Neuroscience,* 10 (1990): 1176–1196.

27. C. L. Colby, J.-R. Duhamel, and M. E. Goldberg, "Oculocentric Spatial Representation in Parietal Cortex," *Cerebral Cortex,* 5 (1995): 470–481; J.-R. Duhamel, F. Bremmer, S. Ben Hamed, and W. Graf, "Spatial Invariance of Visual Receptive Fields in Parietal Cortex Neurons," *Nature,* 389 (1997): 845–848.

28. L. Petit, C. Orssaud, N. Tzourio, F. Crivello, A. Berthoz, and B. Mazoyer, "Functional Anatomy of a Prelearned Sequence of Horizontal Saccades in Man," *European Journal of Neuroscience,* 16 (1996): 3726–3741.

29. W. F. McDaniel, D. M. Compton, and S. R. Smith, "Spatial Learning following Posterior Parietal or Hippocampal Lesions," *NeuroReport,* 5 (1994): 1713–1717. See also our recent findings from a mental locomotor task [O. Ghaem, E. Mellet, F. Crivello, N. Tzourio, B. Mazoyer, A. Berthoz, and M. Denis, "Mental Navigation along Memorized Routes Activates the Hippocampus, Precuneus, and Insula," *NeuroReport* 8 (1997): 739–744].

30. M. N. Shadlen and W. T. Newsome, "Motion Perception: Seeing and Deciding," *Proceedings of the National Academy of Sciences,* 93 (1996): 628–633.

31. J.-R. Duhamel, C. L. Colby, and M. E. Goldberg, "The Updating of the Representation of Visual Space in Parietal Cortex by Intended Eye Movements," *Science,* 255 (1992): 90–92.

32. P. Viviani and A. Berthoz, "Voluntary Deceleration and Perceptual Activity during Oblique Saccades," in R. Baker and A. Berthoz, eds., *Control of Gaze by Brain Stem Neurons* (Amsterdam: Elsevier, 1977), pp. 23–28.

33. M. A. Segraves and M. E. Goldberg, "Functional Properties of Corticotectal Neurons in the Monkey's Frontal Eye Field," *Journal of Neurophysiology,* 58 (1987): 1387–1419.

34. D. Ferrier, *Functions of the Brain.*

35. Penfield and Boldrey, "Somatic Motor and Sensory Representation"; Penfield and Rasmussen, *Cerebral Cortex of Man.*

36. E. Melamed and B. Larsen, "Saccadic Eye Movements in Humans"; P. T. Fox, J. M. Fox, M. E. Raichle, and R. M. Burde, "The Role of the Cerebral Cortex in the Generation of Voluntary Saccades," *Journal of Neurophysiology,* 54 (1985): 348–369.

37. L. Petit, C. Orssaud, N. Tzourio, G. Salamon, B. Mazoyer, and A. Berthoz, "PET Study of Voluntary Saccadic Eye Movements in Humans: Basal Ganglia-Thalamocortical System and Cingulate Cortex Involvement," *Journal of Neurophysiology,* 69 (1993): 1009–1017; Petit et al., "Functional Anatomy," pp. 3726–3741.

38. L. Lang, D. Cheyne, R. Kristeva, R. Beisteiner, G. Lindinger, and L. Deecke, "Three-Dimensional Localization of SMA Activity Preceding Voluntary Movement. A Study of Electric and Magnetic Fields in a Patient with Infarction of the Right Supplementary Motor Area," *Experimental Brain Research,* 87 (1991): 688–695.

39. M. P. Deiber, R. E. Passingham, J. G. Colebatch, K. J. Friston, P. D. Nixon, and R. S. Frackowiak, "Cortical Areas and the Selection of Movement: A Study with Positron Emission Tomography," *Experimental Brain Research,* 84 (1991): 393–402.

40. W. Lang, M. Lang, F. Uhl, C. Koska, A. Kornhuber, and L. Deecke, "Negative Cortical DC Shifts Preceding and Accompanying Simultaneous and Sequential Finger Movements," *Experimental Brain Research,* 71 (1988): 579–587; W. Lang, H. Obrig, G. Lindinger, D. Cheyne, and L. Deecke, "Supplementary Motor Area Activation While Tapping Bimanually Different Rhythms in Musicians," *Experimental Brain Research,* 79 (1990): 504–514.

41. J. Schlag and M. Schlag-Rey, "Evidence for a Supplementary Eye Field," *Journal of Neurophysiology,* 57 (1987): 179–200.

42. B. Gaymard, C. Pierrot-Deseilligny, and S. Rivaud, "Impairment of Sequences of Memory-Guided Saccades after Supplementary Motor Area Lesions," *Annals of Neurology,* 28 (1990): 622–626; B. Gaymard, S. Rivaud, and C. Pierrot-Deseilligny, "Role of the Left

and Right Supplementary Motor Areas in Memory-Guided Saccade Sequences," *Annals of Neurology,* 34 (1993): 404–406.

43. Petit et al., "PET Study," and "Functional Anatomy."

44. S. Y. Musil, C. R. Olson, and M. E. Goldberg, "Visual and Oculomotor Properties of Single Neurons in Posterior Cingulate Cortex of Rhesus Monkey," *Society of Neuroscience Abstracts,* 16 (1990): 1221.

45. P. S. Goldman-Rakic, "Circuitry of Primate Prefrontal Cortex and Regulation of Behavior by Representational Memory," in V. B. Mountcastle, F. Plkum, and D. Humphrey, eds., *Handbook of Physiology. The Nervous System V* (Bethesda: American Physiological Society, 1987), pp. 373–417; S. Funahashi, C. Bruck, and P. S. Goldman-Rakic, "Mnemonic Coding of Visual Space in the Monkey's Dorsolateral Prefrontal Cortex," *Journal of Neurophysiology,* 61 (1989): 1–19.

46. J. P. Joseph and P. Barone, "Prefrontal Unit Activity during a Delayed Oculomotor Task in the Monkey," *Experimental Brain Research,* 67 (1987): 460–468; A. Diamond and P. S. Goldman-Rakic, "Comparison of Human Infants and Rhesus Monkeys on Piaget's AB Task: Evidence for Dependence on Dorsolateral Prefrontal Cortex," *Experimental Brain Research,* 74 (1989): 24–40; C. D. Frith, K. Friston, P. F. Liddle, and R. S. Frackowiak, "Willed Action and the Prefrontal Cortex in Man: A Study with PET," *Proceedings of the Royal Society of London. Series B: Biological Sciences,* 244 (1991): 241–246; G. Di Pellegrino and S. P. Wise, "Visuospatial versus Visuomotor Activity in the Premotor and Prefrontal Cortex of a Primate," *Journal of Neuroscience,* 13 (1993): 1227–1243; A. Bechara, A. R. Damasio, H. Damasio, and S. W. Anderson, "Insensitivity to Future Consequences following Damage to Human Prefrontal Cortex," *Cognition,* 50 (1994): 7–15; M. Petrides, "Impairments on Nonspatial Self-Ordered and Externally Ordered Working Memory Tasks after Lesions of the Mid-dorsal Part of the Lateral Frontal Cortex in the Monkey," *Journal of Neuroscience,* 15 (1995): 359–375.

47. D. Boussaoud and S. P. Wise, "Primate Frontal Cortex: Neuronal Activity following Attentional versus Intentional Cues," *Experimental Brain Research,* 95 (1993): 15–27; Di Pellegrino and Wise, "Visuospatial versus Visuomotor Activity," pp. 1227–1243; R. P. Kesner, "Paired Associate Learning in the Rat: Role of Hippocampus, Medial Prefrontal Cortex, and Parietal Cortex," *Psychobiology,* 21 (1993): 183–192.

48. Goldman-Rakic, "Circuitry of Primate Prefrontal Cortex"; J. M. Fuster, *The Prefrontal Cortex: Anatomy, Physiology, and Neuropsychology of the Frontal Lobe* (New York: Raven Press, 1996); S. Dehaene and J.-P. Changeux, "A Simple Model of Prefrontal Cortex Function in Delayed-Response Tasks," *Journal of Cognitive Neuroscience,* 1 (1990): 244–261.

49. Pierrot-Deseilligny et al., "Role of the Different Frontal Lobe Areas."

50. Israël et al., "Vestibular-Guided Saccades in Man."

51. D. Guitton, H. A. Buchtel, and R. Douglas, "Frontal Lobe Lesions in Man Cause Difficulties in Supressing Reflexive Glance and in Generating Goal-Directed Saccades," *Experimental Brain Research,* 58 (1985): 455–472.

52. K. Sasaki and H. Gemba, "Electrical Activity in the Prefrontal Cortex Specific to No-Go Reaction of Conditioned Hand Movement with Colour Discrimination in the Monkey," *Experimental Brain Research,* 64 (1986): 603–606.

53. J. P. Aggleton, N. Neave, S. Nagle, and A. Sahgal, "A Comparison of the Effects of Me-

dial Prefrontal, Cingulate Cortex, and Cingulum Bundle Lesions on Tests of Spatial Memory: Evidence of a Double Dissociation between Frontal and Cingulum Bundle Contributions," *Journal of Neuroscience,* 15 (1995): 7270–7281; J. Mogensen, T. K. Pedersen, S. Holm, and L. E. Bang, "Prefrontal Cortical Mediation of Rats' Place Learning in a Modified Water Maze," *Brain Research Bulletin,* 38 (1995): 425–434.

54. E. Salzmann, "Attention and Memory Trials during Neuronal Recording from the Primate Pulvinar and Posterior Parietal Cortex (Area PG)," *Behavioral Brain Research,* 67 (1995): 241–253.

55. Frith et al., "Willed Action."

56. S. Park and P. S. Holzman, "Association of Working Memory Deficit and Eye Tracking Dysfunction in Schizophrenia," *Schizophrenia Research,* 11 (1993): 55–61.

57. B. M. Sheliga, L. Riggio, and G. Rizzolatti, "Orienting of Attention and Eye Movements," *Experimental Brain Research,* 98 (1994): 507–522; A. Berthoz, "The Role of Inhibition in the Hierarchical Gating of Executed and Imagined Movements," *Cognitive Brain Research,* 3 (1996): 101–113.

58. W. Lang, L. Petit, P. Hölliger, U. Pietrzyck, N. Tzourio, B. Mazoyer, and A. Berthoz, "A Positron Emission Tomography Study of Oculomotor Imagery," *NeuroReport,* 5 (1994): 921–924.

59. Droulez and Berthoz, "Concept of Dynamic Memory," "A Neural Network Model of Sensoritopic Maps with Predictive Short Term Memory Properties," *Proceedings of the National Academy of Sciences,* 88 (1991): 9653–9657, and "The Dynamic Memory Model and the Final Oculomotor and Cephalomotor Integrator," in H. Shimazu and Y. Shinoda, eds., *The Oculomotor System* (Tokyo: Japan Scientific Societies Press, 1991), pp. 1–19; L. E. Mays and D. L. Sparks, "Saccades Are Spatially, Not Retinocentrically, Coded," *Science,* 208 (1980): 1163–1165.

60. J. Piaget, "Le problème neurologique de l'intériorisation des actions en opérations réversibles," *Archives de psychologie,* 32 (1949): 242–258. Ferrier also became interested in this idea. On page 285 of *Functions of the Brain* he wrote: "In these sentences, and particularly the last, Bain seems to me to have clearly indicated the elements of attention, which I conceive to be a combination of the activity of the motor, and of the inhibitory-motor centres."

11 BALANCE

1. A. Thomas, *Equilibre et équilibration* (Paris: Masson, 1940).

2. Lee and Lishman, "Visual Proprioceptive Control of Stance."

3. Dichgans and Brandt, "Visual-Vestibular Interaction."

4. See Chapter 4.

5. Lee and Aronson, "Visual Proprioceptive Control of Standing."

6. K. F. Hamann, P. P. Vidal, J. M. Sterkers, and A. Berthoz, "A New Test for Postural Disorders: An Application of Visual Stabilization," *Agressologie,* 20B (1979): 129–130.

7. Clément et al., "Adaptation of Postural Control."

8. F. Lestienne, J. Soechting, and A. Berthoz, "Postural Readjustments Induced by Linear Motion of Visual Scenes," *Experimental Brain Research,* 28 (1977): 363–384; J. F. Soechting and A. Berthoz, "Dynamic Role of Vision in the Control of Posture in Man," *Experimental Brain Research,* 36 (1979): 551–561.

9. See Chapter 2.

10. This mental calculation, used in clinical practice during vestibular tests, is assumed to decrease high-level cognitive factors by occupying the subject with a mental task.

11. Sometimes the expression "posturally blind" is used to refer to persons who are insensitive to fluctuations of the visual environment.

12. I. Viaud-Delmon, Y. P. Ivanenko, A. Berthoz, and R. Jouvent, "Sex, Lies, and Virtual Reality," *Nature*, 1 (1998): 15–16.

13. M. Reuchlin, J. Lautrey, C. Marendaz, and T. Ohlman, *Cognition: L'universel et l'individuel* (Paris: Presses Universitaires de France, 1989); M. Denis, *Image et Cognition* (Paris: Presses Universitaires de France, 1989).

14. L. M. Nashner and A. Berthoz, "Visual Contribution to Rapid Motor Responses during Postural Control," *Brain Research*, 150 (1978): 403–407.

15. For a review, see Prochazka, "Sensorimotor Gain Control."

16. C. S. Sherrington, *The Integrative Action of the Nervous System* (New Haven: Yale University Press, 1906).

17. Feldman and Levin, "Positional Frames of Reference"; Bizzi et al., "Posture Control and Trajectory Formation." See Chapter 4.

18. W. Hess, "Teleokinetisches und ereismatisches Kraftsystem in der Biomotorik," *Helvetica Physiologica et Pharmacologica Acta*, 1 (1943): C62–C63.

19. Evarts, "Pyramidal Tract Activity"; E. V. Evarts, C. Fromm, J. Kroller, and V. A. Jennings, "Motor Cortex Control of Finely Graded Forces," *Journal of Neurophysiology*, 49 (1983): 1199–1215.

20. Paillard, "Le corps et ses langages d'espace," and "L'intégration sensori-motrice et idéo-motrice," in M. Richelle, J. Requin, and M. Robert, eds., *Traité de psychologie expérimentale* (Paris: Presses Universitaires de France, 1994), pp. 925–961; A. Riehle and J. Requin, "Monkey Primary Motor and Premotor Cortex: Single-Cell Activity Related to Prior Information about Direction and Extent of an Intended Movement," *Journal of Neurophysiology*, 61 (1989): 534–549.

21. Sechenov, *Selected Works*.

22. J. Massion, V. Gurfinkel, M. Lipshits, A. Obadia, and K. Popov, "Stratégie et synergie: deux niveaux de contrôle de l'équilibre au cours du mouvement. Effets de la microgravité," *Comptes rendus hebdomadaires de l'Académie des sciences (Paris, Série III)*, 314 (1992): 87–92; P. Crenna, C. Frigo, J. Massion, and A. Pedotti, "Forward and Backward Axial Synergies in Man," *Experimental Brain Research*, 65 (1987): 538–548; A. Pedotti, P. Crenna, A. Deat, C. Frigo, and J. Massion, "Postural Synergies in Axial Movements: Short and Long-Term Adaptation," *Experimental Brain Research*, 74 (1989): 3–10; J. Massion, F. Viallet, R. Massarino, and R. Khalil, "La région de l'aire motrice supplémentaire est impliquée dans la coordination entre posture et mouvement," *Comptes rendus hebdomadaires de l'Académie des sciences (Paris, Série III)*, 308 (1989): 417–423.

23. S. Bouisset and M. Zattara, "A Sequence of Postural Movements Precedes Voluntary Movement," *Neuroscience Letters*, 22 (1989): 263–270, and "Biomechanical Study of the Programming of Anticipatory Postural Adjustments Associated with Voluntary Movement," *Journal of Biomechanics*, 20 (1987): 735–742; D. Bazalgette, M. Zattara, N. Bathien, S. Bouisset, and P. Rondot, "Postural Adjustments Associated with Rapid Vol-

untary Arm Movements in Patients with Parkinson's Disease," in M. D. Yahr and K. J. Bergman, eds., *Advances in Neurology,* vol. 45 (New York: Raven Press, 1986), pp. 371–374; Clément et al., "Adaptation of Postural Control."

24. L. Mouchnino, R. Aurenty, J. Massion, and A. Pedotti, "Is the Trunk a Reference Frame for Calculating Leg Position?" *NeuroReport,* 4 (1993): 125–127.

25. J. Babinski, "De l'asynergie cérébelleuse," *Revue de Neurologie,* 7 (1989): 806–816; see recent findings on the physiopathology of this synergy in Bouisset and Zattara, "Biomechanical Study"; Clément et al., "Adaptation of Postural Control"; Crenna et al., "Forward and Backward."

26. Bazalgette et al., "Postural Adjustments," pp. 371–374; F. Viallet, J. Massion, R. Massarino, and R. Khalil, "Coordination between Posture and Movement in Parkinsonism and SMA Lesion," in L. Deecke, J. C. Eccles, and V. B. Mountcastle, eds., *From Neuron to Action* (Berlin: Springer, 1990), pp. 65–70; R. G. Lee, I. Tonolli, F. Viallet, R. Aurenty, and J. Massion, "Preparatory Postural Adjustment in Parkinsonian Patients with Postural Instability," *Canadian Journal of Neurological Sciences,* 22 (1995): 126–135.

27. Massion et al., "La région de l'aire motrice supplémentaire."

28. Y. Gahery and J. Massion, "Coordination between Posture and Movement," in E. V. Evarts, S. P. Wise, and D. Bousfield, eds., *The Motor System in Neurobiology* (Amsterdam: Elsevier Biomedical Press, 1985), pp. 121–125.

29. Y. Paulignan, M. Dufosse, M. Hugon, and J. Massion, "Acquisition of Coordination between Posture and Movement in a Bimanual Task," *Experimental Brain Research,* 77 (1989): 337–348.

30. G. Clément, T. Pozzo, and A. Berthoz, "Contribution of Eye Positioning in the Control of Upside-Down Vertical Posture," *Experimental Brain Research,* 73 (1988): 569–576.

31. H. Head and G. Holmes, "Sensory Disturbances from Cerebral Lesions," *Brain,* 34 (1911): 102–244.

32. V. S. Gurfinkel, E. E. Debreva, and Y. S. Levik, "The Role of the Internal Model in the Perception of Position and Planning of Hand Movements," *Fiziologiia Cheloveka,* 12 (1986): 769–776; V. S. Gurfinkel and Y. S. Levick, "Perceptual and Automatic Aspects of the Postural Body Scheme," in Paillard, *Brain and Space,* pp. 174–162.

33. E. D. Adrian, *The Physical Background of Perception* (Oxford: Clarendon Press, 1947), pp. 93–94.

34. See Chapter 2.

35. A. Berthoz, "Coopération et substitution entre le système saccadique et les réflexes d'origine vestibulaires: faut-il réviser la notion de réflèxe?" *Revue Neurologique,* 145 (1989): 513–526.

36. These experiments were continued by E. Bizzi's group at MIT.

37. Gomi and Kawato, "Adaptive Feedback Control Models."

38. J. Massion and L. Rispal-Padel, "Thalamus: fonctions motrices," *Revue Neurologique,* 142 (1986): 327–336.

39. M. Poncet, J. F. Pellissier, M. Sebahoun, and C. J. Masser, "A propos d'un cas d'autotopagnosie secondaire à une lésion pariéto-occipitale de l'hémisphère majeur," *Encéphale,* 60 (1971): 110–123.

40. M. I. Posner, M. J. Nissen, and W. C. Ogden, "Attended and Unattended Processing

Modes: The Role of Set for Spatial Location," in H. L. Pick and I. J. Saltzman, eds., *Modes of Perceiving and Processing Information* (Hillsdale, N.J.: Erlbaum, 1988), pp. 137–157.

41. A. Sirigu, J. Grafman, K. Bressler, and T. Sunderland, "Multiple Representations Contribute to Body Knowledge Processing," *Brain,* 114 (1991): 629–642.

42. E. Bonda, M. Petrides, S. Frey, and A. Evans, "Neural Correlates of Mental Transformations of the Body-in-Space," *Proceedings of the National Academy of Sciences,* 92 (1995): 11180–11184.

43. M. M. Mesulam, "A Cortical Network for Directed Attention and Unilateral Neglect," *Annals of Neurology,* 10 (1981): 309–325.

12 ADAPTATION

1. A. Gonshor and G. Melvill Jones, "Extreme Vestibulo-ocular Adaptation by Prolonged Optical Reversal of Vision," *Journal of Physiology,* 256 (1976): 381–414.

2. See Figure 2.4.

3. M. Ito, "Cerebellar Control of the Vestibulo-ocular Reflex around the Flocculus Hypothesis," *Annual Review of Neuroscience,* 5 (1982): 275–296; *The Cerebellum and Neuronal Control* (New York: Raven, 1984).

4. F. A. Miles and S. G. Lisberger, "Plasticity in the Vestibulo-ocular Reflex: A New Hypothesis," *Annual Review of Neuroscience,* 4 (1981): 273–299; S. Du Lac, J. L. Raymond, T. J. Sejnowski, and S. G. Lisberger, "Learning and Memory in the Vestibulo-Ocular Reflex," *Annual Review of Neuroscience,* 18 (1995): 409–441.

5. Berthoz and Melvill-Jones, *Adaptive Mechanisms in Gaze Control;* G. Melvill-Jones, D. Guitton, and A. Berthoz, "Changing Patterns of Eye-Head Coordination during Six Hours of Optically Reversed Vision," *Experimental Brain Research,* 69 (1988): 531–544.

6. G. Melvill-Jones, A. Berthoz, and B. Segal, "Adaptative Modification of the Vestibuloocular Reflex by Mental Effort in Darkness," *Experimental Brain Research,* 56 (1984): 149–153.

7. A. Berthoz, "Coopération et substitution entre le système saccadique et les réflexes d'origine vestibulaires: faut-il réviser la notion de réflexe?" *Revue Neurologique,* 145 (1989): 513–526; Viaud-Delmon et al., "Sex, Lies, and Virtual Reality"; I. Viaud-Delmon, Y. Ivanenko, A. Berthoz, and R. Jouvent, "Anxiety and Integration of Visual Vestibular Information Studied with Virtual Reality," *Biological Psychiatry,* in press.

8. Nudo et al., "Use-Dependent Alterations."

9. J.-C. Baudrillard, D. Rousié, H. Foucart, C. Defache, J.-C. Hache, P. Van Tichelen, and J.-M. Lerais, "IRM des canaux semi-circulaires dans les asymétries cranio-faciales et le syndrome cranio-mandibulaire," *Journal de Radiologie,* 76 (1995): 579–585.

10. Ivanenko et al., "The Contribution of Otoliths."

11. Vitte et al., "Recovery from Vestibular Deficits."

13 THE DISORIENTED BRAIN

1. J. Sully, *Illusions: A Psychological Study* (New York: Da Capo Press, 1982), p. 333.

2. K. S. Lashley, *Brain Mechanisms and Intelligence* (Chicago: University of Chicago Press, 1929).

3. I. Krechevsky, "Hypothesis in Rats," *Psychological Reviews,* 39 (1932): 517–532; quotation from p. 528.

4. C. Darlot, P. Denise, J. Droulez, B. Cohen, and A. Berthoz, "Eye Movements Induced by Off-Vertical Axis Rotation (OVAR) at Small Angles of Tilt," *Experimental Brain Research,* 73 (1988): 91–105, and "Motion Perception Induced by Off-Vertical Axis Rotation (OVAR) at Small Angles of Tilt," *Experimental Brain Research,* 73 (1988): 106–114.

5. A. Fessard and A. Tournay, "Quelques données et réflexions sur le phénomène de la post-contraction involontaire," *L'Année psychologique,* 50 (1950): 216–235. This illusion was recently investigated by Lestienne and Gurfinkel.

6. J. P. Roll, J. C. Gilhodes, R. Roll, and F. Harley, "Are Proprioceptive Sensory Inputs Combined into a Gestalt?" in T. Inoue and J. T. MacClelland, eds., *Attention and Performance XVI* (Cambridge, Mass.: MIT Press, 1995), pp. 290–314; Lackner and Levine, "Changes in Apparent Body Orientation."

7. Roll et al., "Are Proprioceptive Sensory Inputs Combined?" pp. 307–308.

8. Rolls and O'Mara, "Neurophysiological and Theoretical Analysis"; O'Mara et al., "Whole Body Motion."

9. The recent findings of Damasio regarding the significance of the prefrontal cortex in relationships between vegetative and cognitive mechanisms and a rereading of the work of the Sokolov school on relationships between the hippocampus and the reticular formation stimulated this speculation.

10. H. Hécaen and J. De Ajuriaguerra, *Méconnaissances et hallucinations corporelles: intégration et désintégration de la somatognosie* (Paris: Masson et Cie, 1952).

11. Shepard, "Ecological Constraints," p. 436.

14 ARCHITECTS HAVE FORGOTTEN THE PLEASURE OF MOVEMENT

1. Baron Georges Eugène Haussmann (1809–1891) was the French city planner whose redesign of Paris between 1853 and 1870 resulted in the broad streets, radiating avenues, and vistas for which the city is famous. [Trans.]

CONCLUSION

1. Bergson, *Matter and Memory,* p. 46.

2. H. Bergson, *Mind-Energy: Lectures and Essays,* trans. H. Wildon Carr (Westport, Conn.: Greenwood, 1975), p. 58.

3. A. Sirigu, J. R. Duhamel, L. Cohen, B. Pillon, B. Dubois, and Y. Abid, "The Mental Representation of Hand Movements after Parietal Cortex Damage," *Science,* 273 (1996): 1564–1568.

WORKS CITED

Abeles, M. *Local Cortical Circuits.* Berlin: Springer, 1982.

Adrian, E. D. *The Physiological Background of Perception.* Oxford: Clarendon Press, 1947.

Aggleton, J. P., N. Neave, S. Nagle, and A. Sahgal. "A Comparison of the Effects of Medial Prefrontal, Cingulate Cortex, and Cingulum Bundle Lesions on Tests of Spatial Memory: Evidence of a Double Dissociation between Frontal and Cingulum Bundle Contributions." *Journal of Neuroscience,* 15 (1995): 7270–7281.

Akbarian, S., O.-J. Grüsser, and W. O. Guldin. "Corticofugal Projections to the Vestibular Nuclei in Squirrel Monkeys: Further Evidence of Multiple Cortical Vestibular Fields." *Journal of Comparative Neurology,* 332 (1993): 89–104.

Amblard, B., A. Berthoz, and F. Clarac, eds. *Posture and Gait: Development, Adaptation, and Modulation.* Amsterdam: Elsevier, 1988.

Amorim, M. A., S. Glasauer, K. Corpinot, and A. Berthoz. "Updating an Object's Orientation and Location during Non-Visual Navigation: A Comparison between Two Processing Modes." *Perception and Psychophysics* 59 (1997): 404–418.

Andersen, R. A. "Visual and Oculomotor Functions of the Posterior Parietal Cortex." *Annual Review of Neuroscience,* 12 (1989): 377–403.

Andersen, R. A., R. M. Bracewell, S. Barash, J. W. Gnadt, and L. Fogasse. "Eye Position Effects on Visual, Memory and Saccade-Related Activity in Area LIP and 7a of the Macaque." *Journal of Neuroscience,* 10 (1990): 1176–1196.

Andersen, R. A., G. K. K. Essick, and R. M. Seigel. "Neurons of Area 7 Activated by Both Visual Stimuli and Oculomotor Behavior." *Experimental Brain Research,* 67 (1987): 316–322.

Andersen, R. A., and V. B. Mountcastle. "The Influence of the Angle of Gaze upon the Excitability of the Light Sensitive Neurons of the Posterior Parietal Cortex." *Journal of Neuroscience,* 3 (1983): 532–548.

Anokhin, P. K. *Biology and Neurophysiology of Conditioned Reflexes and Their Role in Adaptive Behaviour.* Oxford: Pergamon Press, 1974.

Arbib, M. A. "Perceptual Structures and Distributed Motor Control." In *Handbook of Physiology,* sect. 1, *The Nervous System,* vol. 2, part 2, ed. V. B. Brooks. Bethesda: American Physiological Society, 1981, 1449–1480.

———. "Interaction of Multiple Representations of Space in the Brain." In *Brain and Space,* ed. J. Paillard. Oxford: Oxford University Press, 1993, 380–403.

Arbib, M., and S. Amari. "Sensorimotor Transformations in the Brain (with a Critique of the Tensor Theory of the Cerebellum)." *Journal of Theoretical Biology,* 112 (1985): 123–155.

Arbib, M., T. Iberall, and G. Bingham. "Opposition Space as a Structuring Concept for the Analysis of Skilled Hand Movements." *Experimental Brain Research,* 15 (1986): 158–173.

Aristotle. *Psychologie. Opuscules.* Paris: Dumont, 1847.

———. *De Sensu and De Memoria,* trans. G. R. T. Ross. New York: Arno, 1973.

Aubert, H. "Über eine scheinebare Drehung von Objecten bei Neigung des Kopfes nach rechts oder links." *Virchows Archiv*, 20 (1967): 381–393.

Babinski, J. "De l'asynergie cérébelleuse." *Revue de Neurologie*, 7 (1899): 806–816.

Bach-y-Rita, P. "Vision Substitution by Tactile Image Projection." *Nature*, 221 (1989): 963–964.

Baker, R., and A. Berthoz. "Organization of Vestibular Nystagmus in the Oblique Oculomotor System." *Journal of Neurophysiology*, 37 (1974): 195–217.

———. "Is the Prepositus Hypoglossi Nucleus the Source of Another Vestibular Ocular Pathway?" *Brain Research*, 86 (1975): 121–172.

———. *Control of Gaze by Brain Stem Neurons*. Amsterdam: Elsevier, 1977.

Baker, R., M. Gresty, and A. Berthoz. "Neuronal Activity in the Prepositus Hypoglossi Nucleus Correlated with Vertical and Horizontal Eye Movements in the Cat." *Brain Research*, 101 (1976): 366–371.

Barbur, J. L., A. J. Harlow, and L. Weiskrantz. "Spatial and Temporal Response Properties of Residual Vision in a Case of Hemianopia." *Philosophical Transactions of the Royal Society of London. Series B: Biological Sciences*, 343 (1994): 157–166.

Barlow, J. S. "Inertial Navigation as a Basis for Animal Navigation." *Journal of Theoretical Biology*, 6 (1964): 76–117.

Baudrillard, J.-C., D. Rousié, H. Foucart, C. Defache, J.-C. Hache, P. Van Tichelen, and J.-M. Lerais. "IRM des canaux semi-circulaires dans les asymétries cranio-faciales et le syndrome cranio-mandibulaire." *Journal of Radiology*, 76 (1995): 579–585.

Bazalgette, D., M. Zattara, N. Bathien, S. Bouisset, and P. Rondot. "Postural Adjustments Associated with Rapid Voluntary Arm Movements in Patients with Parkinson's Disease." In *Advances in Neurology*, vol. 45, ed. M. D. Yahr and K. J. Bergman. New York: Raven Press, 1986, 371–374.

Bechara, A., A. R. Damasio, H. Damasio, and S. W. Anderson. "Insensitivity to Future Consequences following Damage to Human Prefrontal Cortex." *Cognition*, 50 (1994): 7–15.

Belen'kii, V. E., V. S. Gurfinkel, and Y. I. Pal'tsev. "On Elements of Voluntary Movement Control." *Biofizika*, 12 (1967): 135–141.

Bergson, H. "L'âme et le corps." In *Le Matérialisme actuel*. Paris: Flammarion, 1913.

———. *The Creative Mind*, trans. Mabelle L. Andison. New York: Philosophical Library, 1946.

———. *Mind-Energy: Lectures and Essays*, trans. H. Wildon Carr. Westport, Conn.: Greenwood, 1975.

———. *Matter and Memory*, trans. Nancy Margaret Paul and W. Scott Palmer. New York: Zone Books, 1988.

Beritoff, J. S. *Neural Mechanisms of Higher Vertebrate Behavior*. New York: Brown and Co., 1965.

Bernstein, N. A. "Some Emergent Problems of the Regulation of Motor Acts," *Questions of Psychology*, no. 6 (1957). Reprinted in *Human Motor Actions: Bernstein Reassessed*, ed. H. T. A. Whiting. Advances in Psychology, vol. 17. Amsterdam: North-Holland, 1984, 354–355.

———. *The Coordination and Regulation of Movement*. New York: Pergamon Press, 1967.

Bertenthal, B. I., and R. K. Clifton. "Perception and Action." In *Handbook of Child Psychology*, ed. D. Kuhn and R. Siegler. New York: Wiley, 1996.

Berthoz, A. "Coopération et substitution entre le système saccadique et les réflexes d'origine vestibulaires: faut-il réviser la notion de réflexe?" *Revue Neurologique*, 145 (1989): 513–526.

———. "Reference Frames for the Perception and Control of Movement." In *Brain and Space*, ed. J. Paillard. Oxford: Oxford University Press, 1991, 82–111.

———. "La coopération des sens et du regard dans la perception du mouvement." In *Le Corps en jeu*, ed. O. Aslan. Paris: CNRS Editions, 1993, 17–25.

———, ed. *Multisensory Control of Movement*. Oxford: Oxford University Press, 1993.

———. "Neural Basis of Decision in Perception and in Control of Movement." In *Neurobiology of Decision-Making*, ed. A. R. Damasio, H. Damasio, and Y. Christen. Berlin: Springer, 1996, 83–100.

———. "How Does the Cerebral Cortex Process and Utilize Vestibular Signals?" In *Disorders of the Vestibular System*, ed. R. W. Baloh and G. M. Halmagyi. Oxford: Oxford University Press, 1996, 113–125.

———. "The Role of Inhibition in the Hierarchical Gating of Executed and Imagined Movements." *Cognitive Brain Research*, 3 (1996): 101–113.

———. "Hippocampal and Parietal Contribution to Topokinetic and Topographic Memory." In *The Hippocampal and Parietal Foundations of Spatial Cognition*, ed. N. Burgess, K. J. Jeffery, and J. O'Keefe. Oxford: Oxford University Press, 1999, 381–399.

———. *Leçons sur le corps, le cerveau et l'esprit*. Paris: Odile Jacob, 1999.

Berthoz, A., A. Amorim, S. Glasauer, R. Grasso, Y. Takei, and I. Viaud-Delmon. "Dissociation between Distance and Direction during Locomotor Navigation." In *Wayfinding Behaviour*, ed. R. G. Golledge. Baltimore: John Hopkins University Press, 1999, 328–348.

Berthoz, A., R. Baker, and W. Precht. "Labyrinthine Control of Inferior Oblique Motoneurons." *Experimental Brain Research*, 18 (1973): 225–241.

Berthoz, A., J. Droulez, P. P. Vidal, and K. Yoshida. "Neural Correlates of Horizontal Vestibulo-ocular Reflex Cancellation during Rapid Eye Movements in the Cat." *Journal of Physiology*, 419 (1989): 717–751.

Berthoz, A., W. Graf, and P. P. Vidal. *The Head-Neck Sensory-Motor System*. New York: Oxford University Press, 1991.

Berthoz, A., A. Grantyn, and J. Droulez. "Some Collicular Neurons Code Saccadic Eye Velocity." *Neuroscience Letters*, 72 (1987): 289–294.

Berthoz, A., and A. Guell. "Space Neuroscience Research." *Brain Research Reviews*, 28 (1998): 1–234.

Berthoz, A., I. Israël, P. Georges-François, R. Grasso, and T. Tsuzuku. "Spatial Memory of Body Linear Displacement: What Is Being Stored?" *Science*, 269 (1995): 95–98.

Berthoz, A., I. Israël, T. Vieville, and D. S. Zee. "Linear Head Displacement Measured by the Otoliths Can Be Reproduced through the Saccadic System." *Neuroscience Letters*, 82 (1987): 285–290.

Berthoz, A., and G. Melvill-Jones. *Adaptative Mechanisms in Gaze Control*. Amsterdam: Elsevier, 1985.

Berthoz, A., and S. Metral. "A Torque-Producing Stimulator for the Study of Muscular Response to Variable Forces." *Proceedings of the International Symposium on Biomechanics III.* Basel: Karger, 1973, 158–165.

Berthoz, A., B. Pavard, and L. Young. "Perception of Linear Horizontal Self Motion Induced by Peripheral Vision (Linear-Vection)." *Experimental Brain Research,* 23 (1974): 471–489.

Berthoz, A., and L. Petit. "Les mouvements du regard: une affaire de saccades. Un modèle pour l'étude des circuits de la décision et de l'imagination motrice." *La Recherche,* 289 (1996): 58–65.

Berthoz, A., and T. Pozzo. "Intermittent Head Stabilisation during Postural and Locomotory Tasks in Humans." In *Posture and Gait: Development, Adaptation, and Modulation,* ed. B. Amblard, A. Berthoz, and F. Clarac. Amsterdam: Elsevier, 1988, 189–198.

Berthoz, A., and P. P. Vidal. *Noyaux Vestibulaires et Vertiges.* Paris: Arnette, 1993.

Berthoz, A., P. P. Vidal, and W. Graf, eds. *The Head-Neck Sensory-Motor System.* New York: Oxford University Press, 1991.

Berthoz, A., K. Yoshida, and P. P. Vidal. "Horizontal Eye Movement Signals in Second Order Vestibular Nuclei Neurons in the Alert Cat." *Annals of the New York Academy of Sciences,* 374 (1981): 144–156.

Bienenstock, E., and C. von der Malsburg. "Statistical Coding and Short Term Synaptic Plasticity: A Scheme for Knowledge Representation in the Brain." In *Disordered Systems and Biological Organization,* ed. E. Bienenstock, F. Fogelman, and G. Weisbuch. Berlin: Springer, 1986, 247–272.

Bisiach, E., M. L. Rusconi, and G. Vallar. "Remission of Somatoparaphrenic Delusion through Vestibular Stimulation." *Neuropsychologia,* 29 (1991): 1029–1031.

Bizzi, E., N. Accornero, W. Chapple, and N. Hogan. "Posture Control and Trajectory Formation during Arm Movement." *Journal of Neuroscience,* 4 (1984): 2738–2744.

Bloomberg, J., G. Melvill-Jones, B. Segal, S. McFarlane, and J. Soul. "Vestibular Contingent Voluntary Saccades Based on Cognitive Estimates of Remembered Vestibular Information." *Advances in Oto-Rhino-Laryngology,* 40 (1988): 71–75.

Bonda, E., M. Petrides, S. Frey, and A. Evans. "Neural Correlates of Mental Transformations of the Body-in-Space." *Proceedings of the National Academy of Sciences,* 92 (1995): 11180–11184.

Bootsma, R. J., and C. E. Peper. "Predictive Visual Information Sources for the Regulation of Action with Special Emphasis on Catching and Hitting." In *Vision and Motor Control,* ed. L. Proteau and D. Elliot. Amsterdam: Elsevier, 1992, 285–314.

Bottini, G., E. Paulesu, R. Sterzi, E. Warburton, R. J. Wise, G. Vallar, R. S. Frackowiak, and C. D. Frith. "Modulation of Conscious Experience by Peripheral Sensory Stimuli." *Nature,* 376 (1995): 778–781.

Bottini, G., R. Sterzi, E. Paulesu, G. Vallar, S. Cappa, F. Erminio, R. E. Passingham, C. D. Frith, and R. S. J. J. Frackowiak. "Identification of the Central Vestibular Projections in Man: A Positron Emission Tomography Activation Study." *Experimental Brain Resarch,* 99 (1994): 164–169.

Bouisset, S., and M. Zattara. "Biomechanical Study of the Programming of Anticipatory Postural Adjustments Associated with Voluntary Movement." *Journal of Biomechanics,* 20 (1987): 735–742.

————. "A Sequence of Postural Movements Precedes Voluntary Movement." *Neuroscience Letters*, 22 (1989): 263–270.

Bourdieu, P. *Le Sens pratique*. Paris: Editions de Minuit, 1980.

Boussaoud, D., and S. P. Wise. "Primate Frontal Cortex: Neuronal Activity following Attentional versus Intentional Cues." *Experimental Brain Research*, 95 (1993): 15–27.

Bouveresse, J. *Langage, perception et réalité*. Nîmes: Jacqueline Chambon, 1995.

Bouyer, J. J., M. F. Montaron, J. M. Vahnée, M. Albert, and A. Rougeul. "Anatomical Localization of Beta Rhythms in Cats." *Neuroscience*, 22 (1987): 863–869.

Bradley, D. C., M. Maxwell, R. A. Andersen, M. S. Banks, and K. V. Shenoy. "Mechanisms of Heading Perception in Primate Visual Cortex." *Science*, 273 (1996): 1544–1547.

Brandt, T., J. Dichgans, and E. Koenig. "Differential Effects of Central versus Peripheral Vision on Egocentric and Exocentric Motion Perception." *Experimental Brain Research*, 16 (1973): 476–491.

Brandt, T., and M. Dieterich. "Skew Deviation with Ocular Torsion: A Vestibular Brainstem Sign of Topographic Diagnostic Value." *Annals of Neurology*, 33 (1993): 528–534.

Bridgeman, B. "A Review of the Role of Efference Copy in Sensory and Oculomotor Control Systems." *Annals of Biomedical Engineering*, 23 (1995): 409–422.

Buchanan, J. J., and J. A. S. Kelso. "Posturally Induced Transitions in Rhythmic Multijoint Limb Movements." *Experimental Brain Research*, 94 (1993): 131–142.

Buisseret, P., and E. Gary-Bobo. "Development of Visual Cortical Orientation Specificity after Dark Rearing: Role of Extra-ocular Proprioception." *Neuroscience Letters*, 13 (1979): 259–263.

Buisseret, P., and M. Imbert. "Visual Cortical Cells: Their Developmental Properties in Normal and Dark Reared Kittens." *Journal of Physiology*, 255 (1976): 511–525.

Burnod, Y. *An Adaptative Network, the Cerebral Cortex*. Paris: Masson, 1988.

Buzsáki, G., Z. Horváth, R. Urioste, J. Hetke, and K. Wise. "High Frequency Network Oscillations in the Hippocampus." *Science*, 256 (1992): 1025–1027.

Buzsáki, G., R. Llinás, W. Singer, A. Berthoz, and Y. Christen. *Temporal Coding in the Brain*. Berlin: Spring, 1994.

Cajal, S. R. *Histologie du système nerveux de l'homme et des vertébrés*. Madrid: Raycar, 1972 [original pub. Paris, 1911].

Cappa, S., R. Sterzi, G. Vallar, and E. Bisiach. "Remission of Hemineglect and Anosognosia during Vestibular Stimulation." *Neuropsychologia*, 25 (1994): 722–732.

Celichowski, J., F. Emonet-Denand, Y. Laporte, and J. Petit. "Distribution of Static γ Axons in Cat Peroneus Tertius Spindles Determined by Exclusively Physiological Criteria." *Journal of Neurophysiology*, 71 (1994): 722–732.

Changeux, J.-P. *L'Homme neuronal*. Paris: Fayard, 1983.

Châtelet, G. *Les Enjeux du mobile—Mathématiques, physique, philosophie*. Paris: Le Seuil, 1993.

Chatila, M., C. Milleret, P. Buser, and A. Fougeul. "A 10 Hz 'Alpha-Like' Rhythm in the Visual Cortex of the Waking Cat." *Electroencephalography and Clinical Neurophysiology*, 83 (1992): 217–222.

Chen, L. L., L. Lin, E. J. Green, C. A. Barnes, and B. McNaughton. "Head Direction Cells in the Rat Posterior Cortex. I. Anatomical Distribution and Behavioural Modulation." *Experimental Brain Research*, 101 (1994): 8–23.

Chen, L. L., L. Lin, C. A. Barnes, and B. McNaughton. "Head Direction Cells in the Rat

Posterior Cortex. II. Contributions of Visual and Ideothetic Information to the Directional Firing." *Experimental Brain Research,* 101 (1994): 24–34.

Clément, G., V. S. Gurfinkel, F. Lestienne, M. Lipshits, and K. Popov. "Adaptation of Postural Control in Weightlessness." *Experimental Brain Research,* 57 (1984): 61–72.

Clément, G., and F. Lestienne. "Adaptive Modifications of Postural Attitude in Conditions of Weightlessness." *Experimental Brain Research,* 72 (1988): 381–389.

Clément, G., T. Pozzo, and A. Berthoz. "Contribution of Eye Positioning in the Control of Upside-Down Vertical Posture." *Experimental Brain Research,* 73 (1988): 569–576.

Colby, C. L., J.-R. Duhamel, and M. E. Goldberg. "Oculocentric Spatial Representation in Parietal Cortex." *Cerebral Cortex,* 5 (1995): 470–481.

Columb, Georges [Christophe pseud]. *L'Idée fixe du savant Cosinus.* Paris: Armand Colin, 1939.

Crenna, P., C. Frigo, J. Massion, and A. Pedotti. "Forward and Backward Axial Synergies in Man." *Experimental Brain Research,* 65 (1987): 538–548.

Curthoys, I. S., C. H. Markham, and N. Furuya. "Direct Projection of Pause Neurons to Nystagmus-Related Excitatory Burst Neurons in the Cat Pontine Reticular Formation." *Experimental Neurology,* 83 (1984): 414–422.

Cutting, J. E., and P. M. Vishton. "Perceiving Layout and Knowing Distances: The Integration, Relative Potency and Contextual Use of Different Information about Depth." In *Perception of Space and Motion.* San Diego: Academic Press, 1995, 69–117.

Dai, M. J., I. Curthoys, and G. M. Halmagyi. "A Model of Otolith Stimulation." *Biological Cybernetics,* 60 (1989): 185–194.

Damasio, A. R. "The Brain Binds Entities and Events by Multiregional Activation from Convergence Zones." *Neural Computation,* 1 (1989): 123–132.

———. *Descartes' Error.* New York: Putnam, 1994.

Damasio, A. R., H. Damasio, and Y. Christen. *Neurobiology of Decision-Making.* Berlin: Springer, 1996.

Darlot, C., P. Denise, J. Droulez, B. Cohen, and A. Berthoz. "Eye Movements Induced by Off-Vertical Axis Rotation (OVAR) at Small Angles of Tilt." *Experimental Brain Research,* 73 (1988): 91–105.

———. "Motion Perceptions Induced by Off-Vertical Axis Rotation (OVAR) at Small Angles of Tilt." *Experimental Brain Research,* 73 (1988): 106–114.

Darwin, C. "Origin of Certain Instincts." *Nature,* 179 (1887): 417–418.

———. *The Expression of the Emotions in Man and Animals.* New York: Philosophical Library, 1955.

De Ajuriaguerra, J. *Résumés des cours.* Paris: Collège de France, 1976.

Decéty, J., and D. H. Ingvar. "Brain Structures Participating in Mental Simulation of Motor Behaviour: A Neurophysiological Interpretation." *Acta Physiological Scandinavica,* 73 (1990): 13–34.

Decéty, J., M. Jeannerod, M. Germain, and J. Pastène. "Vegetative Response during Imagined Movement Is Proportional to Mental Effort." *Behavioural Brain Research,* 24 (1991): 1–5.

Decéty, J., M. Jeannerod, and C. Prablanc. "The Timing of Mentally Represented Actions." *Behavioural Brain Research,* 34 (1989): 35–42.

Deecke, L., D. W. F. Schwartz, and J. M. Fredrickson. "Nucleus Ventro-posterior Inferior

(VPI) as the Vestibular Thalamic Relay in the Rhesus Monkey. I. Field Potential Investigation." *Experimental Brain Research*, 20 (1974): 88–190.

Dehaene, S., and J.-P. Changeux. "A Simple Model of Prefrontal Cortex Function in Delayed-Response Tasks." *Journal of Cognitive Neuroscience*, 1 (1990): 244–261.

Deiber, M. P., R. E. Passingham, J. G. Colebatch, K. J. Friston, P. D. Nixon, and R. S. Frackowiak. "Cortical Areas and the Selection of Movement: A Study with Positron Emission Tomography." *Experimental Brain Research*, 84 (1991): 393–402.

De Kleijn, A., and T. Magnus. "Uber die Funktion der Otolithen." *Pflugers Archiv*, 186 (1921): 6–81.

Deniau, J. M., and G. Chevalier. "Disinhibition as a Basic Process in the Expression of Striatal Functions. The Striatonigral Influence on Thalamocortical Cells of the Ventromedial Thalamic Nucleus." *Brain Research*, 334 (1981): 227–233.

Denis, M. "Approches différentielles de l'imagerie mentale." In *Cognition: L'Universel et l'individuel*, ed. M. Reuchlin, J. Lautrey, C. Marendaz, and T. Ohlman. Paris: Presses Universitaires de France, 1989.

———. *Image et cognition*. Paris: Presses Universitaires de France, 1989.

De Waele, C., W. Graf, A. Berthoz, and F. Clarac. "Vestibular Control of Skeletal Geometry." In *Posture and Gait*, ed. A. Berthoz and F. Clarac. Amsterdam: Elsevier, 1988, 423–432.

Diamond, A., and P. S. Goldman-Rakic. "Comparison of Human Infants and Rhesus Monkeys on Piaget's AB Task: Evidence for Dependence on Dorsolateral Prefrontal Cortex." *Experimental Brain Research*, 74 (1989): 24–40.

Dichgans, J., and T. Brandt. "Visual-Vestibular Interaction and Motion and Perception." In *Cerebral Control of Eye Movements and Motion Perception*, ed. J. Dichgans and E. Bizzi. New York: Karger, 1972, 327–338.

———. "Visual-Vestibular Interactions: Effects on Self-Motion Perception and Postural Control." In *Handbook of Sensory Physiology*, vol. 5, ed. H. Leibowitz and H. L. Teuber. Berlin: Springer, 1978, 755–804.

Dichgans, J., C. L. Schmidt, and W. Graf. "Visual Input Improves the Speedometer Function of the Vestibular Nuclei in the Goldfish." *Experimental Brain Research*, 18 (1973): 319–322.

Dieterich, M., and T. Brandt. "Thalamic Infarctions: Differential Effects on Vestibular Function in the Roll Plane (35 Patients)." *Neurology*, 43 (1993): 1732–1740.

Di Pellegrino, G., L. Fadiga, L. Fogassi, V. Gallese, and G. Rizzolatti. "Understanding Motor Events: A Neurophysiological Study." *Experimental Brain Research*, 91 (1992): 176–180.

Di Pellegrino, G., and S. P. Wise. "Visuospatial versus Visuomotor Activity in the Premotor and Prefrontal Cortex of a Primate." *Journal of Neuroscience*, 13 (1993): 1227–1243.

Donders, F. C. "Beitrag zur Lehre von den Bewegungen des menschlichen Auges." In *Hollandischen Beitragen zu den Anatomischen und Physiologischen Wissenschaften*, vol. 1. Amsterdam, 1847, 104–145.

Droulez, J., and A. Berthoz. "Servo-Controlled (Conservative) versus Topological (Projective) Modes of Sensory Motor Control." In *Disorders of Posture and Gait*, ed. W. Bles and T. Brandt. Amsterdam: Elsevier, 1988, 83–97.

———. "The Concept of Dynamic Memory in Sensorimotor Control." In *Motor Control:*

Concepts and Issues, ed. D. R. Humphrey and J.-J. Freund. Chichester: Wiley, 1990, 137–161.

———. "The Dynamic Memory Model and the Final Oculomotor and Cephalomotor Integrator. In *The Oculomotor System*, ed. H. Shimazu and Y. Shinoda. Tokyo: Japan Scientific Societies Press, 1991, 1–19.

———. "A Neural Network Model of Sensoritopic Maps with Predictive Short-Term Memory Properties." *Proceedings of the National Academy of Sciences*, 88 (1991): 9653–9657.

Droulez, J., A. Berthoz, and P. P. Vidal. "Use and Limits of Visual Vestibular Interaction in the Control of Posture. Are There Two Modes of Sensorimotor Control?" In *Vestibular and Visual Control on Posture and Locomotor Equilibrium*, ed. M. Igarashi and F. Owen Black. Basel: Karger, 1985, 14–25.

Droulez, J., and V. Cornilleau-Pérès. "Application of the Coherence Scheme to the Multisensory Fusion Problem." In *Multisensory Control of Movement*, ed. A. Berthoz. Oxford: Oxford University Press, 1993, 485–508.

Droulez, J., and C. Darlot. "The Geometric and Dynamic Implications of the Coherence Constraints in Three-Dimensional Sensorimotor Coordinates." In *Attention and Performance XIII*, ed. M. Jeannerod. Hillsdale, N.J.: Erlbaum, 1989, 495–526.

Duhamel, J.-R., F. Bremmer, S. Ben Hamed, and W. Graf. "Spatial Invariance of Visual Receptive Fields in Parietal Cortex Neurons." *Nature*, 389 (1997): 845–848.

Duhamel, J.-R., C. L. Colby, and M. E. Goldberg. "Congruent Representations of Visual and Somatosensory Space in Single Neurons of Monkey Ventral Intraparietal Cortex (Area VIP)." In *Brain and Space*, ed. J. Paillard. Oxford: Oxford University Press, 1991, 223–236.

———. "The Updating of the Representation of Visual Space in Parietal Cortex by Intended Eye Movements." *Science*, 255 (1992): 90–92.

Du Lac, S., J. L. Raymond, T. J. Sejnowski, and S. G. Lisberger. "Learning and Memory in the Vestibulo-ocular Reflex." *Annual Review of Neuroscience*, 18 (1995): 409–441.

Eccles, J. C., M. Ito, and J. Szentogothai. *The Cerebellum as a Neuronal Machine*. Berlin: Springer, 1967.

Eckhorn, R., R. Bauer, W. Jordan, M. Brosch, W. Kruse, M. Munk, and H. J. Reitboeck. "Coherent Oscillations: A Mechanism of Feature Linking in the Visual Cortex?" *Biological Cybernetics*, 60 (1988): 121–130.

Eichenbaum, H., S. I. Wiener, M. L. Shapiro, and N. J. Cohen. "The Organization of Spatial Coding in the Hippocampus: A Study of Neural Ensemble Activity." *Journal of Neuroscience*, 9 (1989): 2764–2775.

Emonet-Dénand, F., Y. Laporte, P. B. C. Matthews, and J. Petit. "On the Subdivision of Static and Dynamic Fusimotor Actions on the Primary Ending of Cat Muscle Spindle." *Journal of Physiology*, 261 (1977): 827–861.

Ernst, H. *Les Principes de la mécanique* (1894). Cited in R. N. Shepard, "Ecological Constraints on Internal Representation: Resonant Kinematics of Perceiving, Imagining, Thinking, and Dreaming." *Psychological Review*, 91 (1984): 417–447.

Eshkol, N., and A. Wachmann. *Movement Notation*. London: Weidenfeld and Nicolson, 1958.

Etienne, A. S., R. Mowrer, and F. Saucy. "Limitations in the Assessment of Path-Dependent Integration." *Behaviour*, 106 (1988): 81–111.

Evarts, E. V. "Relation of Pyramidal Tract Activity to Force Exerted during Voluntary Movements." *Journal of Neurophysiology,* 31 (1968): 14–27.

———. "Role of Motor Cortex in Voluntary Movements in Primates." In *Handbook of Physiology,* sect. 1, vol. 2, *Motor Control,* ed. V. B. Brooks. Bethesda: American Physiological Society, 1981, 1083–1120.

Evarts, E. V., C. Fromm, J. Kroller, and V. A. Jennings. "Motor Cortex Control of Finely Graded Forces." *Journal of Neurophysiology,* 49 (1983): 1199–1215.

Ewert, J. P. "Neural Mechanisms of Prey-Catching and Avoidance Behavior in the Toad *(Bufo bufo L.)." Brain Behaviour and Evolution,* 3 (1970): 36–56.

———. "Neuroethology of Releasing Mechanisms: Prey-Catching in Toads." *Behavioral and Brain Sciences,* 10 (1987): 337–405.

Faber, D. S., W. S. Young, P. Legendre, and H. Korn. "Intrinsic Quantal Variability Due to Stochastic Properties of Receptor-Transmitter Interactions." *Science,* 258 (1992): 1494–1498.

Feldman, A. G., and M. F. Levin. "The Origin and Use of Positional Frames of Reference in Motor Control." *Behavioral and Brain Sciences,* 18 (1995): 723–744.

Ferrier, D. "Experiments on the Brain of Monkeys." *Philosophical Transactions of the Royal Society of London,* 165 (1875): 433–488.

———. *The Functions of the Brain.* New York: G. P. Putnam's Sons, 1876.

Fessard, A., and A. Tournay. "Quelques données et réflexions sur le phénomène de la post-contraction involontaire." *L'Année psychologique* (1950): 216–235.

Findley, L. J., R. Capildeo, and A. Tremor, eds. *Movement Disorders.* London: Macmillan, 1988.

Flor, H., T. Elbert, S. Knecht, C. Wienbruch, C. Pantev, N. Birbaumer, W. Larbig, and E. Baub. "Phantom-Limb Pain as a Perceptual Correlate of Cortical Reorganization following Arm Amputation." *Nature,* 375 (1995): 482–484.

Flourens, P. *Recherches expérimentales sur les propriétés et les fonctions du système nerveux dans les animaux vertébrés.* Paris: Crévot, 1824.

Fogassi, L., V. Galese, L. Fadiga, G. Luppino, M. Matelli, and G. Rizzolatti. "Coding of Peripersonal Space in Inferior Premotor Cortex (Area F4)." *Journal of Neurophysiology,* 76 (1996): 141–157.

Fox, P. T., J. M. Fox, M. E. Raichle, and R. M. Burde. "The Role of the Cerebral Cortex in the Generation of Voluntary Saccades." *Journal of Neurophysiology,* 54 (1985): 348–369.

Friberg, L., T. S. Olsen, P. Roland, O. B. Paulson, and N. A. Lassen. "Focal Increase of Blood Flow in the Cerebral Cortex of Man during Vestibular Stimulation." *Brain,* 108 (1985): 609–623.

Frith, C. D., K. Friston, P. F. Liddle, and R. S. Frackowiak. "Willed Action and the Prefrontal Cortex in Man: A Study with PET." *Proceedings of the Royal Society of London. Series B: Biological Sciences,* 244 (1991): 241–246.

Frith, U. *Autism: Explaining the Enigma.* Cognitive Development. Oxford: Basil Blackwell, 1989.

Fuchs, A., and W. Becker, eds. *Progress in Oculomotor Research.* Amsterdam: Elsevier, 1981.

Fukuda, T. *Statokinetic Reflexes in Equilibrium and Movement.* Tokyo: University of Tokyo Press, 1983.

Funahashi, S., C. Bruck, and P. S. Goldman-Rakic. "Mnemonic Coding of Visual Space in the Monkey's Dorsolateral Prefrontal Cortex." *Journal of Neurophysiology*, 61 (1989): 1–19.

Fuster, J. M. *The Prefrontal Cortex: Anatomy, Physiology, and Neuropsychology of the Frontal Lobe*. New York: Raven Press, 1996.

Gahery, Y., and J. Massion. "Coordination between Posture and Movement." In *The Motor System in Neurobiology*, ed. E. V. Evarts, S. P. Wise, and D. Bousfield. Amsterdam: Elsevier Biomedical Press, 1985, 121–125.

Gavrilov, V. V., S. I. Wiener, and A. Berthoz. "Whole Body Rotations Enhance Hippocampal Theta Rhythm, Slow Activity in Awake Rats Passively Transported on a Mobile Robot." *Annals of the New York Academy of Sciences*, 781 (1996): 385–398.

Gaymard, B., C. Pierrot-Deseilligny, and S. Rivaud. "Impairment of Sequences of Memory-Guided Saccades after Supplementary Motor Area Lesions." *Annals of Neurology*, 28 (1990): 622–626.

Gaymard, B., S. Rivaud, and C. Pierrot-Deseilligny. "Role of the Left and Right Supplementary Motor Areas in Memory-Guided Saccade Sequences." *Annals of Neurology*, 34 (1993): 404–406.

Georgopoulos, A. P. "Current Issues in Directional Motor Control." *Trends in Neurosciences*, 18 (1995): 506–510.

Georgopoulos, A. P., M. D. Crutcher, and A. B. Schwartz. "Cognitive Spatial Motor Processes. III. Motor Cortical Prediction of Movement Direction during an Instructed Delay Period." *Experimental Brain Research*, 75 (1989): 183–194.

Georgopoulos, A. P., J. F. Kalaska, R. Caminiti, and J. T. Massey. "On the Relations between the Direction of Two-Dimensional Arm Movements and Cell Discharge in Primate Motor Cortex." *Journal of Neuroscience*, 2 (1982): 1527–1537.

Georgopoulos, A. P., J. T. Lurito, M. Petrides, A. B. Schwartz, and J. T. Massey. "Mental Rotation of the Neuronal Population Vector." *Science*, 243 (1989): 234–236.

Georgopoulos, A. P., A. B. Schwartz, and R. E. Kettner. "Neuronal Population Coding of movement Direction." *Science*, 233 (1986): 1416–1429.

Gertsmann, J. "Problems of Imperception of Disease and of Impaired Body Territories with Organic Lesions. Relation to Body Scheme and Its Disorders." *Archives of Neurology and Psychiatry*, 48 (1942): 890–913.

Ghaem, O., E. Mellet, F. Grivello, N. Tzourio, B. Mazoyer, A. Berthoz, and M. Denis. "Mental Navigation along Memorized Routes Activates the Hippocampus, Precuneus, and Insula." *NeuroReport* 8 (1997): 739–744.

Gibson, J. J. *The Senses Considered as Perceptual Systems*. Boston: Houghton, 1966.

———. "The Theory of Affordances." In *Perceiving, Acting, and Knowing*, ed. R. E. Shaw and J. Bransford. Hillsdale, N.J.: Erlbaum, 1977.

Gielen, S. "Muscle Activation Patterns and Joint-Angle Coordination in Multijoint Movements." In *Multisensory Control of Movement*, ed. A. Berthoz. New York: Oxford University Press, 293–312.

Gielen, C. C. A. M., and E. J. Van Zuylen. "Coordination of Arm Muscles during Flexion and Supination: Applications of the Tensor Analysis Approach." *Neuroscience*, 17 (1986): 527–539.

Gladwin, T. *East Is a Big Bird*. Cambridge, Mass.: Harvard University Press, 1970.

Glasauer, S., and H. Mittelstaedt. "Determinants of Orientation in Microgravity." *Acta Astronautica*, 27 (1992): 1–9.

Goethe, Johann Wolfgang von. *Faust*, trans. Randall Jarrell. New York: Farrar, Straus, and Giroux, 1959.

Goldman-Rakic, P. "Circuitry of Primate Prefrontal Cortex and Regulation of Behavior by Representational Memory." In *Handbook of Physiology*, vol. 5, *The Nervous System*, ed. V. B. Mountcastle, F. Plkum, and D. Humphrey. Bethesda: American Physiological Society, 1987, 373–417.

Gomi, H., and M. Kawato. "Adaptive Feedback Control Models of the Vestibulocerebellum and Spinocerebellum." *Biological Cybernetics*, 68 (1992): 105–114.

Gonshor, A., and G. Melvill Jones. "Extreme Vestibulo-ocular Adaptation by Prolonged Optical Reversal of Vision." *Journal of Physiology*, 256 (1976): 381–414.

Goodale, M. A., A. D. Milner, L. S. Jakobson, and D. P. Carey. "Neurological Dissociation between Perceiving Objects and Grasping Them." *Nature*, 349 (1991): 154–156.

Graf, W., and R. Baker. "Adaptive Changes of the Vestibulo-Ocular Reflex in Flatfish Are Achieved by Reorganization of Central Nervous Pathways." *Science*, 221 (1983): 777–779.

Grantyn, A. "How Visual Inputs to the Ponto-Bulbar Reticular Formation Are Used in the Synthesis of Premotor Signals during Orienting." In *Progress in Brain Research*, vol. 80, ed. J. H. J. Allum and M. Hulliger. Amsterdam: Elsevier, 1989, 159–170.

Grantyn, A., and A. Berthoz. "The Role of the Tecto-reticulo-spinal System in Control of Head Movement." In *Control of Head Movement*, ed. G. W. Peterson and F. Richmond. Oxford: Oxford University Press, 1987, 224–244.

Grantyn, A., and R. Grantyn. "Axonal Patterns and Sites of Termination of Cat Superior Colliculus Neurons Projecting in the Tecto-bulbo-spinal Tract." *Experimental Brain Research*, 46 (1982): 243–256.

Grantyn, A., R. Grantyn, V. Robine, and A. Berthoz. "Electroanatomy of Tectal Efferent Connections Related to Eye Movements in the Horizontal Plane." *Experimental Brain Research*, 37 (1979): 149–172.

Grasso, R., C. Assaiante, P. Prévost, and A. Berthoz. "Development of Anticipatory Orienting Strategies during Locomotor Tasks in Children," *Neuroscience and Biobehavioral Review*, 22 (1998): 533–539.

Grasso, R., S. Glasauer, U. Takei, and A. Berthoz. "The Predictive Brain: Anticipatory Control of Head Direction for the Steering of Locomotion." *NeuroReport*, 7 (1996): 1170–1174.

Grasso, R., P. Prévost, Y. Ivanenko, and A. Berthoz. "Eye-Head Coordination for the Steering of Locomotion in Humans: An Anticipatory Synergy." *Neuroscience Letters*, 253 (1998): 115–118.

Gray, C. M., P. König, A. K. Engel, and W. Singer. "Oscillatory Responses in Cat Visual Cortex Exhibit Inter-Columnar Synchronization Which Reflects Global Stimulus Patterns." *Nature*, 338 (1989): 334–337.

Graziano, M., G. Yap, and C. Gross. "Coding of Visual Space by Premotor Neurons." *Science*, 266 (1994): 1054–1057.

———. "The Representation of Extrapersonal Space: A Possible Role for Bimodel-Tactile Neurons." In *The Cognitive Neurosciences*, ed. M. Gazzaniga. Cambridge, Mass.: MIT Press, 1994.

Grüsser, O. J. "Cortical Representation of Head Movement in Space and Some Psychophysical Considerations." In *The Head-Neck Sensory-Motor System*, ed. A. Berthoz, P. O. Vidal, and W. Graf. New York: Oxford University Press, 1991.

Grüsser, O. J., and T. Landis. *Visual Agnosias and Related Disorders*, vol. 12. Vision and Visual Dysfunction. Basingstoke, UK: Macmillan, 1991.

Grüsser, O. J., M. Pause, and U. Schreiter. "Localisation and Responses of Neurones in the Parieto-Insular Vestibular Cortex of Awake Monkeys *(Macaca fascicularis)*." *Journal of Physiology*, 430 (1990): 537–557.

———. "Vestibular Neurones in the Parieto-Insular Cortex of Monkeys *(Macaca fascicularis)*." *Journal of Physiology*, 430 (1990): 559–583.

Guitton, G., H. A. Buchtel, and R. Douglas. "Frontal Lobe Lesions in Man Cause Difficulties in Supressing Reflexive Glance and in Generating Goal-Directed Saccades." *Experimental Brain Research*, 58 (1985): 455–472.

Guldin, W. O., and O. J. Grüsser. "Single Unit Responses in the Vestibular Cortex of Squirrel Monkeys." *Neuroscience Abstracts*, 13 (1987): 1224.

Gurfinkel, V. S. "The Mechanisms of Postural Regulation in Man." In *Physiology and General Biology Reviews*, ed. T. Turpaev. Chur, Switzerland: Harwood Academic Publishers, 1994.

Gurfinkel, V. S., E. E. Debreva, and Y. S. Levik. "The Role of the Internal Model in the Perception of Position and Planning of Hand Movements." *Fiziologiia Cheloveka*, 12 (1986): 769–776.

Gurfinkel, V. S., and Y. S. Levick. "Perceptual and Automatic Aspects of the Postural Body Scheme." In *Brain and Space*, vol. 12, ed. J. Paillard. Oxford: Oxford University Press, 1986.

Hamann, K. F., P. P. Vidal, J. M. Sterkers, and A. Berthoz. "A New Test for Postural Disorders: An Application of Visual Stabilization." *Agressologie*, 20B (1979): 129–130.

Hanneton, S., A. Berthoz, J. Droulez, and J.-J. Slotine. "Does the Brain Use Sliding Variables for the Control of Movements?" *Biological Cybernetics*, 77 (1997): 381–393.

Haustein, W., and H. Mittelstaedt. "Evaluation of Retinal Orientation and Gaze Direction in the Perception of the Vertical." *Vision Research*, 30 (1990): 255–262.

Head, H., and G. Holmes. "Sensory Disturbances from Cerebral Lesions." *Brain*, 34 (1911): 102–244.

Hécaen, H., and J. de Ajuriaguerra. *Méconnaissances et hallucinations corporelles: intégration et désintégration de la somatognosie*. Paris: Masson et Cie, 1952.

Held, R., and A. Hein. "Movement-Produced Stimulation in the Development of Visually Guided Behavior." *Journal of Comparative Physiological Psychology*, 56 (1963): 872–876.

Heliodorus of Emesa. *The Aethiopica*, vol. 5. The Athenian Society's Publications. Athens: Privately printed for the Athenian Society, 1897.

Helmholtz, H. von. *Treatise on Physiological Optics*, trans. J. P. C. Sothall. New York: Dover, 1962.

Henn, V., L. R. Young, and C. Finley. "Vestibular Nucleus Units in Alert Monkeys Are Also Influenced by Missing Visual Field." *Brain Research,* 71 (1974): 144–149.

Hertz, H. *The Principles of Mechanics,* trans. D. E. Jones and J. T. Walley. New York: Dover, 1956.

Hess, W. "Teleokinetisches und ereismatisches Kraftsystem in der Biomotorik." *Helvetica Physiologica et Pharmacologica Acta,* 1 (1943): C62–C63.

Hietanen, J. K., and D. Perrett. "A Role of Expectation in Visual and Tactile Processing within the Temporal Cortex." In *Brain Mechanisms in Perception and Memory: From Neuron to Behavior,* ed. T. E. Ono, L. R. Squire, M. Raichle, D. L. Perrett, and M. Fukuda. Oxford: Oxford University Press, 1993, 83–103.

Hikosaka, O. "Role of the Basal Ganglia in Motor Learning: A Hypothesis." In *Brain Mechanisms in Perception and Memory: From Neuron to Behavior,* ed. T. E. Ono, L. R. Squire, M. Raichle, D. L. Perrett, and M. Fukuda. Oxford: Oxford University Press, 1993, 497–513.

Hikosaka, O., M. Sakamoto, and N. Miyashita. "Effects of Caudate Nucleus Stimulation on Substantia Nigra Cell Activity in Monkey." *Experimental Brain Research,* 95 (1993): 457–472.

Hikosaka, O., and R. H. Wurtz. "Visual and Oculomotor Functions of Monkey Substantia Nigra Pars Reticulata. III. Memory-Contingent Visual and Saccade Responses." *Journal of Neurophysiology,* 49 (1983): 1268–1284.

Hoffman, K. P. "Responses of Single Neurons in the Pretectum of Monkeys to Visual Stimuli in Three Dimensional Space." In *Representation of Three-Dimensional Space in the Vestibular, Oculomotor, and Visual Systems,* ed. B. Cohen and V. Henn. New York: New York Academy of Sciences, 1988, 1–261.

Holst, E. von. "Relations between the Central Nervous System and the Peripheral Organs." *Journal of Animal Behaviour,* 2 (1954): 89–94.

Holst, E. von, and H. Mittelstaedt. "Das Reafferenzprinzip. Wechselwirkungen zwischen Zentralnervensystem und Peripherie." *Naturwissenschaften,* 37 (1950): 464–476.

Humphrey, D. R., and H. J. Freund, eds. *Motor Control: Concepts and Issues.* New York: Wiley, 1991.

Husserl, E. *Idées directrices pour une phénoménologie et une philosophie phénoménologique pure,* vol. 2, *Recherches phénoménologiques pour la constitution.* Paris: Presses Universitaires de France, 1982.

Hyvarinen, J. *The Parietal Cortex of Monkey and Man.* Berlin: Springer, 1982.

Israël, I. "Memory-Guided Saccades: What Is Memorized?" *Experimental Brain Research,* 90 (1992): 221–224.

Israël, I., and A. Berthoz. "Contribution of the Otoliths to the Calculation of Linear Displacement." *Journal of Neurophysiology,* 62 (1989): 247–263.

Israël, I., N. Chapuis, S. Glasauer, O. Charade, and A. Berthoz. "Estimation of Passive Linear Whole Body Displacement in Humans." *Journal of Neurophysiology,* 70 (1993): 1270–1273.

Israël, I., R. Grasso, P. Georges-François, T. Tzuzuku, and A. Berthoz. "Spatial Memory and Path Integration Studied by Self-Driven Linear Displacement." *Journal of Neurophysiology,* 77 (1999): 3180–3192.

Israël, I., S. Rivaud, A. Berthoz, and C. Pierrot-Deseilligny. "Cortical Control of Vestibular

Memory-Guided Saccades." *Annals of the New York Academy of Sciences,* 656 (1992): 472–484.

Israël, I., S. Rivaud, B. Gaymard, A. Berthoz, and C. Pierrot-Deseilligny. "Cortical Control of Vestibular-Guided Saccades in Man." *Brain,* 118 (1995): 1169–1183.

Israël, I., S. Rivaud, C. Pierrot-Desilligny, and A. Berthoz. "Delayed VOR: An Assessment of Vestibular Memory for Self Motion." In *Tutorials in Motor Neuroscience,* ed. J. Requin and J. Stelmach. Dordrecht: Kluwer, 1991, 599–607.

Ito, M. "Cerebellar Control of the Vestibulo-ocular Reflex around the Flocculus Hypothesis." *Annual Review of Neuroscience,* 5 (1982): 275–296.

———. *The Cerebellum and Neuronal Control.* New York: Raven, 1984.

Ivanenko, Y., R. Grasso, I. Israël, and A. Berthoz. "The Contribution of Otoliths and Semi-circular Canals to the Perception of Two-Dimensional Passive Whole-Body Motion in Humans." *Journal of Physiology,* 502 (1997): 223–233.

James, William. *The Principles of Psychology,* vol. 2. New York: Holt, 1890.

Jami, L. "Golgi Tendon Organs in Mammalian Skeletal Muscle: Functional Properties on Central Actions." *Physiological Reviews,* 72 (1992): 623–666.

Janet, P. *Les Débuts de l'intelligence.* Paris: Flammarion, 1935.

Jeannerod, M. "The Posterior Parietal Cortex as a Spatial Generator." In *Brain Mechanisms and Spatial Vision,* eds. D. I. Ingle, J. Jeannerod, and D. Lee. Dordrecht: Martinus Nijhoff, 1985, 279–298.

———. *Neurophysiological and Neuropsychological Aspects of Spatial Neglect.* Amsterdam: Elsevier, 1987.

———. *The Neural and Behavioral Organisation of Goal-Directed Arm Movements.* Oxford: Clarendon Press, 1988.

———, ed. *Attention and Performance XIII.* Hillsdale, N.J.: Erlbaum, 1990.

———. "A Neurophysiological Model for the Directional Coding of Reaching Movements." In *Brain and Space,* ed. J. Paillard. Oxford: Oxford University Press, 1991.

———. "The Representing Brain: Neural Correlates of Motor Intention and Imagery." *Behavioral and Brain Sciences,* 17 (1994): 187–202.

Joliot, M., U. Ribary, and R. Llinás. "Human Oscillatory Brain Activity near 40 Hz Coexists with Cognitive Temporal Binding." *Proceedings of the National Academy of Sciences,* 91 (1994): 11748–11751.

Jones, M. G., and K. E. Spells. "A Theoretical and Comparative Study of the Functional Dependance of the Semi-Circular Canal upon its Physical Dimensions." *Proceedings of the Royal Society of London. Series B: Biological Sciences,* 157 (1963): 403–419.

Joseph, J. P., and P. Barone. "Prefrontal Unit Activity during a Delayed Oculomotor Task in the Monkey." *Experimental Brain Research,* 67 (1987): 460–468.

Kalaska, J. F., R. Caminiti, and A. P. Georgopoulos. "Cortical Mechanisms Related to the Direction of Two-Dimensional Arm Movements: Relations in Parietal Area 5 and Comparison with Motor Cortex." *Experimental Brain Research,* 51 (1983): 247–260.

Kant, Immanuel. *Critique of Pure Reason,* trans. Norman Kemp Smith. New York: St. Martin's Press, 1961.

Karabelas, A. B., and A. K. Moschovakis. "Nigral Inhibitory Termination on Efferent Neurons of the Superior Colliculus. An Intracellular Horseradish Peroxidase Study in the Cat." *Journal of Cognitive Neuroscience,* 239 (1985): 309–329.

Karnath, H. O., K. Christ, and W. Hartje. "Decrease of Contralateral Neglect by Neck Muscle Vibration and Spatial Orientation of Trunk Midline." *Brain*, 116 (1993): 383–396.

Kawano, K., M. Sasaki, and M. Yamashita. "Vestibular Input to Visual Tracking Neurons in the Posterior Parietal Association Cortex of the Monkey." *Neuroscience Letters*, 17 (1980): 55–60.

Kawato, M., K. Furukawa, and R. Suzuki. "A Hierarchical Neural Network Model for Control and Learning of Voluntary Movements." *Biological Cybernetiks*, 57 (1987): 169–185.

Kelso, J. A. S. "Phase Transitions and Critical Behaviour in Human Bimanual Coordination." *American Journal of Physiology*, 14 (1984): R1000–1004.

Kelso, J. A. S., J. D. Delcolle, and G. Schöner. "Action-Perception as a Pattern Formation Process." In *Attention and Performance*, ed. M. Jeannerod. Hillsdale, N.J.: Erlbaum, 1990, 139–169.

Kesner, R. P. "Paired Associate Learning in the Rat: Role of Hippocampus, Medial Prefrontal Cortex, and Parietal Cortex." *Psychobiology*, 21 (1993): 183–192.

Korn, H., and D. S. Faber. "Organisation and Cellular Mechanisms Underlying Chemical Inhibition in a Vertebrate Neuron." In *Molecular and Cellular Interactions Underlying Higher Brain Functions*. Progress in Brain Research, ed. J.-P. Changeux. Amsterdam: North Holland, 1983.

Kosslyn, S. M., C. F. Chabris, C. J. Marsolek, R. A. Jacobs, and O. Koenig. "On Computational Evidence for Different Types of Spatial Relations Encoding: Reply to Cook et al." *Journal of Experimental Psychology: Human Perception and Performance*, 21 (1995): 423–431.

Kosslyn, S. M., C. F. Chabris, C. J. Marsolek, and O. Koenig. "Categorical versus Coordinate Spatial Representations: Computational Analysis and Computer Simulations." *Journal of Experimental Psychology: Human Perception and Performance*, 18 (1992): 562–577.

Krechevsky, I. "Hypothesis in Rats." *Psychological Reviews*, 39 (1932): 517–532.

Lackner, J. R. "Some Contributions of Touch, Pressure, and Kinesthesis to Human Spatial Orientation and Oculomotor Control." *Acta Astronautica*, 8 (1981): 825–830.

———. "Some Proprioceptive Influences on the Perceptual Representation of Body Shape and Orientation." *Brain*, 111 (1981): 281–297.

———. "Orientation and Movement in Unusual Force Environments." *Psychological Science*, 4 (1993): 134–142.

Lackner, J. R., and M. S. Levine. "Changes in Apparent Body Orientation and Sensory Localization Induced by Vibration of Postural Muscles: Vibratory Myesthetic Illusions." *Aviation, Space, and Environmental Medicine*, 50 (1979): 346–354.

Lacquaniti, F., M. Carrozzo, and N. Borghese. "The Role of Vision in Tuning Anticipatory Motor Responses of the Limbs." In *Multisensory Control of Movement*, ed. A. Berthoz. New York: Oxford University Press, 1993, 379–390.

Lacquaniti, F., E. Guigon, L. Bianchi, S. Farraina, and R. Caminiti. "Representing Spatial Information for Limb Movements: Role of Area 5 in the Monkey." *Cerebral Cortex*, 5 (1995): 391–409.

Lacquaniti, F., and C. Maioli. "The Role of Preparation in Tuning Anticipatory and Reflex Responses during Catching." *Journal of Neuroscience*, 9 (1989): 134–148.

Lacquaniti, F., J. Soechting, and C. Terzuolo. "Path Constraints on Point to Point Arm Movements in Three-Dimensional Space." *Neuroscience,* 17 (1986): 313–324.

Land, M. F., and D. N. Lee. "Where We Look When We Steer." *Nature,* 369 (1994): 742–744.

Lang, W., D. Cheyne, R. Kristeva, R. Beisteiner, G. Lindinger, and L. Deecke. "Three-Dimensional Localization of SMA Activity Preceding Voluntary Movement. A Study of Electric and Magnetic Fields in a Patient with Infarction of the Right Supplementary Motor Area." *Experimental Brain Research,* 87 (1991): 688–695.

Lang, W., M. Lang, F. Uhl, C. Koska, A. Kornhuber, and L. Deecke. "Negative Cortical DC Shifts Preceding and Accompanying Simultaneous and Sequential Finger Movements." *Experimental Brain Research,* 71 (1988): 579–587.

Lang, W., H. Obrig, G. Lindinger, D. Cheyne, and L. Deecke. "Supplementary Motor Area Activation While Tapping Bimanually Different Rhythms in Musicians." *Experimental Brain Research,* 79 (1990): 504–514.

Lang, W., L. Petit, P. Hölliger, U. Pietrzyck, N. Tzourio, B. Mazoyer, and A. Berthoz. "A Positron Emission Tomography Study of Oculomotor Imagery." *NeuroReport,* 5 (1994): 921–924.

Laporte, Y., F. Emonet-Dénand, and L. Jami. "The Skeletofusimotor or β Innervation of Mammalian Muscle Spindles." *Trends in Neuroscience,* 4 (1981): 97–99.

Lashley, K. S. *Brain Mechanisms and Intelligence.* Chicago: University of Chicago Press, 1929.

Lee, D. N. "A Theory of Visual Control of Braking Based on Information about Time-to-Collision." *Perception,* 5 (1976): 437–459.

Lee, D. N., and E. Aronson. "Visual Proprioceptive Control of Standing in Human Infants." *Perception and Psychophysics,* 14 (1974): 529–532.

Lee, D. N., and J. R. Lishman. "Visual Proprioceptive Control of Stance." *Journal of Human Movement Studies,* 1 (1975): 87–95.

Lee, D. N., and P. E. Reddish. "Plummeting Gannets: A Paradigm of Ecological Optics." *Nature,* 293 (1985): 293–294.

Lee, D. N., and D. S. Young. "Visual Timing in Interceptive Actions." In *Brain Mechanisms and Spatial Vision,* ed. D. J. Ingle, M. Jeannerod, and D. N. Lee. Dordrecht: Martinus Nijhoff, 1996, 1–30.

Lee, R. G., I. Tonolli, F. Viallet, R. Aurenty, and J. Massion. "Preparatory Postural Adjustment in Parkinsonian Patients with Postural Instability." *Canadian Journal of Neurological Sciences,* 22 (1995): 126–135.

Lestienne, F., J. Soechting, and A. Berthoz. "Postural Readjustments Induced by Linear Motion of Visual Scenes." *Experimental Brain Research,* 28 (1977): 363–384.

Levin, M. F., and A. G. Feldman. "The Role of Stretch Reflex Threshold Regulation in Normal and Impaired Motor Control." *Brain Research,* 657 (1994): 23–30.

Lisman, J. E., and M. A. P. Idiart. "Storage of Short-Term Memories in Oscillatory Subcycles." *Science,* 267 (1995): 1512–1515.

Livingstone, M., and D. Hubel. "Segregation of Form, Color, Movement, and Depth: Anatomy, Physiology, and Perception." *Science,* 240 (1988): 740–749.

Llinás, R. "Possible Role of Tremor in the Organisation of the Nervous System." In *Movement Disorders,* ed. L. J. Findley, R. Capildeo, and A. Tremor. London: Macmillan, 1988, 475–477.

———. "The Intrinsic Electrophysiological Properties of Mammalian Neurons: Insights into Central Nervous System Function." *Science,* 242 (1988): 1654–1664.

———. "The Noncontinuous Nature of Movement Execution." In *Motor Control: Concepts and Issues.* Chichester: Wiley, 1991, 223–242.

———. "Oscillations in CNS Neurons: A Possible Role for Cortical Interneurons in the Generation of 40-Hz Oscillations." In *Induced Rhythms in the Brain,* ed. E. Basar and T. Bullock. Boston: Birkhauser, 1992, 269–283.

Llinás, R., and U. Ribary. "Coherent 40-Hz Oscillation Characterizes Dream State in Humans." *Proceedings of the National Academy of Sciences,* 90 (1993): 2078–2081.

Llinás, R., and C. Sotelo. *The Cerebellum Revisited.* New York: Springer, 1992.

Lobel, E., J. F. Kleine, D. Le Bihan, A. Leroy-Willig, and A. Berthoz. "Functional MRI of Galvanic Vestibular Stimulation." *Journal of Neurophysiology,* 80 (1998): 2699–2709.

Lorente de Nó, R. "Vestibulo-ocular Reflex Arc." *Archives of Neurology and Psychiatry,* 30 (1933): 245–291.

Lorenz, Konrad. *Behind the Mirror: A Search for a Natural History of Human Knowledge,* trans. Ronald Taylor. New York: Harcourt Brace Jovanovich, 1977.

Lotze, R. H. *Medizinische Psychologie oder Physiologie der Seele.* Leipzig: Weidemann, 1852.

Mach, E. *Grundlinien der Lehre von den Bewegungsempfindungen.* Amsterdam: Bonset, 1967.

MacKay, D. G. *The Organization of Perception and Action.* New York: Springer, 1987.

Magnin, M., M. Jeannerod, and P. T. S. Putkonen. "Vestibular and Saccadic Influences on Dorsal and Ventral Nuclei of the Lateral Geniculate Body." *Experimental Brain Research,* 21 (1974): 1–18.

Malsburg, C. von der. "Nervous Structures with Dynamical Links." *Berichte der Bunsengesellschaft für physikalische Chemie,* 89 (1985): 703–710.

Marey, E. J. *Le Mouvement.* Paris: Masson, 1894.

———. *La Machine animale.* Paris: Revue EPS, 1993.

Marr, D. "Simple Memory: A Theory for Archicortex." *Proceedings of the Royal Society of London. Series B: Biological Sciences,* 262 (1971): 23–81.

Massion, J. "Postural Control System." *Current Opinion in Neurobiology,* 4 (1994): 877–887.

Massion, J., V. Gurfinkel, M. Lipshits, A. Obadia, and K. Popov. "Stratégie et synergie: deux niveaux de contrôle de l'équilibre au cours du mouvement. Effets de la microgravité." *Comptes rendus hebdomadaires de l'Académie des sciences (Paris, Série III),* 314 (1992): 87–92.

Massion, J., and L. Rispal-Padel. "Thalamus: fonctions motrices." *Revue Neurologique,* 142 (1986): 327–336.

Massion, J., F. Viallet, R. Massarino, and R. Khalil. "La région de l'aire motrice supplémentaire est impliquée dans la coordination entre posture et mouvement." *Comptes rendus hebdomadaires de l'Académie des sciences (Paris, Série III),* 308 (1989): 417–423.

Mast, F., S. Kosslyn, and A. Berthoz. "Visual Mental Imagery Interferes with Allocentric Orientation Judgements." *Neuroreport,* 110 (1999): 3549–3553.

Matthews, B. L., J. H. Ryu, and C. Bockaneck. "Vestibular Contribution to Spatial Orientation." *Acta Orolaryngologica,* 468 (1989): 149–154.

Mays, L. E., and D. L. Sparks. "Saccades Are Spatially, Not Retinocentrically, Coded." *Science,* 208 (1980): 1163–1165.

McCrea, R. A., K. Yoshida, C. Evinger, and A. Berthoz. "The Location, Axonal Arborization, and Termination Sites of Eye-Movement-Related Secondary Vestibular Neurons Demonstrated by Intra-axonal HRP Injection in the Alert Cat." In *Progress in Oculomotor Research,* ed. A. Fuchs and W. Becker. Amsterdam: Elsevier, 1981, 379–386.

McDaniel, W. F., D. M. Compton, and S. R. Smith. "Spatial Learning following Posterior Parietal or Hippocampal Lesions." *NeuroReport,* 5 (1994): 1713–1717.

McIntyre, J., E. Gurfinkel, M. Lipshits, J. Droulez, and V. Gurfinkel. "Measurement of Human Force Control during Constrained Arm Motion using a Force-Activated Joystick." *Journal of Neurophysiology,* 17 (1995): 1201–1222.

Melamed, E., and B. Larsen. "Cortical Activation during Saccadic Eye Movements in Humans: Localisation by Cerebral Blood Flow Increases." *Annals of Neurology,* 5 (1989): 79–88.

Melvill-Jones, G., A. Berthoz, and B. Segal. "Adaptative Modification of the Vestibulo-ocular Reflex by Mental Effort in Darkness." *Experimental Brain Research,* 56 (1984): 149–153.

Melvill-Jones, G., D. Guitton, and A. Berthoz. "Changing Patterns of Eye-Head Coordination during Six Hours of Optically Reversed Vision." *Experimental Brain Research,* 69 (1988): 531–544.

Merleau-Ponty, M. *Le Visible et l'Invisible.* Paris: Gallimard, 1964.

———. *The Visible and the Invisible,* ed. Claude Lefort, trans. Alphonso Lingis. Northwestern University Studies in Phenomenology and Existential Philosophy. Evanston, Ill.: Northwestern University Press, 1968.

———. *Résumés de cours au Collège de France.* Paris: Gallimard, 1968.

———. *The Essential Writings of Merleau-Ponty,* ed. Alden L. Fisher. New York: Harcourt, Brace, and World, 1969.

———. *La Nature: notes-cours du Collège de France.* Paris: Le Seuil, 1995.

Mesulam, M. M. "A Cortical Network for Directed Attention and Unilateral Neglect." *Annals of Neurology,* 10 (1981): 309–325.

Metcalfe, T., and M. Gresty. "Self-controlled Reorienting Movements in Response to Rotational Displacements in Normal Subjects and Patients with Labyrinthine Diseases." In *Sensing Motion,* ed. D. L. Tomko, B. Cohen, and F. E. Guedry. New York: Annals of the New York Academy of Sciences, 1992.

Michotte, A. *Causalité, permanence et réalité phénoménales.* Paris: Béatrice-Nauwelaerts, 1962.

Miles, F. A., and S. G. Lisberger. "Plasticity in the Vestibulo-ocular Reflex: A New Hypothesis." *Annual Review of Neuroscience,* 4 (1981): 273–299.

Miles, R., K. Toth, A. Gulyas, N. Hajos, and T. F. Freund. "Differences between Somatic and Dendritic Inhibition in the Hippocampus." *Neuron,* 16 (1996): 814–823.

Miller, S., M. Potegal, and L. Abraham. "Vestibular Involvement in a Passive Transport and Return Task." *Physiological Psychology,* 11 (1982): 1–10.

Milner, B. "Visual Recognition and Recall after Right Temporal Lobe Excision in Man." *Neuropsychologia,* 6 (1968): 191–209.

Mishkin, M. "A Memory System in the Monkey." *Philosophical Transactions of the Royal Society. Series B: Biological Sciences,* 98 (1982): 85–95.

Mistlin, A. J., and D. I. Perrett. "Visual and Somatosensory Processing in the Macaque Temporal Cortex: The Role of 'Expectation.'" *Experimental Brain Research,* 82 (1990): 437–450.

Mittelstaedt, H. "The Role of Otoliths in the Perception of the Orientation of Self and World to the Vertical." *Zoologische Jahrbücher der Physiologie,* 95 (1991): 419–425.

———. "Evidence of Somatic Graviception from New and Classical Investigations." *Acta Otolaryngologica,* 115 (suppl. 520) (1995): 186–187.

———. "Somatic Graviception." *Biological Psychology,* 42 (1996): 53–74.

Mittelstaedt, H., and S. Glasauer. "Illusions of Verticality in Weightlessness." *Clinical Investigations,* 71 (1993): 732–739.

Mittelstaedt, H., and M. L. Mittelstaedt. "Mechanismen der Orientierung ohne richtende Aussensreise." *Fortschritte der Zoologie,* 21 (1973): 46–58.

———. "Homing by Path Integration in a Mammal." *Naturwissenschaften,* 67 (1992): 566–567.

Mogensen, J., T. K. Pedersen, S. Holm, and L. E. Bang. "Prefrontal Cortical Mediation of Rats' Place Learning in a Modified Water Maze." *Brain Research Bulletin,* 38 (1995): 425–434.

Moschovakis, A. B., A. B. Karabelas, and S. Highstein. "Structure Function Relationship in the Primate Superior Colliculus. I. Morphological Classification of Efferent Neurons." *Journal of Neurophysiology,* 60 (1988): 232–262.

———. "Structure Function Relationship in the Primate Superior Colliculus. II. Morphological Identity of Presaccadic Neurons." *Journal of Neurophysiology,* 60 (1988): 263–302.

Mouchnino, L., R. Aurenty, J. Massion, and A. Pedotti. "Is the Trunk a Reference Frame for Calculating Leg Position?" *NeuroReport,* 4 (1993): 125–127.

Mountcastle, V. H., J. C. Lynch, A. Georgopoulos, H. Sakata, and C. Acuna. "Posterior Parietal Association Cortex of the Monkey: Command Functions for Operations within Extrapersonal Space." *Journal of Neurophysiology,* 38 (1975): 871–908.

Müller, J. "Über das Aubert'sche Phänomen." *Zeitschrift für Psychologie der Sinnesorgane,* 49 (1916): 109–249.

Munoz, D. P., and R. H. Wurtz. "Fixation Cells in Monkey Superior Colliculus. I. Characteristics of Cell Discharge." *Journal of Neurophysiology,* 70 (1993): 559–575.

Musil, S. Y., C. R. Olson, and M. E. Goldberg. "Visual and Oculomotor Properties of Single Neurons in Posterior Cingulate Cortex of Rhesus Monkey." *Society of Neuroscience Abstracts,* 16 (1990): 1221.

Muybridge, E. *The Human Figure in Motion.* New York: Dover, 1957.

Myashita, Y. "Where Visual Perception Meets Memory." *Annual Review of Neuroscience,* 16 (1993): 245–263.

Nashner, L. M., and A. Berthoz. "Visual Contribution to Rapid Motor Responses during Postural Control." *Brain Research,* 150 (1978): 403–407.

Neisser, U. *Cognition and Reality.* San Francisco: W. H. Freeman, 1976.

Nudo, R. J., G. W. Milliken, W. M. Jenkins, and M. M. Merzenich. "Use-Dependent Alterations of Movement Representations in Primary Motor Cortex of Adult Squirrel Monkeys." *Journal of Neuroscience,* 16 (1996): 785–807.

O'Keefe, J., and L. Nadel. *The Hippocampus as a Cognitive Map.* Oxford: Oxford University Press, 1978.

O'Keefe, J., and M. Recce. "Phase Relationship between Hippocampal Place Units and the EEG Theta Rhythm." *Hippocampus,* 3 (1993): 317–330.

Olivier, E., A. Grantyn, M. Chat, and A. Berthoz. "The Control of Slow Orienting Eye Movements by Tectoreticulospinal Neurons in the Cat: Behavior, Discharge Patterns, and Underlying Connections." *Experimental Brain Research,* 93 (1993): 435–449.

O'Mara, S. M., E. Rolls, A. Berthoz, and R. P. Kesner. "Neurons Responding to Whole Body Motion in the Primate Hippocampus." *Journal of Neuroscience,* 14 (1994): 6511–6523.

Ono, T., B. L. McNaughton, S. Molotchnidoff, E. T. Rolls, and H. Nishigo. *Perception, Memory, and Emotion: Frontiers in Neuroscience.* Cambridge: Pergamon, 1996.

Ono, T., L. R. Squire, M. Raichle, D. I. Perrett, and M. Fukuda, eds. *Brain Mechanisms of Perception and Memory: From Neuron to Behavior.* Oxford: Oxford University Press, 1993.

Orban, G. A., P. Dupont, B. De Bruyn, R. Vogels, R. Vanderberghe, and L. Mortelmans. "A Motion Area in Human Visual Cortex." *Proceedings of the National Academy of Sciences,* 92 (1995): 993–997.

Paillard, J. "Les déterminants moteurs de l'organisation spatial." *Cahiers de psychologie,* 14 (1971): 261–316.

———. "The Multichanneling of Visual Cues and the Organisation of a Visually Guided Response." In *Tutorials in Motor Behavior,* ed. G. E. Stelmach and J. Requin. Amsterdam: North Holland, 1980, 259–279.

———. "Le corps et ses langages d'espace." In *Le Corps en psychiatrie,* ed. E. Jeddi. Paris: Masson, 1982, 53–69.

———. "Posture and Locomotion: Old Problems and New Concepts. Foreword." In *Posture and Gait: Development, Adaptation, and Modulation,* ed. B. Amblard, A. Berthoz, and F. Clarac. Amsterdam: Elsevier, 1988, 5–12.

———. *Brain and Space.* Oxford: Oxford University Press, 1991.

———. "L'intégration sensori-motrice et idéo-motrice." In *Traité de psychologie expérimentale,* ed. M. Richelle, J. Requin, and M. Robert. Paris: Presses Universitaires de France, 1994, 925–961.

Park, S., and P. S. Holzman. "Association of Working Memory Deficit and Eye Tracking Dysfunction in Schizophrenia." *Schizophrenia Research,* 11 (1993): 55–61.

Paulignan, Y., M. Dufosse, M. Hugon, and J. Massion. "Acquisition of Coordination between Posture and Movement in a Bimanual Task." *Experimental Brain Research,* 77 (1989): 337–348.

Pavard, B., and A. Berthoz. "Perception du mouvement et orientation spatial (Revue bibliographique)." *Le Travail humain,* 2 (1976): 207–226.

———. "Linear Acceleration Modifies the Perceived Velocity of a Moving Scene." *Perception and Psychophysics,* 6 (1977): 529–540.

Pavlov, I. P. *Les Réflexes conditionnés.* Paris: Presses Universitaires de France, 1927.

Pedotti, A., P. Crenna, A. Deat, C. Frigo, and J. Massion. "Postural Synergies in Axial Movements: Short and Long-Term Adaptation." *Experimental Brain Research,* 74 (1989): 3–10.

Pellionisz, A. J. "Coordination: A Vector-Matrix Description of Transformations of Over-complete CNS Coordinates and a Tensorial Solution using the Moore-Penrose Generalized Inverse." *Journal of Theoretical Biology*, 101 (1984): 353–375.

Pellionisz, A. J., and R. Llinás. "Brain Modelling by Tensor Network Theory and Computer Simulation. The Cerebellum: Distributed Processor for Predictive Coordination." *Neuroscience*, 4 (1979): 323–348.

———. "Tensorial Approach to the Geometry of Brain Function. Cerebellar Coordination via a Metric Tensor." *Neuroscience*, 5 (1980): 1761–1770.

———. "Space-Time Representation in the Brain. The Cerebellum as a Predictive Space-Time Metric Tensor." *Neuroscience*, 7 (1982): 2949–2970.

Penfield, W. "Vestibular Sensation and the Cerebral Cortex." *Annales d'otorhinolryngologie*, 66 (1957): 691–698.

Penfield, W., and E. Boldrey. "Somatic Motor and Sensory Representation in the Cerebral Cortex of Man as Studied by Electrical Stimulation." *Brain*, 60 (1937): 389–443.

Penfield, W., and T. Rasmussen. *The Cerebral Cortex of Man: A Clinical Study of Localization of Function*. New York: Macmillan, 1957.

Penfield, W., and L. Roberts. *Langage et Mécanismes cérébraux*. Paris: Presses Universitaires de France, 1963.

Peper, L., R. J. Bootsma, D. R. Mestre, and F. C. Bakker. "Catching Balls: How to Get the Hand to the Right Place at the Right Time." *Journal of Experimental Psychology (Human Perception and Performance)*, 20 (1994): 591–612.

Perret, C., and A. Berthoz. "Evidence on Static and Dynamic Fusimotor Actions on the Spindle Response to Sinusoidal Stretch during Locomotor Activity in the Cat." *Experimental Brain Research*, 18 (1973): 178–188.

Perrett, D. I., J. K. Hietanen, M. W. Oram, and P. J. Benson. "Organisation and Function of Cells Responsive to Faces in the Temporal Cortex." *Philosophical Transactions of the Royal Society. Series B: Biological Sciences*, 335 (1992): 23–30.

Perrett, D., A. J. Mistlin, A. J. Chitty, P. A. J. Smith, D. D. Potter, R. Broennimann, and M. H. Harries. "Specialised Face Processing and Hemispheric Asymmetry in Man and Monkey: Evidence from Single Unit and Reaction Time Studies." *Behavioural Brain Research*, 29 (1988): 245–258.

Perrett, D. I., E. T. Rolls, and W. Caan. "Visual Neurons Responsive to Faces in the Monkey Temporal Cortex." *Experimental Brain Research*, 47 (1982): 342.

Peterson, B. W., J. Baker, and A. J. Pellionisz. "Comparison of Spatial Transformation in Vestibulo-ocular and Vestibulo-spinal Reflexes." *Proceedings of the Symposium on the Representation of Three-Dimensional Space in the Vestibular, Oculomotor, and Visual Systems*. Bologna: Barany Society, 1988.

Petit, L., C. Orssaud, N. Tzourio, F. Crivello, A. Berthoz, and B. Mzaoyer. "Functional Anatomy of a Prelearned Sequence of Horizontal Saccades in Man." *European Journal of Neuroscience*, 16 (1996): 3726–3741.

Petit, L., C. Orssaud, N. Tzourio, G. Salamon, B. Mazoyer, and A. Berthoz. "PET Study of Voluntary Saccadic Eye Movements in Humans: Basal Ganglia-Thalamocortical System and Cingulate Cortex Involvement." *Journal of Neurophysiology*, 69 (1993): 1009–1017.

Petrides, M. "Impairments on Nonspatial Self-Ordered and Externally Ordered Working Memory Tasks after Lesions of the Mid-dorsal Part of the Lateral Frontal Cortex in the Monkey." *Journal of Neuroscience,* 15 (1995): 359–375.

Piaget, J. "Le problème de l'intériorisation des actions en opérations réversibles." *Archives de psychologie,* 32 (1949): 242–258.

————. *The Origin of Intelligence in the Child.* London: Routledge and Kegan Paul, 1953.

Pierrot-Deseilligny, C., I. Israël, A. Berthoz, S. Rivaud, and B. Gaymard. "Role of the Different Frontal Lobe Areas in the Control of the Horizontal Component of Memory-Guided Saccades in Man." *Experimental Brain Research,* 95 (1993): 166–171.

Poincaré, H. *The Value of Science,* trans. George Bruce Halsted. New York: Science Press, 1907.

————. *Science and Hypothesis.* New York: Dover, 1952.

Poncet, M., J. F. Pellissier, M. Sebahoun, and C. J. Masser. "A propos d'un cas d'autotopagnosie secondaire à une lésion pariéto-occipitale de l'hémisphère majeur." *Encéphale,* 60 (1971): 110–123.

Posner, M. I., M. J. Nissen, and W. C. Ogden. "Attended and Unattended Processing Modes: The Role of Set for Spatial Location." In *Modes of Perceiving and Processing Information,* ed. H. L. Pick and I. J. Saltzman. Hillsdale, N.J.: Erlbaum, 1988, 137–157.

Pozzo, T., A. Berthoz, and L. Lefort. "Head Stabilisation during Various Locomotor Tasks in Humans. I. Normal Subjects." *Experimental Brain Research,* 82 (1990): 97–106.

Prochazka, A. "Sensorimotor Gain Control: A Basic Strategy of Motor Systems?" *Progress in Neurobiology,* 33 (1989): 281–307.

Proust, Marcel. *Remembrance of Things Past,* trans. C. K. Scott Moncrieff and Terence Kilmartin. London: Chatto and Windus, 1981.

Rademaker, G. G. J. *Réactions labyrinthiques et équilibre.* Paris: Masson, 1935.

Reuchlin, M., J. Lautrey, C. Marendaz, and T. Ohlman. *Cognition: L'universel et l'individuel.* Paris: Presses Universitaires de France, 1989.

Ribot, T. *The Psychology of the Emotions.* Contemporary Science Series. London: W. Scott, 1897.

Riehle, A., and J. Requin. "Monkey Primary Motor and Premotor Cortex: Single-Cell Activity Related to Prior Information about Direction and Extent of an Intended Movement." *Journal of Neurophysiology,* 61 (1989): 534–549.

Rispal-Padel, L., F. Cicirata, and C. Pons. "Cerebellar Nuclear Topography of Simple and Synergistic Movements in the Alert Baboon *(Papio papio)." Experimental Brain Research,* 47 (1982): 365–380.

Rizzolatti, G., and V. Galeze. "Mechanisms and Theories of Spatial Neglect." In *Handbook of Neuropsychology,* ed. F. Boller and J. Gravman. Amsterdam: Elsevier, 1988, 289–313.

Rizzolatti, G., M. Gartilucci, R. M. Camarda, V. Gallex, G. Luppino, M. Matelli, and L. Fogassi. "Neurons Related to Reaching-Grasping Arm Movements in the Rostral Part of Area 6 (Area 6a)." *Experimental Brain Research,* 82 (1990): 337–350.

Rizzolatti, G., L. Riggio, and B. M. Sheliga. "Space and Selective Attention." In *Attention and Performance XV,* ed. C. Umilta and M. Moscovitch. Hillsdale, N.J.: Erlbaum, 1994, 232–265.

Roll, J. P., J. C. Gilhodes, R. Roll, and F. Harley. "Are Proprioceptive Sensory Inputs Combined into a Gestalt?" In *Attention and Performance XVI*, ed. T. Inoue and J. T. MacClelland. Cambridge, Mass.: MIT Press, 1995, 290–314.

Roll, J. P., J. C. Gilhodes, R. Roll, and J. L. Velay. "Contribution of Skeletal and Extraocular Proprioception to Kinaesthetic Representation," in *Attention and Performance XIII*, ed. M. Jeannerod. Hillsdale, N.J.: Erlbaum, 1990, 549–566.

Roll, J. P., R. Roll, and J.-L. Velay. "Proprioception as a Link between Body and Extra-personal Space." In *Brain and Space*, ed. J. Paillard. Oxford: Oxford University Press, 1991, 112–132.

Rolls, E. T. "A Theory of Hippocampal Function." *Hippocampus*, 6 (1996): 601–620.

Rolls, E. T., and S. M. O'Mara. "Neurophysiological and Theoretical Analysis of How the Hippocampus Functions in Memory." In *Brain Mechanisms of Perception: From Neuron to Behavior*, ed. T. Ono. New York: Oxford University Press, 1993, 276–297.

Rolls, E. T., and A. Treves. *Neural Networks and Brain Function*. New York: Oxford University Press, 1998.

Ronsard, P. de. *Poems of Pierre de Ronsard*, trans. and ed. Nicholas Kilmer. Berkeley: University of California Press, 1979.

Rousseau, J.-J. *Emile; Or, on Education*. New York: Basic Books, 1979.

Rubens, A. B. "Caloric Stimulation and Unilateral Visual Neglect." *Neurology*, 35 (1985): 1019–1024.

Sakata, H., and M. Kusunoki. "Organization of Space Perception: Neural Representation of Three-Dimensional Space in the Posterior Parietal Cortex." *Current Opinion in Neurobiology*, 2 (1992): 170–174.

Sakata, H., H. Shibutani, and K. Kawano. "Functional Properties of Visual Tracking Neurons in Posterior Parietal Association Cortex of the Monkey." *Journal of Neurophysiology*, 49 (1983): 1364–1380.

Sakata, H., M. Taira, A. Murata, and S. Mine, "Neural Mechanisms of Visual Guidance of Hand Action in the Parietal Cortex of the Monkey." *Cerebral Cortex*, 5 (1995): 429–438.

Salzman, C., and W. T. Newsome. "Neural Mechanisms for Forming a Perceptual Decision." *Science*, 264 (1994): 231–237.

Salzmann, E. "Attention and Memory Trials during Neuronal Recording from the Primate Pulvinar and Posterior Parietal Cortex (Area PG)." *Behavioral Brain Research*, 67 (1995): 241–253.

Sans, A., and E. Scarfone. "Afferent Calyces and Type I Hair Cells during Development. A New Morphological Hypothesis." *Annals of the New York Academy of Sciences*, 781 (1996): 1–12.

Sartre, J.-P. *The Psychology of Imagination*. Westport, Conn.: Greenwood Press, 1978.

Sasaki, K., and H. Gemba. "Electrical Activity in the Prefrontal Cortex Specific to No-Go Reaction of Conditioned Hand Movement with Colour Discrimination in the Monkey." *Experimental Brain Research*, 64 (1986): 603–606.

Sauvan, S. M., and E. Peterhans. "Neural Integration of Visual Information and Direction of Gravity in the Prestriate Cortex of the Alert Monkey." In *Multisensory Control of Posture*. New York: Plenum, 1995, 43–50.

Schlag, J., and M. Schlag-Rey. "Evidence for a Supplementary Eye Field." *Journal of Neurophysiology*, 57 (1987): 179–200.

Schmidt, R. A. "A Schema Theory of Discrete Motor Skill Learning." *Psychological Review*, 32 (1975): 225–260.

Schöner, G. "A Dynamic Theory of Coordination of Discrete Movements." *Biological Cybernetiks*, 63 (1990): 257–270.

Sechenov, I. M. *Selected Works*. Leningrad: State Publishing House for Biological and Medical Literature, 1935.

Segraves, M. A. "Effects of Frontal Eye Field Stimulation upon Omnipause and Burst Neurons in the Monkey Paramedian Pontine Reticular Formation." *Society of Neuroscience Abstracts*, 18 (1992): 296.10.

Segraves, M. A., and M. E. Goldberg. "Functional Properties of Corticotectal Neurons in the Monkey's Frontal Eye Field." *Journal of Neurophysiology*, 58 (1987): 1387–1419.

Shadlen, M. N., and W. T. Newsome. "Motion Perception: Seeing and Deciding." *Proceedings of the National Academy of Sciences*, 93 (1996): 628–633.

Shallice, T. *Symptômes et modèles en neuropsychologie*. Paris: Presses Universitaires de France, 1993.

Sheliga, B. M., L. Riggio, and G. Rizzolatti. "Orienting of Attention and Eye Movements." *Experimental Brain Research*, 98 (1994): 507–522.

Shepard, R. N. "Ecological Constraints on Internal Representation: Resonant Kinematics of Perceiving, Imagining, Thinking, and Dreaming." *Psychological Review*, 91 (1984): 417–447.

Sherrington, G. S. *The Integrative Action of the Nervous System*. New Haven: Yale University Press, 1906.

———. "Observations on the Sensual Role of the Proprioceptive Nerve Supply of Extrinsic Ocular Muscles." *Brain*, 41 (1918): 332–343.

Shinoda, Y., T. Ohgaki, T. Futami, and Y. Sugiushi. "Vestibular Projections to the Spinal Cord: The Morphology of Single Vestibulo-Spinal Axons." *Vestibulospinal Control of Posture and Movement*, ed. O. Pompeiano and J. Allum. Progress in Brain Research. Amsterdam: Elsevier, 1988, 17–27.

Silberpfennig, J. "Contributions to the Problem of Eye Movements. III. Disturbance of Ocular Movements with Pseudo-hemianopsia in Frontal Lobe Tumors." *Confinia Neurologica*, 4 (1941): 1–13.

Simpson, J. "The Accessory Optic Systems." *Annual Review of Neuroscience*, 7 (1984): 13–14.

Simpson, J. I., and W. Graf. "The Selection of Reference Frames by Nature and Its Investigators," in *Adaptative Mechanisms in Gaze Control. Facts and Theories. Review of Oculomotricity Research*, vol. 1, ed. A. Berthoz and G. Melvill-Jones. Amsterdam: Elsevier, 3–20.

Singer, W. "Search for Coherence: A Basic Principle of Cortical Self-Organization." *Concepts in Neuroscience*, 1 (1990): 1–26.

Sirigu, A., J. R. Duhamel, L. Cohen, B. Pillon, B. Dubois, and Y. Abid. "The Mental Representation of Hand Movements after Parietal Cortex Damage." *Science*, 273 (1996): 1564–1568.

Sirigu, A., J. Grafman, K. Bressler, and T. Sunderland. "Multiple Representations Contribute to Body Knowledge Processing." *Brain*, 114 (1991): 629–642.

Slotine, J.-J. and W. Li. *Applied Nonlinear Control*. Englewood Cliffs, N.J.: Prentice Hall, 1991.

Smith, B. H. "Vestibular Disturbances in Epilepsy." *Neurology,* 10 (1960): 465–469.

Soechting, J. F., and A. Berthoz. "Dynamic Role of Vision in the Control of Posture in Man." *Experimental Brain Research,* 36 (1979): 551–561.

Soechting, J. F., and M. Flanders. "Sensorimotor Representations for Pointing to Targets in Three-Dimensional Space." *Journal of Neurophysiology,* 62 (1989): 582–594.

Soechting, J. F., and B. Ross. "Psychophysical Determination of Coordinate Representation of Human Arm Orientation." *Neuroscience,* 13 (1984): 595–604.

Sokolov, E. N., and O. S. Vinogradova. *Neuronal Mechanisms of the Orienting Reflex.* Hillsdale, N.J.: Erlbaum, 1975.

Squire, L. R., and M. Zola-Morgan. "The Medial Temporal Lobe Memory System." *Science,* 253 (1991): 1380–1386.

Stein, B. E., and M. A. Meredith. *The Merging of the Sciences.* Cambridge, Mass.: MIT Press, 1993.

Sully, J. *Les Illusions des sens et de l'esprit.* Paris: Bibliothèque scientifique internationale, 1883.

Szentagothai, J. "The Elementary Vestibulo-ocular Reflex Arc." *Journal of Neurophysiology,* 13 (1950): 395–407.

Tamura, R., T. Ono, M. Fukuda, and K. Nakamura. "Recognition of Egocentric and Allocentric Visual and Auditory Space by Neurons in the Hippocampus of Monkeys." *Neuroscience Letters,* 109 (1990): 293–298.

———. "Spatial Responsiveness of Monkey Hippocampal Neurons to Various Visual and Auditory Stimuli." *Hippocampus,* 2 (1992): 307–322.

Tanaka, K. "Inferotemporal Cortex and Object Vision." *Annual Review of Neuroscience,* 19 (1996): 109–139.

Taube, J. S., R. U. Muller, and J. B. Ranck Jr. "Head-Direction Cells Recorded from the Postsubiculum in Freely Moving Rats. I. Description and Quantitative Analysis." *Journal of Neuroscience,* 10 (1990): 410–435.

———. "Head-Direction Cells Recorded from the Postsubiculum in Freely Moving Rats. II. Effects of Environmental Manipulations." *Journal of Neuroscience,* 10 (1990): 436–447.

Thier, P., and R. G. Erickson. "Responses of Visual-Tracking Neurons from Cortical Area MST-I to Visual, Eye, and Head Motion." *European Journal of Neuroscience,* 4 (1992): 539–553.

Thier, P., and O. Karnath. *Parietal Lobe Contribution to Orientation in 3D Space.* Berlin: Springer, 1997.

Thomas, A. *Equilibre et équilibration.* Paris: Masson, 1940.

Treves, A., and E. T. Rolls. "Computational Constraints Suggest the Need for Two Distinct Input Systems to the Hippocampal CA3 Network." *Hippocampus,* 2 (1991): 189–199.

———. "What Determines the Capacity of Autoassociative Memories in the Brain?" *Network,* 2 (1991): 371–397.

Trotter, Y., S. Celebrini, B. Stricane, S. Thorpe, and M. Imbert. "Neural Processing of Stereopsis as a Function of Viewing Distance in Primate Visual Cortical Area V1." *Journal of Neurophysiology* 76 (1996): 2872–2875.

Trullier, O., S. Wiener, A. Berthoz, and J. A. Meyer. "Biologically Based Artificial Navigation Systems: Reviews and Prospects." *Progress in Neurobiology,* 51 (1997): 483–544.

Tuohimaa, P., E. Aantaa, K. Toukoniitty, and P. Mäkelä. "Studies of Vestibular Cortical

Areas with Short Living Oxygen 15 Isotopes." *Otorhinolaryngologica*, 45 (1983): 315–321.

Turvey, M. T., and P. N. Kugler. "An Ecological Approach to Perception and Action." In *Human Motor Actions. Bernstein Reassessed*, ed. H. T. A. Whiting. Amsterdam: Elsevier, 1984.

Uhl, F., P. Franzen, G. Lindinger, W. Lang, and L. Deecke. "On the Functionality of the Visually Deprived Occipital Cortex in Early Blind Persons." *Neuropsychologia*, 33 (1994): 256–259.

Umilta, C., C. Mucignat, L. Riggio, C. Barbieri, and G. Rizzolatti. "Programming Shifts of Spatial Attention." *European Journal of Cognitive Psychology*, 6 (1994): 23–41.

Ungerleider, L. G. "Functional Brain Imaging Studies of Cortical Mechanisms for Memory." *Science*, 270 (1995): 769–665.

Vallar, G., E. Lobel, G. Galati, A. Berthoz, L. Pizzamiglio, and D. Le Bihan. "A Fronto-Parietal System for Computing the Egocentric Spatial Frame of Reference in Humans." *Experimental Brain Research*, 124 (1999): 281–286.

Vallar, G., M. L. Rusconi, S. Barozzi, B. Bernardini, D. Ovadia, C. Papagno, and A. Cesarani. "Improvement of Left Visuo-Spatial Hemineglect by Left-Sided Transcutaneous Electrical Stimulation." *Neuropsychologia*, 33 (1994): 73–82.

Vallar, G., R. Sterzi, G. Bottini, S. Cappa, and M. L. Rusconi. "Temporary Remission of Left Hemianesthesia after Vestibular Stimulation. A Sensory Neglect Phenomenon." *Cortex*, 26 (1990): 123–131.

Viallet, F., J. Massion, R. Massarino, and R. Khalil. "Coordination between Posture and Movement in Parkinsonism and SMA Lesion." In *From Neuron to Action*, ed. L. Deecke, J. C. Eccles, and V. B. Mountcastle. Berlin: Springer, 1990, 65–70.

Viaud-Delmon, I., and A. Berthoz. No title. *Current Opinion in Neurobiology*, in press.

Viaud-Delmon, I., Y. P. Ivanenko, A. Berthoz, and R. Jouvent. "Sex, Lies, and Virtual Reality." *Nature*, 1 (1998): 15–16.

———. "Anxiety and Integration of Visual Vestibular Information Studied with Virtual Reality." *Biological Psychiatry*, in press.

Vitte, E., C. Derosier, Y. Caritu, A. Berthoz, D. Hasboun, and D. Soulié. "Activation of the Hippocampal Formation by Vestibular Stimulation: A Functional Magnetic Resonance Study." *Experimental Brain Research*, 112 (1996): 523–526.

Vitte, E., A. Sémont, and A. Berthoz. "Repeated Optokinetic Stimulation in Conditions of Active Standing Facilitates the Recovery from Vestibular Deficits." *Experimental Brain Research*, 102 (1994): 141–148.

Viviani, P. "Motor-Perceptual Interactions: The Evolution of an Idea." In *Cognitive Science in Europe*, ed. M. Imbert, P. Bertelson, R. Kempson, D. Osherson, H. Schnelle, N. Streitz, A. Thomassen, and P. Viviani. Heidelberg: Springer, 1987, 11–39.

Viviani, P., G. Baud-Bovy, and M. Redolfi. "Perceiving and Tracking Kinaesthetic Stimuli: Further Evidence of Motor-Perceptual Interactions." *Journal of Experimental Psychology*, 23 (1997): 1232–1252.

Viviani, P., and A. Berthoz. "Voluntary Deceleration and Perceptual Activity during Oblique Saccades." In *Control of Gaze by Brain Stem Neurons*, ed. R. Baker and A. Berthoz. Amsterdam: Elsevier, 1977, 23–28.

Viviani, P., and T. Flash. "Minimum-Jerk, Two-Thirds Power Law, and Isochrony. Converging Approaches to Movement Planning." *Journal of Experimental Psychology (Human Perception)*, 21 (1995): 32–53.

Viviani, P., and R. Schneider. "A Developmental Study of the Relation between Geometry and Kinematics of Drawing Movements." *Journal of Experimental Psychology (Human Perception)*, 17 (1991): 198–218.

Viviani, P., and N. Stucchi. "Biological Movements Look Uniform: Evidence for Motor-Perceptual Interactions." *Journal of Experimental Psychology (Human Perception)*, 18 (1992): 603–623.

Viviani, P., and C. Terzuolo. "Trajectory Determines Movement Dynamics." *Neuroscience*, 72 (1982): 431–437.

von Uexküll, J. *Streifzüge durch die Umwelten von Tieren und Menschen*. Berlin: Springer, 1965.

———. *Umwelt und Innenwelt der Tieren und Menschen*. Berlin: Springer, 1934.

Walton, K. D., D. Lieberman, A. Llinás, M. Begin, and R. Llinás. "Identification of a Critical Period for Motor Development in Neonatal Rats." *Neuroscience*, 51 (1992), 763–767.

Wexler, M., S. Kosslyn, and A. Berthoz. "Motor Processes in Mental Rotation." *Cognition*, 68 (1998): 77–94.

Wiener, S. I., V. A. Korshunov, R. Garcia, and A. Berthoz. "Inertial, Substratal, and Landmark Cue Control of Hippocampal CA1 Place Cell Activity." *European Journal of Neuroscience*, 7 (1995): 2206–2219.

Wiener, S. I., C. A. Paul, and H. Eichenbaum. "Spatial and Behavioral Correlates of Hippocampal Neuronal Activity." *Journal of Neuroscience*, 9 (1989): 2737–2763.

Wurtz, R. E., C. Duffy, and J. P. Roy. "Motion Processing for Guiding Self-Motion." In *Brain Mechanisms of Perception and Memory*, ed. T. Ono, L. R. Squire, M. Raichle, D. I. Perrett, and M. Fukuda. Oxford: Oxford University Press, 1993.

Wurtz, R. E., and M. E. Goldberg. "Activity of Superior Colliculus in Behaving Monkey. III. Cells Discharging before Eye Movements." *Journal of Neurophysiology*, 35 (1972): 575–586.

Yasui, S., and L. R. Young. "Perceived Visual Motion as Effective Stimulus to Pursuit Eye Movement System." *Science*, 190 (1975): 906–908.

Yoshida, K., R. A. McCrea, A. Berthoz, and P. P. Vidal. "Interneurones inhibiteur de la saccade oculaire horizontale étudiée chez le chat éveillé à l'aide d'injections intra-axonique de peroxydase." *Comptes rendus hebdomadaires de l'Académie des sciences (Paris, Série III)*, 290, D53 (1980): 636–638.

Yoshida, K., A. Berthoz, P. P. Vidal, and R. A. McCrea. "Eye Movement Related Activity of Identified Second-Order Vestibular Neurons in the Cat." In *Progress in Oculomotor Research*, ed. A. Fuchs and W. Becker. Amsterdam: Elsevier, 1981, 371–378.

Yoshida, K., R. A. McCrea, A. Berthoz, and P. P. Vidal. "Properties of Immediate Premotor Inhibitory Burst Neurons Controlling Horizontal Rapid Eye Movements in the Cat." In *Progress in Oculomotor Research*, ed. A. Fuchs and W. Becker. Amsterdam: Elsevier, 1981, 71–81.

Young, L., M. Shelhamer, and S. Modestino. "M.I.T./Canadian Vestibular Experiments in Weightlessness." *Experimental Brain Research*, 64 (1986): 299–307.

CREDITS

Figure 1.2. Redrawn from N. A. Bernstein, "Some Emergent Problems of the Regulation of Motor Acts," *Questions of Psychology* 6 (1957).

Figure 1.3. Redrawn from R. Llinás and U. Ribary, "Coherent 40-Hz Oscillation Characterizes Dream State in Humans." *Proceedings of the National Academy of Sciences, USA* 90 (1993): 2078–2081; Fig. 4, p. 2080. Copyright © 1993 National Academy of Sciences, U.S.A.

Figure 1.4. Redrawn from R. A. Schmidt, "A Schema Theory of Discrete Motor Skill Learning." *Psychological Review,* 32 (1975): 236. Copyright © 1975 by the American Psychological Association. Adapted with permission.

Figure 3.1. Redrawn from M. Livingstone and D. Hubel, "Segregation of Form, Color, Movement, and Depth: Anatomy, Physiology, and Perception." *Science,* 240 (1988): 742. Copyright © 1988 American Association for the Advancement of Science. Reprinted with permission.

Figure 3.6. Redrawn from A. Grantyn, "How Visual Inputs to the Ponto-Bulbar Reticular Formation Are Used in the Synthesis of Premotor Signals during Orienting." In *Progress in Brain Research,* vol. 80, ed. J. H. J. Allum and M. Hulliger. Amsterdam: Elsevier, 1989, 159–170. Reprinted with the permission of the author and Elsevier Science.

Figure 3.7. Adapted from a figure in M. Graziano and C. Gross. *Experimental Brain Research* 97: 100. Copyright © Springer-Verlag, 1993. Reprinted with the permission of the author and Springer-Verlag.

Figure 4.2. Redrawn from M. Graziano and C. Gross, "Multiple Pathways for Processing Visual Space." In T. Inui and J. L. McClelland (eds.), *Attention and Performance* XVI. Cambridge, Mass.: MIT Press, p. 199. Reprinted with the permission of the authors and MIT Press.

Figure 4.3. Redrawn from M. A. Arbib, "Perceptual Structures and Distributed Motor Control." In *Handbook of Physiology,* sec. 1, *The Nervous System,* vol. 2, pt. 2, ed. V. B. Brooks. Bethesda: American Physiological Society, 1981, p. 1468. Reprinted with permission of the American Physiological Society.

Figure 5.1. Redrawn from L. R. Squire and M. Zola-Morgan, "The Medial Temporal Lobe Memory System." *Science* 253 (1991): 1381. Copyright © 1991 American Associaion for the Advancement of Science. Reprinted with permission.

Figure 5.2. European Space Agency (ESA) documents.

Figure 5.3. Redrawn from S. R. Cajal, *Histologie du système nerveux de l'homme et des vertébrés.* Madrid: Raycar, 1972 [Paris, 1911].

Figure 5.4. Redrawn from E. T. Rolls, "A Theory of Hippocampal Function." *Hippocampus, 6* (1996): 601–620. Copyright © 1996. Reprinted by permission of Wiley-Liss, Inc., a subsidiary of John Wiley & Sons, Inc.

Figure 6.3. Redrawn from P. Viviani and T. Flash, "Minimum-Jerk, Two-Thirds Power Law, and Isochrony. Converging Approaches to Movement Planning." *Journal of Experimental Psychology (Human Perception),* 21 (1995): 40. Copyright © 1995 by the American Psychological Association. Adapted with permission.

Figures 7.1 and 7.2. Redrawn from W. Graf and R. Baker, "Adaptive Changes of the Vestibulo-Ocular Reflex in Flatfish Are Achieved by Reorganization of Central Nervous Pathways." *Science,* 221 (1983): 777–778. Copyright © 1983 American Association for the Advancement of Science. Reprinted with permission.

Figure 7.3. Redrawn from T. Fukuda, *Statokinetic Reflexes in Equilibrium and Movement.* Tokyo: Igaku-Shoin, 1981. Copyright © 1981 Igaku-Shoin. Adapted with permission.

Figures 10.2 and 10.3. Redrawn from O. Hikosaka, Figs. 30.5 and 30.9 in "Role of the Basal Ganglia in Motor Learning." In *Brain Mechanisms and Memory: Cells, Systems, and Circuits,* ed. J. L. Gough, N. M. Weinberger, and G. Lynch. New York: Oxford University Press, 1992. Copyright © 1992 by Oxford University Press, Inc. Used with permission of Oxford University Press, Inc.

Figure 11.1. From F. Lestienne, J. Soechting, and A. Berthoz, "Postural Readjustments Induced by Linear Motion of Visual Scenes." *Experimental Brain Research,* 28 (1977): 363–384. Copyright © 1977 Springer-Verlag. Reprinted with permission.

Figure 11.2. Redrawn from J. Massion, "Postural Control System." *Current Opinion in Neurobiology,* 4 (1994): 877–887; Fig. 1, p. 878. Copyright © 1994 Elsevier Science. Reprinted with permission.

Figure 11.3. Redrawn from C. Darwin, *The Expression of the Emotions in Man and Animals.* New York: Philosophical Library, 1955.

Figure 13.1. Redrawn from LPPA and NASA documents.

INDEX

Page numbers in *italics* refer to figures.

Brain (continued)
229–32; versus mind, 1–2; modulation of sensory information by, 29, 35–36; predictive power of, 57; processing of information from sensory receptors, 5–6, 263; processing of movement by, 22, 23; as a tensor, 48–49; use of memory by, 6
Brain imaging, 231–232
Braking (collision prevention), 168–172
Brandt, T., 52
Broom experiment, 223
Bruyère, A., 256, 259

Capture: calculations to perform, 22–23; perceptions in, 165–166; and tracking of targets, 62–64
Caricature, 132
Catching a ball, 170–171, 172–173
Cemetery illusion, 247–248
Centralist theory, 31, 263
Cerebellum: and body schema, 229–230; development of, 212; function of, 64, 65, 164, 201–203, 225, 265; inhibitions in, 193–194; multisensory convergences in, 59, 82
Cerebral imagery, 175
Chakyar community (Kerala), 189
Chance, 255
Châtelet, G., 2
Children: development of movements in, 179, 214; teaching speed sports to visually impaired, 51
Chronography, 140
Churchland, P., 48
Cingulate cortex, 209–210
Cognition, 1–2; and language, 4
Cognitive sciences, multidisciplinary nature of, 11
Coherence: and autism, 93–96, 189, 264; definition of, 56, 57–59, 94–95, 264; between seeing and hearing, 77–89; and spatial neglect, 74–76; and unity of perception, 90–93, 114. See also Perception
Collateral axon branches, 156–157
Colliculus. See Contralateral colliculus; Inferior colliculus; Superior colliculus
Collision prevention (braking), 168–172
Comparator, 13–16, 14, 19
Compensation, 239–241

Composite variables, principle of, 138
Conditioned reflexes, 11
Consecutive optokinetic nystagmus, 55–56
Constructivism, 11
Continuity, in perception, 92–93
Contralateral colliculus, 79
Coordination, of movements, 141
Cordo, P., 223
Coriolis accelerations, 251
Coriolis forces, 6, 119
Cornilleau-Pérès, V., 92
Corollary discharge, theory of, 183
Cortical vision, 50
Cutaneous (skin) receptors, 26, 27, 29–31, 83–86, 105–106. See also Touch
Cybernetic paradigm, 139, 174, 183
Cyon, E., 36

Damasio, A. R., 7
Dargassies, Saint Anne, 186
Darlot, C., 91–92
Darwin, Charles, 32, 118, 139, 188, 232
Dead reckoning, 118, 120, 124
De Ajuriaguerra, J., 83, 186
De animalium motu (Aristotle), 139
Decision making, 7, 13, 95
Deecke, L., 208
Degrees of freedom, control of number of, 141–147, 142, 154
Déjerine, J., 217
Descartes, René, 3, 11
Dichgans, J., 52, 67
Directional sensitivity, 32–33
Displacement, 67, 92; of attention, 192, 211; and balance, 225; memory of, 120–122, 125. See also Acceleration; Velocity
Distance perception, 76–77
Donders, F. C., 150
Dorsal motor nucleus, 65
Dorsal terminal nucleus (DTN), 64
Dorsolateral pontine nucleus (DLPN), 63
Dorsolateral thalamus, 74
Dorsomedial pontine nucleus (DMPN), 63
Dove prisms, 235, 236
Droulez, J., 91–92, 213
Duchenne de Boulogne, 217
Dynamic illusions, 244–246
Dynamic memory, 212–214

Gravito-inertial receptor system, 33
Gravity: detection of, 38, 41; as frame of reference, 100–102; illusions caused by, 244–248; internalization of, 173–174; and prediction of movement, 172; receptors in stomach, 104–105; role in perception, 73
Graziano, M., 109
Grillner, S., 162
Gropius, W., 257
Gross, C., 85–86, 109, 231
Grüsser, O. J., 72, 98
Gurfinkel, V. S., 22, 105, 107, 219, 224, 228, 229

Hallucinations, 253–254, 264
Hartline, H. K., 35
Haussmann, Baron G. E., 258
Head: restriction of, 143–144; stabilization of, 101, *102*, 111–112, 143, *184*; tilt of, 38–42, 72–73, 92
Head, H., 227, 230
Hearing, 25, 77–82
Hein, A., 11, 58
Held, R., 11, 58
Heliodorus of Emesa, 186–187
Helmholtz, H. von, 9, 21, 25, 33, 77, 195, 243
Hemianesthesia, 75
Hemianopia, 75
Hemispatial neglect, 74
Heraclitus of Ephesus, 3
Hertz, H., 137
Hikosaka, O., 201
Hippocampal formation, 73, 74
Hippocampus: anatomy of, *127*; function of, 126–136, 190, 191, 251, 264
Holmes, G., 227
Husserl, E., 16, 87, 132, 168
Hypothesis, best possible, 243
Hyvarinen, J., 98

Idiotropic vector, 104–105
Illusions, 6; of agoraphobia, 253; definition of, 242, 243; of dissociation of the body, 70; of epilepsy, 70; and hallucinations, 253–254, 264; of moon size, 244; and theory of coherence, 91; types of, 244–254; of vection, 52–53; of velocity inversions, 53; of ventriloquism, 58; of vertigo, 69–70, 253; of waterfalls, 55, 252

Imagined movement, 211–212
Immobility, sense of, 41
Implicit frames of reference, 112–114
Inertia, overcoming, 3, 32, 33
Inertial cues, 32, 33
Inertial navigation, 119–120
Inferior colliculus, 78, 80
Inferior olive, dorsal motor nucleus of, 65
Inferior temporal (IT) area, 130
Inhibition, of sensations, 87–88, 192–194
Inhibitory second-order vestibular neurons, 193
Inner ear receptors. *See* Vestibular (inner ear) receptors
Interactive frame of reference, 97
Interneurons, 197
Inverse problem, 177–179
Involuntary post-contraction, 248–249
Ischemia, 227–228
Isochrony, principle of, 31, 147, 152
Ito, M., 116, 193, 236

Jackson, J. Hughlings, 126
James, William, 9–10, *10*
Janet, P., 10, 241
Jeannerod, M., 11, 22, 110, 196
Jerk: definition of, 151; detection of, 33, 34–35, 46; minimalization of, 151–152, *152*, 162
Joint receptors, 26, 27, 31–32
Julesz stereograms, 76
Jung, C., 52

Kalman filters, 174–175
Kandinsky, V., 257
Kant, Immanuel, 5
Kathakali, 188–189
Kawato, M., 178
Kelso, J. A. S., 172
Kepler, J., 140
Kinematic analysis, 140
Kinematics, internalization of, 176
Kinesthesia, 5, 25, 120
Koenderink, J. J., 168
Kohnstamm's illusion, 248–249
Korn, H., 194
Kornhuber, A., 208
Kosslyn, S. M., 110
Krechevsky, I., 243

Postsubiculum, function of, 73
Posture: in absence of gravity, 103–104, 219; and emotions, 139, 188, 232; as first expression of movement, 137; as readiness to move, 190; stabilization of, 43–44, 64, 105–106. *See also* Balance
Power gripping, 160
Pozzo, T., 225
Precision gripping, 160
Predators. *See* Capture
Prediction, 57; and dynamic memory, 212–214; memory for, 115–117, 264–265; in ocular pursuit, 151; of trajectories, 132–134, 151, 185. *See also* Anticipation; Motor prediction
Prefrontal cortex (PFC), 121, 206, 210–211, 264
Prehistoric cave art, 134–136
Preperceptions, 20, 132
Prepositus hypoglossi nucleus, 46, 47
Presynaptic inhibitory mechanisms, 30, 32
Prochazska, A., 88
Proprioception, description of, 26, 27–32
Proust, Marcel, 91, 132
Prouvé, Jean, 257
Purkinje, 183, 195
Putnam, Hilary, 261
Pyramidal cells, 128
Pyramidal cortical neurons, 31, 132–133

Rademaker, G. G. J., 43, 105
Ramón y Cajal, S., 155
Readiness potential, 208, 223
Recce, M., 133
Receptors. *See* Sensory receptors
Reciprocal innervation, 193
Re-excitation (recurring excitation loop), 47
Reference frames. *See* Frames of reference
Reflex actions, 10, 11–13
Regularity, 255
Relative frames of reference, 110–112
Renshaw neurons, 193
Representation, 12, 21–22
Residual vision, 51
Response, 11. *See also* Reflex actions
Retina: function of, 60, 61

Retinal ganglia, 64
Reuchlin, M., 221
Ribot, T., 7
Righting reflexes, 106
Rigidity, in perception, 92, 93, 99
Rizzolatti, G., 20, 200, 212
Robotic movement, theory of nonlinear control of, 31
Rolls, E. T., 130
Rondot, Pierre, 217
Ronsard, Pierre de, 138
Rotational memory, 120–122
Rotations: detection of, 42–43, 46–47, 176. *See also* Mental rotation
Rougeul-Buser, A., 16
Rousié, D., 239
Rousseau, J.-J., 9

Saccades: cortical control of, 204–211; definition of, 62, 192, 194, 195–196; inhibition of, 195, 199–203; production of, 197–199
Saccadic inhibition, 195, 199–203
Saccule: description of, 26, 27, 38; function of, 33, 38–41
Sailors' illusion, 252–253
Sakata, H., 257
Sartre, J.-P., 83
Schema, definition of, 17–18, 18, 20
Schizophrenia, 211, 264
Schmidt, R. A., 17–18, 18, 167
Schöner, G., 172
Sechenov, I. M., 58, 224
Second-order vestibular neurons, 68, 70, 159, 159
Seeing. *See* Sight
Segregation, of sensory signals, 42–43
Selfishness, 265
Self-motion, perception of, 52–56, 218
Semicircular canals: description of, 26, 27, 32–33; function of, 33–38, 44, 66–67, 92; geometry of, 33–38, 60, 65, 71, 100–101; location of, 143
Semiology, qualitative vs. quantitative, 217
Senses: ambiguity of, 90–91; function of, 25; number of, 5, 262; revision of meaning of, 263–264. *See also* Hearing; Sight; Smell; Taste; Touch
Sensory conflict, 251

Sensory receptors: adaptation of, 233–234; configurations of, 5, 37, 107–109, *108*, 130; location of, *26, 27*; processing information from, 19–20

Sensory thalamus, multisensory convergences in, 59

Sensory thresholds, 190

Servo systems, theory of, 47, 49, 174, 183

Set (*Einstellung*), 223–224

Shadlen, M. N., 62

Shepard, R. N., 132, 175–177, 253–254

Sherrington, G. S., 11, 27, 47, 216, 223, 227

Short-term memory, 127–128

Sight, 25; and hearing, 77–82; segregation of sensory signals in, 42–43; in sense of movement, *26, 33*, 50–56, 60–64; and touch, 83–89. *See also* Gaze

Simulation: definition of, 22–24, 37; of muscle contractions, 28–29

Skeletal geometry, 143–144

Skin. *See* Cutaneous (skin) receptors; Touch

Slotine, J.-J., 31, 138, 171

Smell, 25

Sokolov, E. N., 16, 190

Somatoparaphrenia, 75

Somatotopy, 208

Sound. *See* Hearing

Space: extrapersonal, 98–99, 170; impaired perception of, 74–76; personal, 98

Spatial (cognitive) maps, 128–130

Spatial memory: definition of, 116–117; neural basis of, 73–74, 126–136; types of, 117–126

Spatial neglect, 74–76, 100

Spatial representation, 70

Spatial selection, 200

Stabilized vision, 219

Stationary locomotion, 165, 182

Stein, B. E., 78, 82

Steps, initiation of, 225

Stimulus, neuronal model of, 190–191

Stimulus-response paradigm, 50, 139, 180

Strategies, definition of, 154–155, 224, 263

Subcortical vision, 50

Subjective vertical, 41, 104, 218–219

Substantia nigra pars reticulata (SNpr), 200, 211

Substitution, and adaptation, 234–238

Subthalamic nucleus (STN), 201

Sully, J., 242

Superior colliculus (SC): description of, 78; function of, 61, 77–81; and inhibitions, 199–200; multisensory convergences in, 59, 80–81

Superior medial temporal sulcus, 176

Superior temporal sulcus (STS), 87

Supplemental eye field (SEF), 209

Supplementary motor area (SMA), 208–209, 225

Symmetry: perception of, 73, 228–229, 240; preference for, 93

Synergies: coordination of, 162–164; definition of, 44, 154–155, 164

Synergy, principle of, 154–155, 164, 224

Szentagothai, J., 46

Tactile perception. *See* Touch

Targets, tracking of, 62–64

Taste, 25

Tecto-reticulo-spinal neurons (TRSNs), *79*, 199–200

Temporal encoding, theory of, 264

Temporal selection, pauses in, 199

Temporal "windows," 82

Temporonasal direction (orientation), 64

Tensor: brain as, 48–49; definition of, 48

Territorialism, 265

Teuber school, 11

Thalamo-cortical circuitry, 15–16, *17*

Thalamus (Th): and body schema, 230; function of, 73, 166–167; inhibitions in, 193

Thales of Miletus, 256

Thermoreceptors, 30

θ rhythm, 132–133

Thomas, André, 101, 217

Tilt, of head, 38–42, 72–73, 92

Time, and coherence, 90

Time shifts, 81–82

Time τ-to-collision, 168–169

Toads, 166–168

Topographic memory, 117–118

Topokinetic (topokinesthetic) memory, 125

Touch: description of, 25, 29–31; and perception of space, 105–107; and sight, 83–89

Trajectories, prediction of, 132–134

Transfer functions, 217

Translational memory, 122–124

Translations, detection of, 42–43, 176